Studies in Modern History

General Editor: **J. C. D. Clark**, Joyce and Elizabeth Hall Distinguished Professor of British History, University of Kansas

Titles include:

James B. Bell
THE IMPERIAL ORIGINS OF THE KING'S CHURCH IN EARLY AMERICA, 1607–1783

Jonathan Clark and Howard Erskine–Hill (*editors*)
SAMUEL JOHNSON IN HISTORICAL CONTEXT

Benard Cottret (*editor*)
BOLINGBROKE'S POLITICAL WRITINGS
The Conservative Enlightenment

Richard R. Follet
EVANGELICALISM, PENAL THEORY AND THE POLITICS OF CRIMINAL LAW REFORM IN ENGLAND, 1808–30

Andrew Godley
JEWISH IMMIGRANT ENTREPRENEURSHIP IN NEW YORK AND LONDON, 1880–1914

William Anthony Hay
THE WHIG REVIVAL
1808–1830

Philip Hicks
NEOCLASSICAL HISTORY AND ENGLISH CULTURE
From Clarendon to Hume

Mark Keay
WILLIAM WORDSWORTH'S GOLDEN AGE THEORIES DURING THE INDUSTRIAL REVOLUTION IN ENGLAND, 1750–1850

William M. Kuhn
DEMOCRATIC ROYALISM
The Transformation of the British Monarchy, 1861–1914

Kim Lawes
PATERNALISM AND POLITICS
The Revival of Paternalism in Early Nineteenth-Century Britain

Marisa Linton
THE POLITICS OF VIRTUE IN ENLIGHTENMENT FRANCE

Nancy D. LoPatin
POLITICAL UNIONS, POPULAR POLITICS AND THE GREAT REFORM ACT OF 1832

Karin J. MacHardy
WAR, RELIGION AND COURT PATRONAGE IN HABSBURG AUSTRIA
The Social and Cultural Dimensions of Political Interaction, 1521–1622

Robert J. Mayhew
LANDSCAPE, LITERATURE AND ENGLISH RELIGIOUS CULTURE, 1660–1800
Samuel Johnson and Languages of Natural Description

Marjorie Morgan
NATIONAL IDENTITIES AND TRAVEL IN VICTORIAN BRITAIN

James Muldoon
EMPIRE AND ORDER
The Concept of Empire, 800–1800

W. D. Rubinstein and Hilary Rubinstein
PHILOSEMITISM
Admiration and Support for Jews in the English-Speaking World, 1840–1939

Julia Rudolph
WHIG POLITICAL THOUGHT AND THE GLORIOUS REVOLUTION
James Tyrrell and the Theory of Resistance

Lisa Steffen
TREASON AND NATIONAL IDENTITY
Defining a British State, 1608–1820

Lynne Taylor
BETWEEN RESISTANCE AND COLLABORATION
Popular Protest in Northen France, 1940–45

Anthony Waterman
POLITICAL ECONOMY AND CHRISTIAN THEOLOGY SINCE THE ENLIGHTENMENT
Essays in Intellectual History

Doron Zimmerman
THE JACOBITE MOVEMENT IN SCOTLAND AND IN EXILE, 1746–1759

Studies in Modern History
Series Standing Order ISBN 0–333–79328–5
(outside North America only)

You can receive future titles in this series as they are published by placing a standing order. Please contact your bookseller or, in case of difficulty, write to us at the address below with your name and address, the title of the series and the ISBN quoted above.

Customer Services Department, Macmillan Distribution Ltd, Houndmills, Basingstoke, Hampshire RG21 6XS, England

Political Economy and Christian Theology Since the Enlightenment

Essays in Intellectual History

A. M. C. Waterman
Fellow of St John's College, Winnipeg
and Professor of Economics
University of Manitoba

palgrave
macmillan

First published 2004 by
PALGRAVE MACMILLAN
Houndmills, Basingstoke, Hampshire RG21 6XS and
175 Fifth Avenue, New York, N. Y. 10010
Companies and representatives throughout the world

PALGRAVE MACMILLAN is the global academic imprint of the Palgrave Macmillan division of St. Martin's Press, LLC and of Palgrave Macmillan Ltd. Macmillan® is a registered trademark in the United States, United Kingdom and other countries. Palgrave is a registered trademark in the European Union and other countries.

ISBN 1–4039–3913–6

This book is printed on paper suitable for recycling and made from fully managed and sustained forest sources.

A catalogue record for this book is available from the British Library.

Library of Congress Cataloging-in-Publication Data

Waterman, Anthony Michael C.
Political economy and Christian theology since the Enlightenment: essays in intellectual history / A.M.C. Waterman.
p. cm. – (Studies in modern history)
Includes bibliographical references and index.
ISBN 1–4039–3913–6
1. Economics–Religious aspects–Christianity–History of doctrines. 2. Economics–Great Britain–History. I. Title. II. Studies in modern history (Palgrave Macmillan (Firm))

BR115.E3W35 2004
261.85–dc22 2004052589

10 9 8 7 6 5 4 3 2 1
13 12 11 10 09 08 07 06 05 04

Printed and bound in Great Britain by
Antony Rowe Ltd, Chippenham and Eastbourne

Contents

List of Tables and Figures

Acknowledgements

The author wishes to acknowledge with thanks permission of the copyright holders to reprint the following previously published works of his in this volume:

Cambridge University Press: for "The Nexus Between Theology and Political Doctrine in Church and Dissent," pp. 193–218 in K. Haakonssen (ed.), *Enlightenment and Religion: The Case of Rational Dissent* (1996). (Chapter 3)

The Australian Journal of Politics and History: for "The Grand Scheme of Subordination: The Intellectual Roots of Tory Doctrine," pp. 121–33 in *The Australian Journal of Politics and History*, vol. 40, Special Issue, *Ideas and Ideologies: Essays in Memory of Eugene Kamenka* (1994). (Chapter 4)

Cambridge University Press: for "A Cambridge Via Media in Late Georgian Anglicanism," pp. 419–36 in *Journal of Ecclesiastical History*, vol. 42, no. 2 (1991). (Chapter 5)

The Southern Economic Association: for "Economics as Theology: Adam Smith's *Wealth of Nations*," pp. 907–21 in *Southern Economic Journal*, vol. 68, no. 4 (2002). (Chapter 6)

Taylor & Francis Books Ltd (Routledge): for "The Beginning of Boundaries: the Sudden Separation of Political Economy from Christian Theology," pp. 41–63 in Guido Erreygers (ed.), *Economics and Interdisciplinary Exchange* (Routledge, 2001). (Chapter 7)

Kluwer Academic/Plenum Publishers: for "Whately, Senior and the Methodology of Classical Political Economy," pp. 41–60 in H. Geoffrey Brennan and A. M. C. Waterman (eds), *Economics and Religion: Are They Distinct?* (1994). (Chapter 8)

Edward Elgar Publishing Ltd: for "Peasants, Population and Progress in Malthus and Chalmers," pp. 31–53 in Evelyn L. Forget and Richard A. Lobdell (eds), *The Peasant in Economic Thought: "A Perfect Republic"* (1995). (Chapter 9)

The Review of Social Economy: for "Property Rights in John Locke and in Christian Social Teaching," pp. 97–115 in *Review of Social Economy*, vol. 40, no. 2 (1982) (Chapter 10); and for "The Intellectual Context of Rerum Novarum," pp. 465–82 in *The Review of Social Economy*, vol. 49, no. 4 (1991). (Chapter 11)

The Journal of Markets and Morality: for "Market Social Order and Christian Organicism in Centesimus Annus," pp. 220–33 in *The Journal of Markets and Morality*, vol. 2 no. 2. (1999). (Chapter 12)

Kluwer Academic/Plenum Publishers: for "Social Thinking in Established Protestant Churches," pp. 51–67 in James M. Dean and A. M. C. Waterman (eds) *Religion and Economics: Normative Social Theory* (1999). (Chapter 13)

The American Journal of Economics and Sociology and Blackwell Publishers for: "'New Political Economies' Then and Now: Economic Theory and the Mutation of Political Doctrine," pp. 13–51 in *American Journal of Economics and Sociology*, vol. 61, no. 1 (2002). (Chapter 14)

The author is also grateful to Evelyn L. Forget and Richard A. Lobdell, editors of The *Peasant in Economic Thought* listed above, for permission to include "Peasants, Population and Progress in Malthus and Chalmers" in this collection.

Preface

Except for the first two, all of the essays in this book have already appeared in print, either as articles or as chapters in books. Some were written for historians, some mainly for theologians and one or two almost entirely for economists. They have been brought together at the suggestion of Jonathan Clark, who appears to believe that a certain vague coherence may be found lurking behind their bewildering heterogeneity. My own belated attempt to identify some common theme is reported in Chapter 1.

Because of their widely differing provenance I have a large number of colleagues and friends to acknowledge, and it would seem best to thank them in relation to the specific chapter or chapters in which they helped me.

Jonathan Clark who commissioned this book, and Donald Winch who for twenty-five years has patiently, faithfully and repeatedly born with my amateurish attempts to write intellectual history, have given me comments and advice on the introductory Chapter 1, each responding to my importunity under great pressure of time.

Chapter 2 was written for a seminar series on "The Enlightenment Project" (a title I now repent) that I organized in 1999–2000 at the Institute for the Humanities at the University of Manitoba. It has therefore circulated in draft for many years. In addition to the participants at that seminar, especially R. A. Lebrun, I am grateful to Jonathan Clark, Knud Haakonssen, J. G. A. Pocock and Gordon Schochet for helpful comments and advice; also to Natalie Johnson for valuable research assistance, and to whom I am now even more indebted for the vast labor of preparing this collection for publication.

Chapter 3 was originally prepared for a conference on "Rational Dissent" organized by Knud Haakonssen at the Australian National University, the proceedings of which were edited by him and published as *Enlightenment and Religion* (1996). I was much stimulated and helped by all the other participants and especially by Knud himself and Sandy Stewart, and also profited greatly from later discussion with James E. Bradley. Chapter 4, which overlaps with Chapter 3 to some extent, was also written for a conference at that most hospitable of universities: on that occasion to mark the retirement of, and say farewell to, the late Eugene Kamenka, then gravely sick. Eugene attended every session, entertained several of us to dinner a few days later with his usual generous hospitality: and died within a week of that party. It was both a great honour and a moving experience thus to pay tribute to one whose friendship I had enjoyed intermittently for nearly thirty years. The proceedings were published in *The Australian Journal of Politics and History* (1994) and I am indebted to the other participants especially Peter Munz, and also to Mark Francis, Knud Haakonssen and Mary Kinnear for their comments. I am also deeply grateful for the financial support of the Australian National University, which has appointed me to visiting fellowships in the Research School of Social Sciences no fewer than four times since 1991.

The research for Chapter 5, published in *The Journal of Ecclesiastical History* (1991) was funded by the Social Sciences and Humanities Research Council of Canada during my tenure of a bye-fellowship at Robinson College, Cambridge and I am grateful to Jonathan Clark, Graham Cole, John Gascoigne, Mary Kinnear, James Raven and Stephen Taylor for discussion of work-in-progress. Chapter 6 was written during a subsequent visit to Robinson College where I and my wife have been made so welcome on many occasions over the years. It was presented to the Southern Economic Association meetings in Atlanta at the invitation of Robert A. Nelson, and published in *Southern Economic Journal* (2002). I am grateful to Bob for encouraging this work, to some extent complementary to his own though not quite so much as he may have expected, and to Vivienne Brown, Knud Haakonssen, the late Paul Heyne, Emma Rothschild and Donald Winch for their criticism and advice.

If this book has any point and any story to tell at all, that story hinges on Chapter 7. Written, like Chapter 6, whilst I was at Robinson College in 1997–98, it was presented to the fourth annual European Conference on the History of Economic Thought (ECHE 98) at the University of Antwerp on the theme of "Exchange at the Boundaries". I am obliged to Guido Erreygers, one of the conference co-organizers, who edited a selection of the papers presented as *Economics and Interdisciplinary Exchange* (2001), for his encouragement and sound editorial advice, and to the other participants for their comments.

Chapter 8, like Chapter 5, develops further certain themes originally announced in my *Revolution, Economics and Religion* (1991): in this case, the seminal contribution of Richard Whately to the vexed question of the epistemic relation between political economy and Christian theology. It was written during tenure of a research fellowship at the Australian National University as a chapter in *Economics and Religion:*

Are They Distinct? (1994) edited by Geoffrey Brennan and me. In addition to Geoffrey himself I am indebted to Knud Haakonssen, Mary Kinnear, Colin Roger, and to two of the other contributors, Ian Steedman and the late Barry Gordon, economist, athlete, wine-lover and friend who died suddenly as the book was going to press.

Chapter 9 originated in an interdisciplinary Arts Faculty seminar at the University of Manitoba organized by my departmental colleagues Evelyn Forget and Richard Lobdell, who published a selection of the papers as *The Peasant in Economic Thought* (1995). My thanks to them and the other participants, especially Robert O'Kell, for encouragement and comments.

Chapters 10 and 11 originally appeared in *Review of Social Economy* (1982, 1991) and Chapter 12 in *The Journal of Markets and Morality* (1999). All three deal with the problem of how to relate traditional Christian social doctrine to the "new science" of political economy. The research for Chapter 10 was financed by the Christendom Trust, the British Council and the University of Manitoba during my tenure of the Maurice Reckitt Fellowship in Christian Social Thought at the University of Sussex (1979–80), where I began my interdisciplinary studies in intellectual history and where I first met Donald Winch. Norman Cameron, Don McCarthy, Murdith McLean, James Tully and the late Roy Vogt made valuable comments on work-in-process. Chapter 11 was commissioned for a special issue of *Review of Social Economy* to commemorate the inauguration of modern "Catholic Social Teaching" in the Leo XIII's famous encyclical, *Rerum Novarum* (1891). The first draft was presented to the theological discussion group at St John's College, Winnipeg: subsequent drafts at a joint La Trobe-Monash-Melbourne University history seminar; the Humanities Research Centre of the Australian National University; New College, University of New South Wales; and the Australian Catholic University of Australia, Queensland, where it was delivered as a public lecture. Many valuable comments were picked up *en route* and I am especially grateful for those of the late Paul Heyne, R. A. Lebrun, Michael Novak, John Passmore and T. K. Rymes. Chapter 12 considers John-Paul II's own tribute to *Rerum Novarum,* in which I was helped by the comments of Kenneth Minogue.

Chapter 13, like Chapter 8, was originally written as a chapter of a book edited by myself and another, in this case my St John's College and economics departmental colleague James M. Dean. *Economics and Religion: Normative Social Theory* (1999) is a sequel to its similarly named predecessor. My chapter was written during 1997–98 at

Robinson College and I profited from conversations and correspondence with my co-editor and with Geoffrey Brennan, Tony Cramp, Geoff Harcourt, the late Paul Heyne and the Rt Revd Stephen Sykes, then Bishop of Ely.

The last chapter was originally written for the 2000 annual conference of the Department of Economics of the University of Manitoba. The theme of the conference that year was "Political Economy of the 'New Right'" and I attempted to situate a local and contemporary debate in larger historical context. I am much obliged for helpful criticism and comment from all participants especially Fletcher Baragar and James Dean, from two of our guests on that occasion, Geoffrey Brennan and Jim Stanford, from Mark Blaug and Nancy Folbre, and from Laurence Moss who published a revised version in *American Journal of Economics and Sociology* (2002).

The reader should be warned, if he or she does not already suspect it, that because of the variety of publications for which, and circumstances in which, these papers were written, a certain amount of overlapping of material and even of text will be discovered. I hope before long to write a real book about the greatest of all mutations in political thought: the gradual replacement of the rhetoric of Christian theology by that of political economy during the nineteenth and twentieth centuries. In that book all redundancies will be eliminated and at least some of the many empty spaces left by these few essays filled in. Meanwhile I can only hope that any who read this collection will look upon the essays as I have suggested in chapter 1 that they might: that is, as a series of snapshots in which something very much like the right-hand side of one picture may sometimes be seen in the left-hand side of another.

1
Political Economy *and* Christian Theology

"Political economy" and Christian theology coexisted happily in the intellectual world of eighteenth-century Britain. During the nineteenth century they came to be seen as incompatible, even mutually hostile. In the twentieth century they went their separate ways and are no longer on speaking terms. These essays are snapshots of the history of this estrangement.

They also illustrate a profound intellectual mutation characteristic not only of Britain but of many other industrialized countries. Until well into the nineteenth century the political thinking of all save a tiny minority of Europeans and Americans was deeply influenced, if not wholly formed, by the theology of Western Christianity. By the end of the twentieth century political economy, latterly called "economics," had almost entirely replaced theology as the normative framework of political discussion.

How this came to be, and how it is related to the estrangement of economics from theology, are larger questions than this book can answer. But it may at least serve to reassert the central importance of this mutation in the intellectual history of modern political thought. And it may make some modest contribution to a much-needed explanation. Some of the essays throw out hints or suggestions. One or two rule out some obviously wrong answers.

In this Introduction I shall begin by saying a little more about the ousting of theology by economics. The following section reviews some of the more obvious explanations, and the third considers the attention given to theology and to political economy in recent intellectual historiography. In the last section I attempt to explain how and why the two should be combined.

1.1 Economic problems of a "Christian polity"

"Christianity, I need scarcely say, is recognized as part of the Law of Canada as it is of the British Empire." So declared the Minister of Justice of the Dominion of Canada in the Committee of the Whole on the Lord's Day Bill of 1906.[1] There was no dissent. Though this doctrine was struck down in England eleven years later the feeling persisted in many parts of the English-speaking world that society was "Christian" or at the very least, God-fearing. As late as 1960 the Parliament of Canada affirmed "that the Canadian Nation is founded upon principles that acknowledge the supremacy of God."[2]

Formal commitments of this kind were accompanied by a large literature, not yet extinct, devoted to various aspects of the study of *Christian Polity*[3] and *The Return of Christendom*[4] Shortly before the Second World War the most eminent member of the Christendom group, the American poet T. S. Eliot, was invited by the Master and Fellows of Corpus Christi College, Cambridge, to deliver public lectures on this theme. First published in October 1939 as *The Idea of a Christian Society*, they were reprinted eight times by 1954. J. M. Keynes's *General Theory of Employment, Interest and Money* received eight printings between 1936 and 1951. Far exceeding either in circulation, Archbishop William Temple's *Christianity and Social Order* (1942) was one of the most influential books of its age. "The impact of William Temple on my generation was immense," reported Edward Heath, late Prime Minister of the United Kingdom, in his Foreword to the 1976 reprinting:

> ... he propounded with lucidity and vigour his understanding of the Christian ethic in its application to contemporary problems which engrossed us all. Perhaps it was this, more than anything, which appealed to a generation anxiously searching for principles by which to govern its conduct, both public and private. In short, for us he was relevant – when very few others were.[5]

An Oxford economist writing in 1960 described Temple's book as 'one of the foundation piers of the Welfare State.'[6]

There are still living in many parts of the United States, Britain and the old dominions survivors of a generation brought up before 1939 who remember clearly the culture of what was almost universally described in those places as "a Christian country," and which in regions such as Ontario or the South Island of New Zealand lasted until well into the 1950s. But few traces now remain, and in retrospect it appears that the great mutation had begun long before: some time between Walter Bagehot's years as editor of *The Economist* (1860–77) and the First World War at the latest. Here too, the story of the Canadian Lord's Day bill of 1906 is instructive.

"In a Christian community," the Roman Catholic Archbishops of Canada had asked the Select Committee, "is it not preferable that the production of material wealth ... should be restricted, rather than the nation should suffer by the degradation and de-christianizing of its members?"[7] Despite overwhelming support for the bill in the lower house, Parliament eventually answered "No." Representatives of manufacturing and trading interests pointed out to that to shut down blast furnaces and oil refineries every seventh day would be ruinous. Railroad and shipping companies refused to immobilize their trains and ships in mid-journey for twenty-four hours on the Lord's Day. Amendments and compromises were duly made, and the teeth drawn from the bill.[8] By the end of the nineteenth century industrial development, not only in Canada but throughout in the capitalist world, had so raised the opportunity cost of a "Christian polity" that legislators were unwilling to incur that cost on behalf of their constituents. Those among them, therefore, who cared for ideas were increasingly willing to attend to a secular, utilitarian rhetoric gradually constructed by economists and others over the previous hundred years, which purported to justify that unwillingness. Arguments for the maintenance or modification of public policy based on political economy soon came to trump arguments based on Christian theology.

1.2 Explanations of the supplanting of theology by economics

None of the foregoing is intended to suggest that ideas are only the froth racing past on the turbulent surface of "real" or "objective" or "material" "forces." Yet material circumstances do matter. Marx[9] and his followers may have exaggerated the extent to which "the mode of production in material life determines the general character of the social, political and spiritual processes of life." But it would be foolish to deny that the things we have to do to earn our living, and the general economic character of the society in which we do those things, may affect the way we think about life in general.

By means of industrialization, population growth and concentration into great cities, the omnicompetent "bourgeoisie" have not only "rescued a considerable part of the population from the idiocy of rural life:"[10] they have produced high and rising incomes for most, ever-improving medical care, greatly extended life expectancy, and a global network of trade and communications. It seems unlikely that the Christian theological framework of political discourse would have been so very largely displaced by economic considerations if we still dwelt in *The World We Have Lost* consisting largely of rural villages each of a few hundred souls,[11] with eighteenth-century diet, sanitation and dentistry. A powerful side-effect of population growth and mobility moreover, has been a growing need for religious pluralism in order to keep the peace between contiguous neighborhoods of mutually hostile immigrants, unbridgeably divided by "culture" based on creed. In the United States, a far higher proportion of the population is reported as practicing religious (specifically Christian) beliefs than in any other English-speaking country. Yet paradoxically that nation is the most secular, both in its political culture and in the private attitudes and beliefs of its citizens.[12] For when nothing is allowed to count politically save those things on which all are agreed, the merely economic wins by default.

But "internal," purely intellectual considerations, though not as important perhaps as Keynes supposed,[13] are part of any story in which what happens in human societies is the outcome of individual, rational decisions. The objective circumstances did not change by magic. They changed because men and women sought purposefully to realize their private ends. Those ends were shaped, at least in part, by the conversation in educated circles relating to religious, philosophical and political questions. It is "the intellectual historian's characteristic role" to "eavesdrop" upon such conversation, in awareness "that what is overheard has also to be ... "translated" for the benefit of the contemporary reader."[14] Such "translation" is only possible in awareness of the context which gives meaning to the utterances of those we try to overhear and which may provide at least some clue to their intended "illocutionary force."[15] Context includes not only the external, objective circumstances but also the inner logic of the conversation itself. Whatever else Adam Smith was intending to do in writing *Wealth of Nations*, in part at least he was correcting and generalizing François Quesnay's conception of the surplus. By attending to that part of the conversation seemingly shaped by criticism of other participants' ideas, the intellectual historian may contribute – however narrowly and humbly – to an explanation of what happened and why, which is the point of all historiography.[16]

An "internal," intellectual–historiographic account of the replacement of the rhetoric of Christian theology by that of political economy has yet to be produced; though an economist has lately suggested that modern economics is actually the current manifestation of Western religion and is the lawful heir of a putatively defunct Christianity: a *Reaching for Heaven on Earth* that grew out of the liberal-protestant "social gospel" in nineteenth-century America.[17] But at least a great deal has been written about the intellectual downfall of theology during "the Enlightenment" since Sir Leslie Stephen's (1881) pioneering account of eighteenth-century ideas, and a textbook orthodoxy has been defined:

> Modern science allowed us to dispense with the redundant hypothesis of a divine creator. Modern philosophy put paid to teleology, miracles and theodicy. The world was thus made safe for Comte's "social science," which John Stuart Mill could then import and domesticate in backward, reactionary Britain.[18] By late-Victorian times a fully rational politics was at last possible. The Fabian Society and the London School of Economics were up and running.

This edifying tale is still told in some circles, but it begins with a caricature of eighteenth-century thought that few historians now find useful. "Why then," my reader may well ask, "do you yourself use the word "Enlightenment" in your title?" The answer is simply that this is what many prospective readers, even among the learned, have come to accept, with or without the absurdity of a "Project" tacked on. I owe it to them to declare that I find the concept of "the Enlightenment" of no more taxonomic value than such other textbook labels as "Modern" times, "the Renaissance," "the Reformation," or "the Industrial Revolution." When I do use it, as in Chapter 2, I mean to imply – even if I do not always employ – inverted commas.

It is especially important to be aware that eighteenth-century science and philosophy had very little, if anything, to do with the replacement of theology by economics in the late-nineteenth century. Only in Roman Catholic countries, specifically France, did science and philosophy tend to undermine theology. In protestant countries, above all in England and Scotland, science and philosophy were successfully co-opted in support of Christian orthodoxy. David Hume was an outlier. His attack on Design was supposed to have been adequately dealt with by William Paley (1802), thanks to whom the teleological argument remained the rational foundation of Victorian religion down to the

1860s.[19] His criticism of miracles was turned against him by Richard Whately (1819), who showed that if we apply Hume's epistemological criteria to recent history we should be obliged to conclude that Napoleon Buonaparte could never have existed. Insofar as the decline and fall of theology is more than merely a response to economic and demographic changes, the intellectual reasons are to be looked for in the second half of the nineteenth century.

It was the latter part of the decade of the 1860s, Keynes[20] conjectured, "which will ... be regarded by the historians of opinion as the critical moment at which Christian dogma fell away from the serious philosophical world of England, or at any rate of Cambridge." Two celebrated controversies produced this result.

J. S. Mill's devastating attack on Henry Mansell's Bampton Lectures of 1858[21] dragged out into the open the "Problem of Evil" that eighteenth-century apologists and early nineteenth-century natural theologians had largely succeeded in sweeping under the rug. If God is all-powerful then He must be wicked. If His idea of wickedness and "evil" differs from our own – which is what Mansell had suggested – then it is morally repugnant to worship Him. There is and can be no satisfactory theodicy. Christianity, in common with other monotheistic religions, is and must be intellectually incoherent.

The publication of Darwin's *Origin of Species* in 1859, closely followed by Herbert Spencer's *First Principles* (1862) did the damage to teleology that Hume's *Dialogues* had failed to effect, bringing the entire edifice of Paley's theological and social theory to the ground. Paley's *Moral and Political Philosophy* (1785) had been continually reprinted on both sides of the Atlantic from its first appearance in 1785 down to the late 1860s. Until the American Civil War it was the most widely used text-book of politics in the United States.[22] *Moral and Political Philosophy* is based upon a "theological utilitarianism" that finds its justification in holy scripture (which Paley defended in *Horae Paulinae* [1790] and *Evidences of Christianity* [1794]), the ultimate credibility of which rests upon *Natural Theology* [1802], specifically the Argument from Design. Hume's Scotch philosophy had merely bored the pragmatic and unmetaphysical English. But something tangible and visible – something *scientific* – was a different matter. By seeming to show the absence of "Design" in the appearance of "Nature," the theory of biological evolution deprived Victorian religion of its chief intellectual support; and thereby deprived Victorian *politics* of its grounding in Christian theology. All of a sudden Paley's reputation slumped. His books were reprinted no more. "Metaphysical agnosticism,

Evolutionary progress, and – the one remnant still left of the intellectual inheritance of the previous generation – Utilitarian ethics joined to propel the youthful mind in a new direction."[23]

1.3 Intellectual historiography of Christian theology, and of political economy

Religion and theology have never been altogether neglected in English intellectual history. The first large-scale history of ideas in Britain, Stephen's *English Thought in the Eighteenth Century* (1881) was chiefly concerned with the effect of Deism upon Christian orthodoxy and the auxiliary, "indirect" part played by "philosophy" – conceived schizophrenically as somehow detached from, and independent of, theology – in bringing about "intellectual chaos."[24] Lord Acton (1968) and Neville Figgis (1907, 1922) wrote extensively about continental European political ideas in the context of medieval and early modern Catholicism. Basil Willey's series of "background" studies for seventeenth, eighteenth and nineteenth-century English literature[25] do ample justice to the Christian theological context of that literature. Sir Ernest Barker's *Social and Political Thought in Byzantium*[26] describes "a climate of thought" in which "theology naturally became the queen regnant of all knowledge."

More recent authorities in what one might label the "Cambridge" tradition of intellectual history, though heavily concentrated on the evolution of Anglophone political-philosophic thought, have recognized the importance of Christian theology at least for the early modern period. This is most evident in the treatment of John Locke's political theory by Laslett (1960), John Dunn (1969), Richard Tuck (1979), and James Tully (1980). It is "thoroughly unrealistic and occasionally unhistorical" to regard Locke as a bourgeois apologist for the doctrine of "possessive individualism."[27] His theories of property and political obligation can only be understood in light of his assumption that Man is "God's workmanship"[28] and in the context of seventeenth-century religious controversy, in which Locke joined with a will for more than three decades. Quentin Skinner's[29] grand "survey of the transition from the medieval to modern political theory," though pointing to "the development of a modern, naturalistic and secular view of political life," shows the central importance of theological doctrines old and new in fifteenth and sixteenth-century political struggles. J. G. A. Pocock (1980, 1985) has pioneered a revisionist view of the "Enlightenment" in Britain that recognizes the intellectual leadership of Establishment theologians. This view has lately been developed by Brian Young, who has shown that "the boundaries between England's Enlightenment and Counter-Enlightenment are decidedly permeable."[30] Attention has also been directed to the theological ideas of English Protestant Dissent and their relation to the "language of liberty."[31]

Yet much (though by no means all) of this valuable literature retains some flavour of Leslie Stephen's tendentious assumption that the "modern, naturalistic and secular view of political life" was in place by the end of the eighteenth century at the latest: or as Stephen himself put it with heroic anachronism, that "Hume's scepticism completes the critical movement of Locke. It marks of one the great turning points in the history of thought."[32]

This is not a matter of secularist bias, whether or not such bias may have existed in the minds of some. Sir Ernest Barker, for example, had no anti-theological axe to grind. But his widely used textbook on *Political Thought in England, 1848 to 1914* (1928) contains only few and scattered references to Christianity and theology. It is rather, a matter of analytical method.

Intellectual historians, taking a leaf out of the economists' book, have postulated perfect rationality in their agents and perfect information in the implicit social system. Immediately a new "thought" is discovered and found to be true, it must of course be instantly adopted and universally disseminated. Therefore "turning points" exist and may be identified. The anomalous fact that most highly educated and intelligent men and woman in Britain and elsewhere continued both to accept and further develop the theological presuppositions of a Christian society for two more centuries after Hume's "great turning point" (and almost a century after Keynes's "critical moment") is somebody else's business to explain.

What one might loosely characterize as a "social history of ideas," which attempts to deal at a lower level of abstraction with the messy and complicated reality of their reception, transmission and application by the many, has begun to appear in response to this need. The most distinguished example, perhaps, is Boyd Hilton's *The Age of the Atonement* (1988) which traces in great detail the varied ways in which a heterogeneous cast of "evangelicals" influenced, or failed to influence, the climate of social and economic opinion in the nineteenth century. Meanwhile conventional, textbook versions of British political history before 1832 have been challenged by Jonathan Clark (1985) who has argued persuasively that Britain remained a "confessional state" at least until that date. His claim that theology still mattered a lot in this period has been enlarged by other historians.[33] The continuing importance of theology in Victorian political and intellectual culture has been attested to by many recent authors in a wide variety of ways.[34]

When we turn from the treatment of religion and theology to that of political economy the story is quite different. In part the reason for this is obvious.

Religious beliefs and theological ideas determined the intellectual culture of Christian Europe. They shaped and pervaded political controversy in the sixteenth century, which is usually taken as the *terminus a quo* of that so-called "Modern" period which engrosses the attention of most English-speaking historians and their readers. However strange and uncongenial it may now appear, there is no escape from theological wrangling if one wishes to understand how present-day political ideas came into being. But although "economic thought" of some kind is as old as civilization, *œconomie politique* was obscurely born in seventeenth-century France as a humble and humdrum set of recipes for managing the realm as a privately owned estate of Louis XIV. It is true that by the end of that century Pierre de Boisguilbert had elucidated its scientific content by situating it in the Augustinian theodicy he had learned as a student at the Jansenist college of Port-Royal.[35] But few were aware of his work and it was not until Adam Smith (1976) produced *Wealth of Nations* nearly a century later that even a handful of the educated public outside France began to pay much attention to the new science. So far as the Anglo-American world is concerned at any rate, it was only after Dugald Stewart's Edinburgh lectures of 1800–1801 and the first publication of the *Edinburgh Review* in 1802 that "political economy" started to appear on the agenda of political debate.[36]

There is also a less obvious reason, arising from the nature of political economy itself. For although the field of inquiry denoted by that title is, and was intended from the first to be, deeply involved in the art of government, it is based on a core of scientific theory that is abstract, technical, difficult to master (or even understand) without much disciplined application; and therefore repulsive to all save its practitioners. Sir Leslie Stephen[37] did include a long and tolerably well-informed chapter on political economy in his history of eighteenth-century thought that does justice to Smith's French predecessors, though seemingly unconnected with "thought" of any other kind. But after Stephen, intellectual historians, many of whom appear to have their natural habitat in political philosophy, have until lately kept at a prudent distance.

Where real historians have feared to tread, economists themselves – or at any rate the "incompetent or retired" practitioners[38] among whom the author of this book must be numbered – have rushed in to fill the vacuum. For more than a century an "in-house" literature of what is now called "History of Economic Thought" has thriven, never more so than today despite the fact that in many American and British universities undergraduate students of economics are no longer required to learn the history of their discipline. Some of this literature is of great distinction: James Bonar (1885) on Malthus, Keynes's (1972, part II) brilliant "Lives of Economists," Joseph Schumpeter's (1954) classic *History of Economic Analysis*, and Jacob Viner's (1972, 1978) studies of the relation between religious and economic ideas come immediately to mind. But few economists now evince the literary genius of a Keynes, or even the encyclopedic learning of a Schumpeter. Moreover much of what gets written, both in the textbooks and also in the journals, is not really history at all.

Paul Feyerabend[39] analysed the process by which "the *history* of a science becomes an inseparable part of the science itself ... essential for its further *development* as well as giving *content* to the theories it contains at any particular moment." But it is clear from Feyerabend's own examples[40] that by "history" he means the pseudo-history useful only to scientists, and which bears much the same relation to genuine history as a topological diagram of the London Underground bears to a true, scale map with all the usual features surveyed in. I have elsewhere defended the proper use of this pseudo-history – labeled "History of Economic Analysis" (HEA) to distinguish it from "Intellectual History" (IH) – and argued that although it is not a true map it may be useful as a cartographic tool.[41]

The true history to which HEA might be a useful adjunct is of course a genuine intellectual history of political economy that obeys Skinner's[42] methodological canon: "no agent can ... be said to have meant or done something which he could never be brought to accept as a correct description of what he had meant or done." Because IH (of political economy), like HEA, can only be successfully attempted by those who are inward with the arcane discipline of economics, its practitioners have been few. Each of the four major authors identified above, especially Keynes and Viner, are early examples. But since their day eminent economists have found less and less time to reflect upon the history of their science. On those few occasions when – as in the examples of Paul Samuelson (e.g. 1977), Takashi Negishi (1989), Michio Morishima (1989) and Robert Dorfman (e.g. 2001) – they have felt able to take a holiday from the intricate mathematical technicalities of modern economic theorizing, they have generally preferred to translate the work of their pre-mathematical forerunners into their own theoretical idiom; which is the most sophisticated and useful form of HEA.[43]

However, simply because there are now far more economists alive and well than at any previous time, more specialization and division of labor is possible. There is at last a very small, but growing body of genuinely historical scholars who are also fully trained as economists. Senior among these is Donald Winch, himself a pupil of Viner, most of whose books over the past four decades deal with the related themes of "classical political economy," Adam Smith and the "Scottish Enlightenment."[44] The latest in this series, *Riches and Poverty: An Intellectual History of Political Economy in Britain, 1750–1834* (1996), may be taken to illustrate the state of the art in this field.

1.4 Putting the two together

Few as those historians of ideas still are who have been trained in economics, those trained both in economics and in theology are fewer still. Viner's work is almost *sui generis* and even he acquired his detailed knowledge of church history and theology only late in life. Yet a genuine and complete intellectual history of Anglophone political ideas over the past three centuries requires not only a sensitive awareness of the part played by each; it requires an informed understanding of the relation between these two fields of inquiry and of the way in which that relation has changed.

It may not be irrelevant to remark that although political economy and Christian theology have been intimately related, especially during the eighteenth and nineteenth centuries, they are in principle distinct enterprises. Each is able to get on quite well without the other. With due caution against extending the analogy too far, we might illustrate both the intimacy and the distinctness with the example of music.

Human beings made music long before the coming of Christianity. They make music under all other religions. And after the emergence in Europe of a dominant, all-embracing theological framework of civilized life, folk music persisted in relative independence of ecclesiastical control. Yet for more than a thousand years, both in Byzantium and in the West, music was an essential – or at any rate very important – part of the liturgy. The church paid the piper, and to a very large extent the church called the tune. Some forms of music were prescribed, some tolerated and some forbidden. The authorities occasionally intervened to restrain the natural tendency of composers to elaborate their work for purely musical reasons at whatever cost in intelligibility or decency. But in tolerant seasons the church, as chief or only patron, fostered the development of an international "high culture" of music in places as diverse as St Mark's Venice, the *Thomaskirche* at Leipzig, and the Chapel Royal of the Tudor and Stuart courts. Theological considerations were never absent, for the art of music like all other art existed to glorify God in the orthodox manner. Hence for example the interval of the augmented fourth, *diabolus in musicum,* was forbidden because its painfully audible subversion of diatonic harmony was evident mischief of the Great Enemy of divine order.

As large-scale patronage passed from church and court first to the nobility, then to the bourgeoisie and finally to the state, much of this changed. Yet composers continued to set liturgical and other sacred texts, and many still do today when properly paid. The most sophisticated occasionally recall the ancient prohibitions with powerful symbolic effect. Benjamin Britten's *War Requiem,* which superimposes anti-war poems of Robert Owen on the *missa pro defunctis,* opens with the augmented fourth and continually exploits it in order to portray the diabolical nature of modern war. At three crucial points in the work a bi-tonal sequence employing that interval is resolved upon a major triad in token of reconciliation and peace in Christ. Meanwhile, popular music goes its way, as it always has, without much regard if any to high art.

We may view the changing relation between political economy and theology, I suggest, in somewhat the same way. Whatever the religious and political regime, those who have much to gain and much to lose have always thought economic thoughts, and always will. Whenever circumstances permit speculative and scientific inquiry, a few superior minds will ponder the "secret concatenation" by which the getting and spending of the multitude is unwittingly coordinated. But during the centuries of "Christendom" the church regulated as much of that getting and spending as seemed necessary to achieve certain theologically determined ethical goals; and some of its best brains began the investigation of economic processes as an adjunct to casuistry. As the power of the church declined after the sixteenth century the ecclesiastical regulation of economic life fell into desuetude. And by the beginning of the eighteenth century the central scientific conception of modern economics – "spontaneous order" in a competitive, market economy – had emerged from the womb of Catholic theology. One hundred years later political economy began to go its separate way. But economists and theologians remained in conversation throughout the nineteenth century, and as late as Jevons and Wicksteed some of the former at any rate tried hard to maintain the compatibility of political economy with divine order. Even at the present time all hope of reconciling the two disciplines is not yet extinguished. Meanwhile economic thought of a popular kind, the conventional wisdom of bankers and businessmen embodied in the financial pages of the newspapers, has gone on pretty much as always, gloriously indifferent to the lofty speculations of a Samuelson or an Arrow.

As in the case of music, the relation of Christian theology to economic thought, even to the latter-day variety we call "political economy," has been long-lasting, profoundly influential and still not quite at an end. But as in the case of music, that relation is contingent and accidental. There would still be some music and some political economy, however crude and feeble, even had Christendom never existed. What then is the point of a history of the relation between the two, and how might such history be attempted?

If I am right in my belief that economics has gradually ousted theology from the discourse of politics over the past two centuries, and if I am also right in thinking that this is the most significant intellectual mutation ever to occur in the political life of the civilized world, then the point of such a history is obvious. We want to know when it happened, why it happened and how it happened.

Objective, material circumstances are clearly of high importance in any such explanation. But there is plenty of work for the intellectual historian to do. To begin with, we must explain the drastic decline in the authority of Christian theology among the learned in late-Victorian times; though this can be done with little if any reference to political economy. More importantly, we have to explain why it was *political economy* that replaced theology. This will take us back to the beginning of the nineteenth century when, according to Arnold Toynbee, "the bitter argument between economists and human beings" had its origin.[45] Purely intellectual changes in the scientific content of political economy profoundly changed the hitherto benign and maternal relation between theology and economics, eventually causing the former to indict the latter for juvenile delinquency. A complete history of this episode would attend to the conversation economists were having among themselves about such dull and tedious matters as rent, diminishing returns and the effect of technical progress on the distribution of income. It would also attend to the conversation economists were having with the theologically minded, both among their own number and elsewhere, about the implications of these strange new ideas for our view of human life on this earth and our understanding of divine providence. The full story would include, among other things, an account of the mid-Victorian attack on a reformist alliance of political economists and Evangelical philanthropists launched by a succession of literary Romantics, some of whom believed themselves to be informed by at least approximately Christian considerations.

That story has yet to be written. This book is nothing but a straw in the wind. Apart from advertising the need for a complete history, its chief value may lie in the more-or-less chronological series of snapshots or sketches of successive stages in the replacement of theology by economics.

Chapter 2 suggests a new explanation for the puzzling fact that "Enlightenment" in England and other protestant countries produced none of the anticlerical and anti-theological polemic characteristic of its effect in France. Chapters 3 and 4 explore and explain (with some overlapping) the theological framework of political thought in England during the "long eighteenth century." Chapter 5 revisits Leslie Stephen's grievously inaccurate account of intellectual life in Cambridge after the failure of the Feathers Tavern petition.

Political economy makes its first appearance in chapter 6, in which *Wealth of Nations* is read as natural theology. Chapter 7 describes in outline the sudden separation of political economy from Christian economy in the aftermath of T. R. Malthus's first *Essay on Population*. If the argument of this chapter survives criticism, it might become an important element in a complete history of the replacement of theology by economics. Chapter 8 shows how the separatist methodology of modern economics emerged as an unintended consequence of the attempt by Whately and Nassau Senior to protect the Establishment from assault by the Philosophic Radicals. And Chapter 9 explains how the theologically flavored political economy of Malthus and Thomas Chalmers undermined, many decades in advance, a key component of Karl Marx's explanation of the inevitable collapse of capitalism.

Chapters 10, 11 and 12 attend to various aspects of the "Catholic Social Teaching" inaugurated in 1891 by Leo XIII's famous encyclical, *Rerum Novarum*. In Chapter 10 I examine the well-known claim that Leo XIII smuggled Locke's theory of property rights into papal doctrine. Chapter 11 considers the intellectual context of *Rerum Novarum* and attempts to explain why that celebrated proclamation contains both an attack on socialism and a grave warning against capitalism. Chapter 12 considers *Centesimus Annus*, John Paul II's centenary tribute to the pioneering work of his great predecessor. It is shown that there is a serious conceptual dissonance in that work, for an altogether novel affirmation of the benefits of capitalism and a market economy is superimposed on an underlying stratum of traditional anti-capitalist rhetoric. In each of these chapters, which like Chapters 3 and 4 contain some unavoidable overlapping, I have tried to understand and appreciate the belated efforts of the most powerful and influential of all Christian churches to come to terms with the new science of political economy.

In Chapter 13 I turn to established protestant churches, particularly the Church of England in which, partly because of a distinguished succession of theologically minded economists, a *modus vivendi* was reached with political economy a century-and-a-half before the Church of Rome. Towards the end of the nineteenth century however, Anglican theologians and prelates had begun to assume the anti-economistic bias of the Romantics and Christian socialists in their midst, and twentieth-century attempts to combine economics and theology in normative social theory have created serious intellectual difficulties.

As Chapters 2 to 5 contain no reference to political economy, so the final Chapter 14 makes no mention of theology. For from the vantage point of 2000, which is the terminus *ad quem* of that study, economics fills the political landscape and Christian theology has become all but invisible. With hindsight we can see that the process of replacement began long ago, by 1776 at the latest. For a period of roughly eighty years (c. 1870–1950) the inner logic of economic theorizing worked to create a widespread assumption among the educated that a "scientific" socialism could replace capitalism peacefully, retaining all of its economically benign features whilst avoiding its waste, inefficiency and social injustice. During the last fifty years, in abstraction from external circumstances and purely as a result of continuing criticism, refinement and modification, virtually the whole of this pleasing vision has been abandoned.

2
Why the English "Enlightenment" Was Different

At one time "the" Enlightenment was taken by many to have been deliberately, or at any rate implicitly, hostile to religion in general and to Christianity in particular. As late as 1967 this supposedly unitary phenomenon could still be described by a well-known author as "the rise of modern paganism."[1] Yet as long ago as 1932, Ernst von Cassirer had warned us that "if we attempt to test this traditional view by concrete historical facts, we soon come to entertain the gravest doubts and reservations."[2] In Cassirer's judgment, the "strongest intellectual forces of the Enlightenment do not lie in its rejection of belief but rather in the new form of faith which it proclaims, and in the new form of religion which it embodies."[3] And by 1981 at the latest it had become clear that the evident irreligion of the French *philosophes* was quite untypical of the educated classes in protestant nations of the Atlantic world. For "in England, Scotland, Germany, Holland and English North America, 'enlightenment' found a home *within* the Christian churches."[4]

It is widely accepted that an important part of that projected renewal of faith of which Cassirer wrote was an attempt to purge religion of "superstition" and "enthusiasm." This was understood at the time as a typical act of reformation. "That *the corruption of the best things produces the worst,* is grown into a maxim, and is commonly proved, among other instances, by the pernicious effects of *superstition* and *enthusiasm,* the corruptions of true religion."[5] The pernicious effects of "enthusiasm" that Hume had in mind were the many assaults upon law and order in sixteenth and seventeenth century Europe committed by "furious and violent" religious sectaries: "the *anabaptists* in GERMANY, the *camisars* (sic) in FRANCE, the *levellers* and other fanatics in ENGLAND, and the *covenanters* in SCOTLAND."[6]

The pernicious effects of "superstition" were chiefly the religious persecutions which it licensed: the Inquisition, the wars of religion, St Bartholemew's massacre; and Calvin's betrayal of Servetus which echoes in many an Enlightenment text.

Hume was silent about "true religion." It has generally been supposed that "the enlightened" were at best Arian or Socinian; more probably deist like Voltaire, or infidel like Holbach and Diderot. Yet many English thinkers of the first rank remained in communion with their Church throughout the century, and a recent study has warned us that "the boundaries between England's Enlightenment and its Counter-Enlightenment are decidedly permeable."[7] It is my object in this paper to throw new light upon this characteristic feature of the English Enlightenment, described by Margaret C. Jacob and J. G. A. Pocock as conservative, clerical and "magisterial."[8] Was the English case peculiar because "the strong modernising drive that we identify with the Enlightenment was integral to the preservation of the establishment in state and Church"? Knud Haakonssen has lately summarized Pocock's conclusions in this way, and there can be no doubt that the evolution of the English constitution after 1688 does much to explain the differences between France and England in the eighteenth century.[9]

It is not my purpose to contest this view. I seek only to supplement it with another explanation too obvious to have attracted much notice. During the *ancien régime* both France and England were confessional states.[10] But France was papist and England protestant. In each country the exclusion of superstition and enthusiasm was central to "the Enlightenment project." In each country the eucharistic doctrine of transubstantiation was a paradigm of superstition. In papist France, the clergy of the Church of Rome was inhibited from wholehearted participation in Enlightenment by formal commitment of its members to that doctrine. In protestant England, the clergy of the Church of England was free from such constraint. Even the most orthodox cleric could join with a clear conscience in the extirpation of superstition and enthusiasm. The former was identified with popery: the latter with those obnoxious sects which had endangered the peace of the Church in the previous century, and more recently, with Methodism.

The differential consequences for Enlightenment culture are apparent. In France, the intellectual climate was generally anti-clerical and increasingly anti-Christian. In England, the intellectual climate was created almost entirely by the philosophical, theological and political concerns of the most powerful minds – both lay and clerical – within the established church. The seeming exception of Rational Dissent can be ruled out. The characteristic thought of that small but important *élite* was constructed, to a large extent, as an ideological defence against the theological justification of Establishment polity.[11] Edward Gibbon is a solitary outlier: and even he "insists on considering himself *bon protestant – pratiquant, si non croyant*."[12]

In what follows I shall first review the use of the term "enthusiasm" in eighteenth-century polemic. Next, I shall do the same for "superstition." Finally, I shall show how the attempt to exclude each could be regarded in that age as wholly consistent with Anglican orthodoxy.

2.1 Enthusiasm

Samuel Johnson defined three senses of "enthusiasm" in his *Dictionary*:[13] first, as "A vain belief of private revelation; a vain confidence of divine favour or communication;" secondly, as "Heat of imagination; violence of passion, confidence of opinion;" and thirdly, as "Elevation of fancy;" "exaltation of ideas." The first sense is illustrated from John Locke: "Enthusiasm is founded neither on reason nor on divine revelation, but rises from the conceits of a warmed or overwheening brain."[14] It is this primary sense in which the word is used in Enlightenment religious discourse, though the second and third meanings sometimes seem to overlap.

Louis de Cahusac's long article on *ENTHOUSIASME* in volume V of *L'Encyclopédie*[15] treats the word as a term of Philosophy and Belles Lettres (as in Johnson's third sense) and though acknowledging its more common usage (as in Johnson's second sense) as "*une espece de fureur*," attempts to rescue it for the psychology of artistic creation. However this exposition is followed by Mallet's shorter article on *ENTHOUSIASTES* as a term of Ecclesiastical History originally applied to various ancient heretics; according to Theodoret, "*parce qu'étant agités du démon, ils croyoient avoir de véritables inspirations:*"

> *On doûne encore ajourd'hui le nom* **d'Enthousiastes** *aux Anabaptistes, au Quakers ou Trembleurs, qui se croyent remplis d'une inspiration divine, & soûtiennent que la sainte Ecriture doit être expliquée par les lumieres de cette inspiration.*

This usage corresponds with that used in most English writing of the period. However, the singular form, *ENTHOUSIASTE*, when separated from its sense as in Belles Lettres, "*se prend souvent en mauvaise part pour désigner un fanatique.*" And when *FANATIQUE* and *FANATISME* are consulted, we discover that "*Le Fanatisme n'est donc que la superstition mise en action.*" Alexandre Deleyre was the author of this famous article,[16] and unlike the pious and orthodox Mallet, was one of the most violently anti-religious of the Encyclopedists. He eventually became a Jacobin, voted for the execution of Louis XVI, and survived the Terror.[17] Deleyre's willingness to blur the boundary between enthusiasm and superstition may be regarded as a part of his polemical strategy and untypical of the Encyclopedists as a whole.

Hume had noted that whereas "superstition is favourable to priestly power," enthusiasts have always "expressed great independence in their devotion; with a contempt of forms, ceremonies and traditions."[18] In papist countries therefore, enthusiasm was likely to be rare: in protestant countries, more abundant. The "camisars in France" were protestant: "the *enfants terribles* of the Huguenot party" whose violence resembled that of the Scotch covenanters. They were exiled from France by 1710 and known in England – where they were repudiated by the local Huguenots – as "the French Prophets."[19] And though Ronald Knox devoted a considerable portion of his pioneering study to the Jansenists and Molinists, each party of which was a movement within the French Church, only the Jansenist convulsionaries of Saint-Médard exhibited that mass hysteria and contempt for decency and order supposed by the enlightened to be characteristic of enthusiasm.[20] In Britain however, outbursts of enthusiasm had been frequent throughout the seventeenth century, and by 1739 had reappeared as a consequence of the field preaching of Wesley and Whitefield.[21] Therefore, notwithstanding Bishop Lavington's *Enthusiasm of the Methodists and Papists Compared,*[22] the general tendency of English writers was to employ the term to denote the pathological excesses only of protestant zeal since the Reformation.

The earliest attempt by an Anglican writer to investigate the theological claims of enthusiasm is that of the *émigré* son of the illustrious Isaac Casaubon, Meric Casaubon (1599–1671), who obtained preferment in the Church of England in the 1620s.[23] Casaubon's *Treatise Concerning Enthusiasme* (1655) attempts to distinguish true or "supernaturall" enthusiasm – genuine possession by supernatural powers – from "naturall" *Enthusiasme,* meaning "an extraordinary, transcendent, but natural fervency, or pregnancy of the soul, producing strange effects, apt to be mistaken for supernaturall."[24] Writing during the Interregnum whilst deprived of his benefices, Casaubon was circumspect in criticizing the enthusiastic propensities of many of Cromwell's supporters, but he did include a passage in the second edition (1656) which questioned the

Quakers' claim to genuine inspiration.[25] Henry More's more popular *Enthusiasmus Triumphantus*,[26] which like Casaubon's work acknowledged the possibility of a true enthusiasm, appeared in the same year.

After Casaubon's and More's the most important treatment of enthusiasm by English writers are those by Locke (1695) and Shaftesbury.[27] Though the former makes no direct allusion to the English enthusiasts of the time, its readers would have recognized the unflattering account that Locke supplied as typical of the first generation of Quakers, Levellers and Independents. Locke was chiefly concerned with epistemology in *Human Understanding*, and chapter xviii of Book IV deals with faith and reason as grounds of human knowledge:

> *Nothing that is contrary to, and inconsistent with the clear and self-evident Dictates of Reason, has a Right to be urged, or assented to, as a matter of Faith, wherein Reason hath nothing to do.* Whatsoever is divine *Revelation*, ought to overrule all our Opinions, Prejudices, and Interests: Such Submission as this of our *Reason* to *Faith*, takes not away the Land-marks of Knowledge...

It is essential to distinguish the two, for *If the boundaries be not set between Faith and Reason, no Enthusiasm, or extravagancy in Religion can be contradicted.*[28] The chapter which follows treats of enthusiasm itself as a mere pretence at knowledge. Claims to divine knowledge must be tested by "unerring Rules to know whether it is from GOD or no."

> Either "the Truth imbraced is consonant to the *Revelation* in the written word of GOD; or the Action conformable to the dictates of right *Reason* or Holy Writ".[29]

Locke's former pupil, the third Earl of Shaftesbury, wrote his *Letter Concerning Enthusiasm* in 1708, in the immediate aftermath of the Camisard disturbances in London. Like More and the other Cambridge Platonists, but unlike Locke, Shaftesbury was more interested in the second and third senses of Johnson's definition than in the first. He claimed "in some measure" to have "justified enthusiasm" and subscribed himself "your enthusiastic friend."[30] Enthusiasm "is wonderfully powerful and extensive;" "the hardest thing in the world to know fully and distinctly;" "even atheism is not exempt from it." "Nor can divine inspiration, by its outward marks, be easily distinguished from it. For inspiration is a real feeling of the Divine Presence, and enthusiasm a false one."[31] Elsewhere, however, Shaftesbury acknowledges "a kind of enchantment or magic in that which we call enthusiasm."[32] And both in *The Moralists* (1709) and in his answers to criticism of the *Letter* in *Miscellaneous Reflections* (1711) he argues for "a fair and plausible enthusiasm" in the "transports of poets, the sublime of orators, the rapture of musicians."[33] This sense of "enthusiasm" is congruent with that which concerned Cahusac in the *Encylopédie*. The esthetic, rather than epistemological, direction of Shaftesbury's thought led to his relative neglect in England; and to a correspondingly greater attention in Germany.[34]

For in England, a convincing refutation of the knowledge-claims of religious enthusiasts was of high ideological importance. No "European society (not even Scotland) had journeyed so far through the fires of enthusiasm in the seventeenth century, or seen how clearly it could become a revolutionary assault on the governing institutions."[35] Therefore it was Locke, rather than Shaftesbury, who dominated the English usage of "enthusiasm" and its cognates for the next century and more. In the trifurcation of meaning recognized by Johnson's *Dictionary*, the third sense was relegated to the specialized treatment of esthetics, seldom a popular topic in England in any age. The second, etymologically degraded, sense now well-nigh universal, seems to have been in popular use before the end of the century. Boswell twice employs the word in its present-day sense in his *Life of Johnson*.[36] And in his *Journal of a Tour to the Hebrides* he records a conversation on 16 August 1773 in which Johnson mildly derides him for this vulgarity.[37]

It was Johnson's first sense, as illustrated from Locke, that informed "the polemic against enthusiasm" that "persisted in the speech and thought of the governing classes in England for perhaps a century and a half."[38] There appears, alas, to be no evidence for the oft-mentioned gravestone of a certain clergyman, recording that he had "served his Maker for forty years without the smallest sign of enthusiasm." But as late as 1815 the Revd Richard Graves, Regius Professor of Divinity in Trinity College Dublin, could append to the printed copy of his first prælection a list of books recommended to ordinands which included one by himself, written in 1798, "Exonerating the Apostles and Evangelists of the Charge of Enthusiasm."[39]

2.2 Superstition

Johnson's *Dictionary* defines three groups of usage of "superstition:" first, as "Unnecessary fear or scruples in religion; observance of unnecessary and uncommanded rites or practices; religion without morality;" secondly, as "False religion; reverence of beings not proper objects of reverence; false worship;" and thirdly, "Over-nicety, exactness too scrupulous." The second sense, which became the most important in Enlightenment polemic, is illustrated from the New Testament: "They had certain questions against him of their own superstitions."[40]

Louis de Jaucourt's article in volume XV of *L'Encyclopédie*[41] is an amplification, not to say an exaggeration, of the second of these usage-groups. "*En effet, la superstition est un culte de religion, faux, mal dirigé, pleins de vaines terreurs, contraire à la raison & aux saines idées, qu'on doit avoir de l'être suprème.*" Jaucourt had been raised a clandestine protestant and had at one time intended to be a pastor. But his opinions were much influenced by Voltaire, and though less hostile to protestant than to papist religion, he seems to have become a deist at most.[42] As an alternative definition he proposes "*la superstition est cette espece d'enchantement ou de pouvoir magique, que le craints exerce sur notre ame.*" Voltaire had noted that "Superstition born in Paganism, adopted by Judaism, infested the Christian Church from the earliest times;" and claimed that "it is the fundamentals of the religion of one sect which is considered as superstition by another sect."[43] It is clear that Voltaire wished to suggest that superstition is not merely a corruption of true religion, but integral to all religion. Jaucourt, who claimed that superstition is worse than atheism, does not go quite as far as this. But he takes as his motto an aphorism by Aulius Gellius which he translates as "*il faut être pieux, & se bien garder de tomber dans la* ***superstition***." "*Religiosum*" in Jaucourt's source means "superstition" only in Johnson's first sense. But no *philosophe* was ever deterred from his polemical purpose by mere etymology. Hence Jaucourt's tendentious and challenging exordium: superstition is "*tout excès de la religion en general.*"

There is an obvious difference of flavor between the French account of superstition as supplied by Voltaire and Jaucourt, and the English as supplied by Samuel Johnson. Now Johnson was both pious and orthodox, and might be thought an apologist. So let us round up the usual suspects. Hume had come close to Voltaire's position by his remark that "superstition is a considerable ingredient in almost all religions"[44] but typically kept his options open with that "almost." Anthony Collins's *Discourse of Free-Thinking* asserted that there is "no just Remedy to this universal Evil [of superstition] but *Free-Thinking*;" but went on to count as superstitious "any Doctrines not taught by the *Church of England*;" and to defend "free-thinking" as consistent with the work of the recently formed Society for the Propagation of the Gospel.[45] Gibbon seems to have perceived the history of Christianity as a quasi-cyclical temporal process in which the enthusiasm of one period gradually subsides into superstition, only to recur at a later time in reaction to the excesses of the latter.[46] None of these three departed from the letter, at any rate, of Johnson's first two usage-groups. It is difficult to find any other English-speaking author of the eighteenth century understand, by superstition, "all excess of religion in general."

There can be little doubt that this marked difference, not so much in the meaning as in the illocutionary force of "superstition," has a lot to do with the fact that France was oppressively papist throughout the eighteenth century, whereas Great Britain was overwhelmingly and gratefully protestant. And in both nations, superstition was strongly identified with Romanism. Moreover, whether fairly or not, the public image of the church was very different in the two countries.

For though the French church was undoubtedly well served by many faithful pastors and efficient bishops,[47] the scandalous exceptions were too obvious to be ignored. It is perhaps symbolic that one of the most famous of the encyclopedists, Jean le Rond d'Alembert, should have been the bastard son of a lapsed nun who later became rich as the mistress of a cardinal. Voltaire's articles on "adultery," "bishops" and "faith" in his *Philosophical Dictionary*[48] – to mention but a few – are typical of the joyful relish indulged by many of the *philosophes* for any sexual immorality among the clergy. In this they were taught by Bayle, whose biographical article on St Augustine dealt chiefly with the carnal excesses of his subject in early manhood.[49]

Bishops and other dignitaries of the Church of England on the other hand – though less zealous and efficient, perhaps, than of France – were almost uniformly respectable and dull. What of Scotland and other protestant nations? Adam Smith, who, as William Wilberforce noted, "cannot be suspected of having been misled by religious prejudices,"[50] assures us with perfect sincerity, "there is scarce perhaps to be found any where in Europe a more learned, decent, independent, and respectable set of men, than the greater part of the presbyterian clergy of Holland, Geneva, Switzerland, and Scotland."[51] The contrast in tone with Voltaire's sneering prurience could hardly be greater.

In France, therefore, there was a constant tendency to equate superstition not merely with popery but with "religion in general." This finds its extreme form in the anonymous but well-known works of the Baron von Holbach. His celebrated *Système de la Nature*, an explicitly atheist work, treats "superstition" and "religion" as synonyms in its chapter on "Immortality of the Soul."[52] Five years before, Voltaire and d'Alembert were made aware of the posthumous testimony of an obscure curé of Etrepigny, Jean Meslier (1678–1733), who – acknowledging his remorse for teaching "pious lies" – had bequeathed to his parishioners in 1732 a manuscript on "Superstition in All Ages" showing that "all religion is but a castle in the air; that Theology is but ignorance of natural causes reduced to a system."[53] The equation of superstition and religion in France, it would seem, was not confined to a narrow circle of the *élite* in Mme d'Holbach's *salon*.

But in England and Scotland there was no such equation. To a considerable extent, "superstition" was employed as a code-word for popery. The more eirenical were sometimes ready to make some allowances even for that. An influential Cambridge divine, Daniel Waterland (1683–1740), observed "that of the two Extremes, Profaneness and Superstition, the latter is the safest for any to lean to."[54] There is no resemblance between Waterland's charitable opinion and the pragmatic defense of superstition offered by the politically radical, French, Gabriel Bonnot de Mably. For Mably considered religion and superstition from a purely utilitarian standpoint.[55]

The association of superstition and popery has its origin in the protestant Reformation. The most influential bearer of this doctrine in France at the beginning of the eighteenth century was the Huguenot exile Pierre Bayle (1647–1706), whose *Dictionnaire Historique et Critique* (1697) fascinated the leading European thinkers of the next two generations and inspired both the *Encyclopédie* and Voltaire's *Dictionnaire Philosophique*.[56] For Bayle, according to a recent commentator, "the crux of Catholic (sic) superstition, as always, is the *hoc est corpus meum*, the basis of the doctrine of transubstantiation."[57] But Bayle often focuses on Mariology and on hagiology in general, and more fundamentally, on the differences between Rome and the protestant churches as to the proper method of interpreting scripture.

The crudest identification of (papist) superstition with transubstantiation is found in Voltaire's *Tombeau du Fanatisme* which he prudently gave to the world as "An Important Study by Lord Bolingbroke:"

> The *hocus pocus*, or the transubstantiation, which name is ridiculous, became established little by little, after having been unknown in the first centuries of Christianity. We can estimate what veneration a priest or a monk attracted who made a god with four words, and not only one god, but as many as he wished.

Voltaire pointed out in a footnote that the "juggler's trick" that we call "hocus pocus" is derived from *hoc est corpus meum*.[58] In his long article on *TRANSSUBSTANTIATION* in volume XVI of the *Encyclopédie* the criticism is, of course, far less direct. The "expression" is novel, protestants reject it unanimously, and of three "systems" which attempt to explain how transubstantiation occurs, all are defective. The four magic words occur in other French attacks on (Romanist) superstition. The "Christ-worshippers," wrote the Curé Meslier:

> attribute to themselves the power of forming Gods out of dough, and of making as many as they want. For, according to their principles, they have only to say four words over ... these little images of paste, to make as many Gods of them as they desire.[59]

Meslier's identification of Christianity with the Romanism and its characteristic superstition that he himself had rejected is typical of French authors in this period. Among English authors it is to be found only in those such as Toland and Gibbon who, repenting them of the popish doctrines they once had believed, threw out the baby of Christianity with the bathwater of superstition.

2.3 The exclusion of superstition and enthusiasm from establishment religion

It is my contention that within Establishment culture throughout the eighteenth century, enlightenment was generally held to be compatible with orthodoxy. By "Establishment culture" I mean the complex of assumptions, attitudes, interests and commitments that Jonathan Clark[60] has shown united all those in England – and to a considerable extent in Scotland – whose assent to the doctrinal standard of the Church admitted them to public life. By "enlightenment" in this context I mean the Western European campaign to purge religion of enthusiasm and superstition. By "orthodoxy" I mean the body of Christological and soteriological doctrine summarized in the ancient, Trinitarian creeds and integral to the *Book of Common Prayer*.[61] And as I stated at the outset, it is my hypothesis that this putative state of affairs, so very different from that which prevailed in France at the time, may to a large extent be explained by the different force of the word "superstition" in the two countries. Before going further, therefore, I ought to say more about orthodoxy in the Georgian Church. For at least since Sir Leslie Stephen's influential work, is has often been supposed that enlightenment and Anglican orthodoxy were mutually exclusive, and that by the end of the eighteenth century the "intellectual party" in the Church was "Socinian in all but name."[62]

No doubt Newton's natural theology was the most powerful intellectual influence on English religion in the eighteenth century; and no doubt Newton himself (a Fellow of Trinity College) was a secret unitarian, or rather, in Keynes's colorful phrase, a "Judaic monotheist of the school of Maimonides."[63] No doubt John Locke's *Letter Concerning Toleration* (1689) and his *Reasonableness of Christianity* (1695) were feared by some to tend towards Socinianism or even deism. No doubt the works of two of Locke's younger friends, John Toland's *Christianity Not Mysterious* (1696) and Anthony Collins's *Discourse of Free-Thinking* (1713), promoted deism. No doubt it was *Scripture-Doctrine of the Trinity* (1712) written by the influential Newtonian, Samuel Clarke (1675–1729), that set in motion six decades of controversy and self-doubt leading to the Feathers Tavern Petition of 1772; and the loss to Dissent of several conscientious and distinguished divines after the rejection of that petition by Parliament.

Yet – insofar as it makes any sense to say such a thing – the Church of England seems to have been more orthodox at the end of the eighteenth century than it had been at the beginning. In part this was achieved by the work of those members of "the intellectual party" who chose to retain their benefices after 1772, and who felt morally obliged to defend conformity to their former colleagues. I have elsewhere shown[64] that after the mid-1770s the leaders of this party, Richard Watson (1737–1816), William Paley (1743–1805) and John Hey (1734–1815), constructed a characteristically Cambridge *via media*: not between Rome and Geneva, but between *both* Rome *and* Geneva on the one hand (as representing "blind attachment to system") and on the other, "Bolingbroke, Voltaire, Helvetius and Hume" (as representing "the contagion of Infidelity"). The so-called "School of Paley" managed to be at the same time both enlightened and conservative, and was increasingly respected in orthodox circles after the French Revolution threw up new causes of fear and doubt. Paley's books remained on the reading lists for more than a century, and Hey's subtle, rationalistic defense of "mystery" in religion – eclipsing Toland's naive reductionism – influenced the first generation of Tractarians.

But to an even greater extent perhaps, the persistence and gradual strengthening of orthodoxy in the Georgian Church was due to the well-known ability of Establishment culture to digest or co-opt the most unpalatable intrusions, harnessing their vitality, laying claim to their prestige, and filtering out any harmful or dangerous ingredients. Hence, for example, Newtonian science was not a threat but an opportunity. By the time the freshman Robert Malthus went up to Cambridge in 1784 to prepare for ordination, Newton's *Opticks* and *Principia* were required for Honors, and Maclaurin's famous textbook was prescribed reading. Ordinands must read Newton, Maclaurin informed them, because "natural philosophy" demonstrates the existence and attributes of God, leads to belief in an afterlife and a view of this life as a state of probation, and thus prepares us for the study of scripture.[65] Locke too was co-opted. His *Reasonableness of Christianity* – which Watson recommended to his pupils – was based upon the supposition, made explicit in the *Essay on Human Understanding,* of the epistemological sovereignty of Holy Scripture in all matters of religion. Members of the Cambridge "intellectual party" shared this belief wholeheartedly; and made it the basis of their continuing willingness to "speak of the Trinity" and to assent to seemingly mysterious doctrinal formularies.[66] There was a feeling that the robustness of orthodoxy in England, by contrast with France, was a happy consequence of British customs and institutions. "The freedom of enquiry," Richard Watson pointed out to undergraduates, "which has subsisted in this country in the present century, has eventually been of great service to the cause of Christianity."[67] By 1797 the impeccably orthodox, Cambridge-educated layman William Wilberforce could claim that "the melancholy proofs of our depravity" rest "on the same solid basis as the sublime philosophy of Newton;" and could enroll Newton and Locke as exemplars of Christian belief in his best-seller which launched the evangelical movement in the Church of England.[68]

There can be no doubt of the sincerity with which the orthodox joined in the Enlightenment project of excluding enthusiasm and superstition from religion.

In the case of enthusiasm, this was more than simply the need to provide an ideological defense against aggression from protestant sectaries. There was, to be sure, a well-tended folk memory of the 1640s and 1650s. And many a torpid, claret-drinking parson was unpleasantly disturbed no doubt by the undeniably Christian zeal of the Weslyan revival. Ronald Knox was correct to dismiss polemic such as Lavington's as comically misguided. But something far more important was at issue, something fundamental, indeed, to Anglican theological self-understanding: the ultimate authority of holy scripture in all matters of faith. The knowledge-claims of enthusiasm were not just politically subversive: they were blasphemous, and they cut at the root of protestant apologetic as against the Church of Rome.

It was this consideration that animated Joseph Butler in his famous interview with John Wesley in 1738.

> *[Butler]* Mr. Wesley, I will deal plainly with you. I once thought Mr. Whitefield and you well-meaning men. But I can't think so now. For I have heard more of you – matters of fact, sir. And Mr. Whitefield says in his *Journal*, "There are promises still to be fulfilled in me." Sir, the pretending to extraordinary revelations and gifts of the Holy Ghost is a horrid thing, a very horrid thing.[69]

It is "horrid" because it undermines several of the thirty-nine Articles of Religion (articles VI, VIII, XVII, XIX, XX, XXI, XXII, XXVIII) to which Whitefield and Wesley had assented, Articles which, if true, would justify the otherwise inexcusable separation of the Church of England from communion with Rome. It is "very horrid" because it blasphemes the uniqueness and sufficiency of the revelation of God in Christ. On this crucial point the latitudinarians Locke and Tillotson were at one with the Non-Juring Hickes, with the two great converts from Dissent, Secker and Butler, with the high-church Waterland and the low-church Watson, with Simeonite evangelicals, with the Hackney Phalanx and with all the Clapham Sect. Even Wesley himself agreed, and attempted to evade the charge.[70] All of this, especially among those whose formation owed much to Cambridge, was entirely congruent with – though by no means caused by – the iconic status of Sir Isaac Newton throughout that century. For as Pocock has remarked, "The subject of a Newtonian universe was not likely to find God and spirit immanent in the universe or in his own body and perceptions; he would be free from the smallest tincture of enthusiasm."[71]

In the case of superstition, narrowly understood to be "a system of error and usurpation" fabricated by the Church of Rome,[72] there was at least as much agreement among Anglicans of all varieties, including the most scrupulous. The identification of popery with superstition was routine.[73] Richard Watson had reminded his Cambridge students that the thesis *Pontifex Romanus est ille Antichristus quem futurum Scriptura praedixit* is a "primary pillar of the reformed faith."[74] In 1801 the young John Bird Sumner (1780–1862), later to be Queen Victoria's second Archbishop of Canterbury, won the Hulsean prize for his essay which considered "how far the present declining state of the Papacy is an accomplishment of the Prophecies concerning the *Beast with seven heads and ten horns* of the Revelations, and the *Antichrist* or *Man of Sin* of St. Paul."[75]

Pocock has defined "superstition" concisely as "the worship of the godhead in objects perceived by the senses" and suggests that "its ultimate form is probably the Catholic (sic) doctrine of transubstantiation."[76] There seems little doubt that this view of the matter would have been held by all or most Anglicans, lay and clerical, in that age. Article XXVIII of the Church of England states that "Transubstantiation (or the change of the substance of Bread and Wine) in the Supper of the Lord, cannot be proved by holy Writ; but is repugnant to the plain words of Scripture, overthroweth the nature of a Sacrament, and hath given occasion to many superstitions." The Declaration imposed by the Test Act of 1673 ran: "I, *A. B.*, do declare that I believe that there is not any transubstantiation in the Sacrament of the Lord's Supper, or in the elements of bread and wine, at or after the consecration thereof by any person whatsoever." All who sought public office were bound by these standards until the nineteenth century. The object, of course, was to exclude Roman Catholics from public life from fear that they might seek to overthrow the Protestant Succession. That fear that was sharply reawakened by the Jacobite rebellion of 1745. A sermon of the time by Robert Wallace proclaimed "Ignorance and Superstition a Source of Violence and Cruelty, and in particular the Cause of the present Rebellion."[77]

Almost by definition, a protestant was bound to concur in the repudiation of transubstantiation and there is no evidence that any found it hard to do so. The heavy-handed Watson (putting thoughts in Calvin's mind) supposed it "almost impossible that the Scriptures could be so far perverted as to afford the *Romanists* any handle for their doctrine of Transubstantiation, or that the understanding of any human being could have been so far debased, or rather so utterly annihilated, as to believe in it for a moment."[78] His more sophisticated colleague in "the intellectual party," John Hey, concerned to establish the validity of assenting to unintelligible propositions, "would not condemn a Romanist who, as one of the people, gave a verbal assent to it [transubstantiation], merely in submission to authority, if he did not pretend to understand it."[79] By the same token, Hey agreed with Voltaire that the refusal of the nuns of St Cyran to sign the anti-Jansenist propositions was unreasonable.[80]

Anglicans of a low-church persuasion were sometimes inclined to confuse transubstantiation with *any* doctrine of the "real presence" of Christ in the sacrament. Thus Joseph Milner (1744–97), in his "inaccurate and uncritical"[81] *History of the Church of Christ,* having noted that "every true Roman-catholic lays immense stress on the doctrine of transubstantiation" observed that "popish divines were well aware that the doctrine of the real presence is the very foundation of their religion; and that if it once be taken away, there is an end both of their dignity and their power."[82] But thus to imply that the Church of England does not itself have a "doctrine of the real presence" is to put a Zwinglian gloss on the Anglican formularies which the texts can not support. This was perfectly understood by Daniel Waterland, one of the most learned Cambridge divines of the century, whose *Review of the Doctrine of the Eucharist* (1737) sets out and analyses the entire range of Christian belief on the matter. Better than any other document of this period, his judgment on transubstantiation shows clearly how it was that the most pious and orthodox protestants could be "enlightened" in their desire to exclude superstition:

> To the Romanists, who plead warmly for *the very body and blood* in the Eucharist, we make answer, that we do receive *the very body and blood* in it, and through it, as properly as a man *receives* an *estate*, and becomes *possessed* of an *inheritance* by any deeds or conveyances: and what would they have more? Will nothing satisfy, except the *wax* and *parchment* be transubstantiated into *terra firma*, or every *instrument* converted into *arable*? Surely, that is pushing points too far, and turning things most serious into perfect ridicule.[83]

If Enlightenment in France was hostile to religion whereas in England it was not, this is not at all because there was not as much desire in the one country as in the other to exclude enthusiasm and superstition. It seems rather, as Pocock has remarked, that "In England there was ... simply no *infâme* to be crushed."[84]

3
Theology and Political Doctrine in Church and Dissent

It has lately been argued by Jonathan Clark that theological heterodoxy, in particular the Socinian heresy, is "conceptually basic to radicalism in the *ancien régime* sense":

> If Christ were not a person of the Trinity, the Catholic doctrine of the Atonement, held Socinians, was meaningless; if man was not in need of redemption, original sin did not descend by inheritance, and man must be assumed to be both fundamentally benevolent and capable of ordering his own affairs in all respects ... A consequence of a denial that Christ exercised divine authority was that He could not institute a priesthood descending by apostolic succession and exercising its mediatory powers by virtue of that divine right: the Anglican clergy were thus on a par, in point of authority, with Dissenting clergy (or even with the private individual.) If even the Church could not claim divine institution, the State was still more obviously secular. "No bishop, no king" was once more a relevant challenge, if mankind was free to amend or reject its ecclesiastical and political hierarchy in the name of reason, conscience or utility.[1]

Despite certain minor infelicities – there is no one "Catholic doctrine of the Atonement;" "fundamentally benevolent" is not the equivalent of "originally innocent" – this is a useful synopsis of the nexus between theology and political doctrine in both Church and Dissent. However, neither Clark nor any other author that I am aware of has provided the complete statement of which the quoted passage is a summary It is therefore my intention in this chapter to supply such a statement with the object of throwing light upon two related questions: just what was it, exactly, that English Protestant Dissent "dissented" from; and in what sense, if any, may we say that the Rational Dissenters were "rational" and the orthodox Dissenters were not?

I shall argue that there existed from before 1662 until sometime after 1800 a coherent and widely accepted body of establishment social theory rooted in Catholic ecclesiology ultimately biblical in origin, transmitted unimpaired to the post-Reformation Church of England, and expressed most authoritatively in the *Book of Common Prayer.* Tory or High-Church ecclesiastical and civil polity was fully consistent with, if not actually entailed by, the ecclesiology of the Prayer Book. But that ecclesiology is itself a consequence of Pauline soteriology: and Pauline soteriology depended upon the central Christological conception, only defined adequately in the fifth century, of Christ as being both fully human and fully divine. The doctrine of the Holy Trinity can he regarded as a corollary of the Incarnation. Hence there is a clear conceptual nexus running from the Trinity and orthodox Christology through Pauline soteriology, Catholic ecclesiology, post-Reformation ecclesiastical and civil polity, to the political doctrines arising from and conveniently labelled by the "grand scheme of subordination."

Now if it could be shown convincingly both that orthodox Christology (and only orthodox Christology) implied the principle of subordination and also that the principle of subordination (and only the principle of subordination) implied orthodox Christology we should have a convenient method of demonstrating Clark's claim that theological heterodoxy was "conceptually basic" to "radical" or "democratic" political doctrines in the eighteenth century (or indeed at any other time). For if heterodoxy be defined as that which is not orthodox, and "democratical principles" as a denial of the principle of subordination, then the logical equivalence of orthodoxy and subordination would imply the logical equivalence of heterodoxy and democratical principles.

This result would immediately suggest a criterion of "rationality" for Dissent. Dissenters who clung to Christological orthodoxy whilst rejecting establishment polity and the principle of subordination would be clearly self-contradictory and therefore "irrational." Those who rejected the entire establishment package, as did many of those known to historians as Rational Dissenters, would be internally consistent and therefore "rational."

A case might be made that orthodox Christology and Catholic ecclesiology entail each other. There is no possibility, however, of establishing a similar relation between ecclesiology and ecclesiastical polity, hence there can be no logical equivalence of orthodoxy with the principle of subordination. Though establishment doctrine is coherent and well-formed its theological foundations do not *uniquely* determine a particular set of political principles. Hence it is and was possible in principle to deduce a different, somewhat more "democratical" *polity* from the same *ecclesiology* and so to exonerate orthodox Dissenters from the charge of irrationality.

What this means is that some support may be found for Clark's claim that heterodoxy is "conceptually basic" to democratical principles, but little if any for his stronger claim that such principles are, in any strictly logical sense, "entailed by Socinianism."[2] The most that we can say, though it is by no means negligible, is that during the period from 1662 to the repeal of the Test Act in 1828 all Dissenters confronted, and were to a significant extent excluded from participation in, an *ancien régime* based upon the unity of Church and state; that there existed a generally understood and widely accepted rationale of this *régime* which deduced its key political doctrine of subordination from theological first principles; and that some of the bolder spirits among that intellectual elite known as Rational Dissent took a short way with establishment by denying the entire body of establishment doctrine – a sufficient though not a *necessary* condition of their consistency in rejecting both orthodox Christology and the principle of subordination.

By *orthodox Christology* I shall mean the belief, defined in the Nicene and the Athanasian creeds, that Jesus is of "one substance" (*homo-ousios*) with God the Father; yet also "Substance of his Mother, born in the world; Perfect God, and Perfect Man." The doctrine of the Holy Trinity, celebrated at length in the Athanasian creed, if not exactly a corollary, is dependent upon this definition: all varieties of Christian (not merely *Christological*) heterodoxy deny both *homo-ousios* and, therefore, the Holy Trinity in one way or another.

By *the principle of subordination* I shall mean the belief that "Without order there is no living in public society, because the want thereof is the mother of confusion, whereupon division of necessity followeth, and out of division inevitable *destruction* ... And if things or persons be ordered this doth imply that they are distinguished by degrees. For order is a gradual disposition." The underlying conception of society is strictly organicist: "the very Deity itself both keepeth, and requireth for ever this to be kept as a law, that whensoever there is a coagmentation of many, the lowest be knit to the highest by that which being interjacent may cause each to cleave unto other, and so all to continue one."[3]

The nexus between the two conceptions is complex and I have attempted to illustrate it in Figure 3.1. Orthodox Christology, I maintain, is a soteriological conception which gives rise to Catholic ecclesiology and is implied by it. Catholic ecclesiology was transmitted unimpaired at the Reformation to the Tudor and Jacobean Church; it informs the *Book of Common Prayer* and other Anglican formularies and authorities. These in turn supply the rationale of Anglican ecclesiastical polity and Restoration civil polity. The two latter are seventeenth-century manifestations of the principle of subordination.

In constructing my argument I shall focus upon the definitive document of the post-Reformation Church of England, the *Book of Common Prayer*. First I shall show that it transmits a Catholic ecclesiology that implies and is implied by orthodox Christology. Secondly I shall argue that the organicism which is realized in establishment ecclesiastical and civil polity is presupposed and articulated by the Prayer Book and other Anglican sources *as a corollary of the Catholic ecclesiology they transmit*. In a final section I shall consider some of the implications of these ideas for Dissenting doctrines in the eighteenth century.

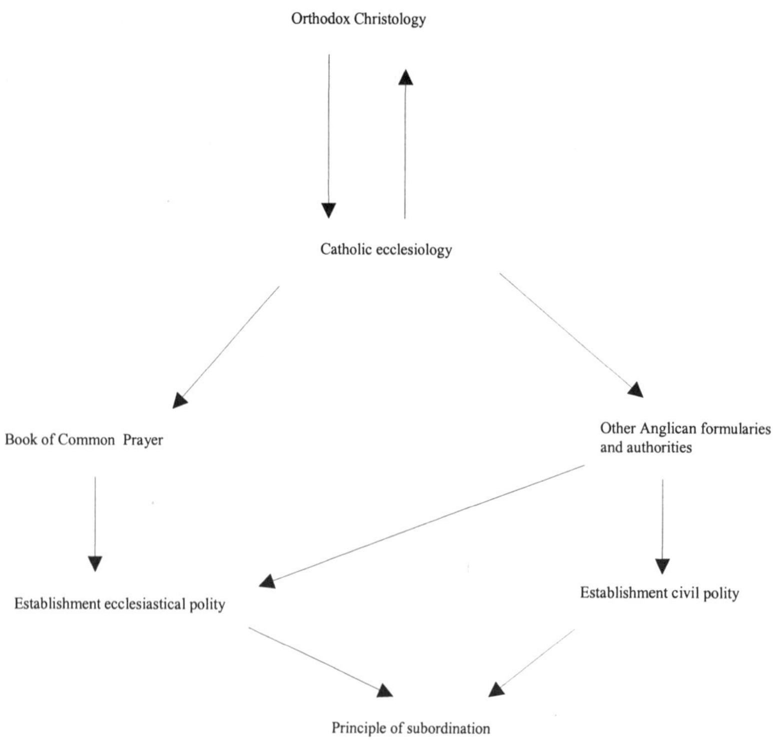

Figure 3.1 The relation between orthodox Christology and the principle of subordination

3.1 Ecclesiology of the *Book of Common Prayer*

A dissenter is defined by that from which he dissents. The historical phenomenon known as English Protestant Dissent, or simply "Dissent," is therefore determined, or at any rate inaugurated, by the Act of Uniformity (May, 1662). Under this Act all clerics were required to meet three tests by the Feast of St Bartholomew (17 August) following: to read publicly the daily offices of the revised *Book of Common Prayer* and make "unfeigned assent and consent" to all that the Prayer Book contained; to renounce rebellion and the Solemn League and Covenant (1643); and to have received episcopal ordination. Though implementation of the Act was confused and uncertain,[4] some 1,700 clerics, nearly 20 percent of the whole, either resigned their benefices or were eventually ejected under its provisions.[5] English (as distinct from Scottish and Irish) Presbyterian support for the Covenant had never been very strong[6] and rebellion was wholly discredited. Hence it was the *Book of Common Prayer,* together with the episcopacy it presupposed, that was the chief sticking point for those nonconforming ministers and their lay supporters who comprised the first generation of English Dissent.

The 1662 recension of the Prayer Book differed only slightly from the Jacobean and Elizabethan versions, which in turn were very conservative revisions of the second Edwardine book of 1552. Most of the significant changes in 1662, especially those made to the Ordinal, were intended to strengthen the Laudian (and indirectly pre-Reformation) interpretation of its rites which Dissenters and their Puritan forebears denied with such rancour. For, since Elizabethan times, a Protestant party in the Church of England had regarded the liturgy as "unperfect Boke, culled and picked out of that Popishe dunghill the Portuise and Masseboke, full of all abominiations."[7] And as late as 1773 the Earl of Chatham defended the Dissenters' zeal in a famous antithesis: "They contend for a scriptural and spiritual worship; we have a Calvinistic Creed, a Popish liturgy and Arminian clergy."[8]

What Chatham and the Elizabethan Puritans meant by "popish" in this context is that the *Book of Common Prayer* consists almost entirely of translations – with substantial editing and modification – of pre-Reformation service books, chiefly of the Sarum use. Though the government's message to the Devon revels of 1549 was less than perfectly ingenuous it contained a core of truth: "It seemeth to you a new service, and indeed is none other but the old; the self-same words in English, which were in Latin, saving a few things taken out."[9]

Whether or not Lord Chatham was correct about the "Calvinistic Creed" and the "Arminian clergy," it was the liturgy rather than these which determined the character and ethos of the Church of England. For creeds and their interpretation are a matter of doubtful disputation by experts. Though there were plenty of Anglican divines, who interpreted the Articles and other formularies in a Catholic sense they could always be ignored or dismissed by the Dissenters as "Romanizers" and in many cases countered by opposing passages from the writings of Low-Church (or "Latitudinarian") divines such as Hoadley or Chillingworth. But the Prayer Book was something that every church-going English man, woman and child – that is to say, the vast majority of the population – heard continually from earliest infancy to the last hours of their lives. Common people with no education at all knew large portions of it by heart simply by hearing it so often. There is evidence to suggest that the Prayer Book was not popular before the 1640's.[10] It was probably its suppression during the Interregnum, and its replacement by the detested ranting of the Puritans, that endeared it to the general public. It was used clandestinely during the 1650's and joyfully welcomed back at the Restoration. Thereafter it became enshrined in the affections of all Churchmen and its (genuine) excellencies much exaggerated. "Our incomparable liturgy" was regarded by ten generations of Englishmen altogether ignorant of Latin and Oriental rites as the *ne plus ultra* of Christian worship. Novelists like George Eliot and Thomas Hardy reveal its influences upon the speech, and therefore the thinking, of country folk until well into the nineteenth century. Its idiom saturated the language and remained normative for literary culture down to the Second World War.

The *Book of Common Prayer* both presupposes and inculcates what I shall call a "Catholic" ecclesiology: by which I mean a conception of the Church not as a voluntary, human society of like-minded believers but as a divine society, transcending space and time, uniting the living and the dead, joined with angelic creatures in the everlasting praise of God:

> To thee all Angels cry aloud; The Heavens, and all the Powers therein.
> To Thee Cherubin, and Seraphin: continually do cry ...
> The glorious company of the Apostles ....
> The goodly fellowship of the Prophets ...
> The noble army of Martyrs ...
> The Holy Church throughout all the world: doth acknowledge thee.[11]

The "Holy Church throughout all the world," described in the communion office as "Christ's Church militant here in earth" is the visible and temporal portion of a much larger society. And "although in the visible Church the evil be ever mingled with the good,"[12] it is the means whereby the faithful in this life are united with Christ, and through him with all who now stand in the presence of God:

> Therefore with Angels and Archangels,
> and with all the company of heaven,
> we laud and magnify thy glorious Name;
> evermore praising thee ...[13]

An individual enters the Church by faith and baptism, the latter usually in infancy. Initiation is supernatural. The celebrant prays that the child may receive "that thing which by nature he cannot have: that he may be baptized with Water and the Holy Ghost, and received into Christ's Holy Church, and be made a lively member of the same." Baptism is a ritual drowning after which the child is "born again" to new life in union with Christ." The celebrant gives thanks, "Seeing now ... that this child is regenerated, and grafted into the body of Christ's church."[14] The initiative comes from God, who has chosen to "adopt" the child, who in baptism is "made a member of Christ, the child of God, and an inheritor of the kingdom of heaven."[15] The concluding prayer combines these themes:

> We yield thee hearty thanks, most merciful Father, that it hath pleased thee to regenerate this infant with thy Holy spirit, to receive him for thine own child by adoption, and to incorporate him into thy Holy Church.[16]

The Church is variously referred to in the *Book of Common Prayer* as the "ark" (Baptism), the "spouse" of Christ (Matrimony, Ordering of Priests), Christ's "flock" (Baptism, Ordering of Priests, Consecration of Bishops), the "Lord's family" (Ordering of Priests), the "household" of God (Collects: Epiphany V, Trinity XXII), an edifice "built upon the foundation of the Apostles and Prophets, Jesus Christ himself being the head corner-stone" (Collect: SS Simon and Jude), and "the fellowship of Christ's Religion" (Collect: Easter III). But by far the most frequently used image, and that which controls the meaning of all the others, is that of the Church as the "mystical body" (or simply "body") of Christ. God has "knit together" his "elect in one communion and fellowship, in the mystical body" of "Christ our Lord" (Collect: All Saints'). In the Eucharist the worshippers know that they are "very members incorporate in the mystical body of thy Son, which is the blessed company of all faithful people" (Communion Office). The homily provided for the marriage service contains the strongest statement of the quasi-organic unity between Christ and the faithful:

> He that loveth his wife loveth himself: for no man ever yet hated his own flesh, but nourisheth and cherisheth it, even as the Lord the Church: for *we are members of his body, of his flesh, and of his bones.*[17]

It is evident that the ultimate source of these conceptions is the soteriology and related ecclesiology of St Paul (e.g. I Cor. 12: 12–27; Rom. 12: 4–5; Eph. 4: 4–16, etc.).

In a classic study of Pauline theology J. A. T Robinson has shown how the concepts of *Sarx* (translated as "flesh") and *Soma* (translated as "body") are used by the Apostle to explain, or at any rate suggest, how the human race is "saved" by the death and resurrection of Christ. Man is made "the image and likeness of God" (Gen. 2: 26–7). But natural human existence is a state of radical separation from God, symbolised and explained by the Fall (Gen. 2: 17; 3: 3,4; Rom. 5: 12–21). But the effect of the original disobedience of Adam is (potentially) reversed by the perfect obedience of Christ, the "last Adam" (I Cor. 15: 45), and this is made known in the resurrection:

> Christ being raised from the dead dieth no more: death hath no more dominion over him.
> For in that he died, he died unto sin once: but in that he liveth, he liveth unto God. (Rom. 6: 9–10; Proper Anthems for Easter Day)

It is the body of Christ *and only the body of Christ* which rises from the dead, becomes "the first-fruits of them that slept" (I Cor. 15: 20) and inaugurates the new life of restored unity between God and man.[18]

Now "as in Adam all die, even so," St Paul promises (I Cor. I5: 22), "*in Christ* shall all be made alive" (my italics). The rest of the human race participate in the victory of Christ over death and the reconciliation of God and man *precisely insofar as they may become "members" of the body of Christ.* By faith and baptism they "die" to the old life we inherit from Adam and rise to the new life we share with Christ:

> Grant, O Lord, that as we are baptized into the death of thy blessed Son our Saviour Jesus Christ, so by continual mortifying our corrupt

> affections we may be buried with him: and that through the grave, and gate of death, we may pass to our joyful resurrection.[19]

It follows that the Church is the extension of the life of the Risen Christ. "In you," St Paul reminded his Corinthian converts, "the evidence for the truth of Christ has found confirmation" (I Cor. I: 6, *NEB*). His spirit will guide the Church "into all truth" (John 16: 13); where two or three gather in his name, Christ will be in the midst (Matt. 18: 20); the Eucharist is a perpetual *anamnesis* of his death and resurrection (I Cor. II: 26; Matt. 26: 29) wherein the faithful are "partakers of his most blessed Body and Blood."[20] We may summarize the argument so far as follows: humans find "salvation" through union with Christ; union with Christ means union with the Church, his (mystical) body; hence follows the patristic slogan, *Salus extra ecclesiam non est.*[21]

As mediator between God and man, Christ must be fully united with each. For as St Athanasius put it in a memorable aphorism, God "was made man, that we might be made God."[22] A Christology that ignores or minimizes the divinity of Christ (e.g. Ebionite, Arian, etc.), or that ignores or minimizes his humanity (e.g. gnostic, Apollinarian, etc.), tends to undermine the Christian gospel of salvation. So does a Christology which, though recognizing both the divine and the human nature of Christ, fails to maintain the unity of his person (e.g. Nestorian, Eutychian, etc.). Orthodox Christology – a Christology, that is to say, that is consistent with the evangelical understanding of salvation in Christ – requires a clear statement of union of God and man in Christ which maintains both natures in one person. This was first supplied in the creed of the Council of Nicaea (325) which repudiated the Christology of Arius and vindicated that of St Athanasius. Christ is "of one substance with the Father" *(homo-ousios to Patri:* in the Western Church *consubstantialem patri)* and "was made Man" *(kai enanthropesanta).* The term *enanthropesanta,* not found in the Bible, was specially coined "to express the *permanent* union of God with Human Nature,"[23] but it proved inadequate against Apollinarian and Nestorian deviations and a more careful definition was required at the Council of Chalcedon (451). The so-called creed of St Athanasius, roughly contemporaneous with Chalcedon, greatly amplified the Christological clauses. Christ is:

> God, of the Substance of the Father, begotten before the worlds: and Man, of the Substance of his Mother, born in the world;
> ... Who although he be God and Man: yet he is not two, but one Christ
> ... not by confusion of Substance: but by unity of Person.[24]

It is a relatively short step from a clear statement of the unity of the two natures in the person of Christ to a fully articulated doctrine of the Holy Trinity. The earliest Christian baptismal formula was Trinitarian (Matt. 28: 19) and the earliest (baptismal) creeds included the three fold affirmation of belief in Father, Son and Paraclete.[25] Once the terminology of "person" and "substance" had been worked out in relation to Christ it could be applied symmetrically to each member of the Trinity:

> Whosoever will be saved: before all things it is necessary that he hold the
> Catholick Faith ...
> And the Catholick Faith is this: that we worship one God in Trinity, and Trinity
> in Unity;
> Neither confounding the *Persons:* nor dividing the *Substance* ...[26]

Orthodoxy means "right (correct, etc.) *worship* (praise, etc.)." It is impossible to regulate the way the faithful *think*. But in the tradition inherited by the Church of England it was held to be proper that the unity of the Church should find expression in a form of worship that represents as well as is humanly possible the (essentially mysterious) relation between God and man revealed in Jesus Christ.

It need hardly be said that the *Book of Common Prayer*, like all Catholic liturgy, is replete with Trinitarian references and ascriptions. The daily psalms and canticles begin and end with the antiphon *Gloria Patri;* one or more of the three creeds are recited at least twice daily; many collects and other prayers end with a Trinitarian ascription; the Litany begins with a Trinitarian invocation; the *Quicunque vult* (which Newman regarded not so much as a creed as a hymn of praise to the Holy Trinity) is appointed for Christmas, Easter, Ascension, Pentecost and other great feasts; the feast of the Holy Trinity is celebrated in its traditional place in the Western Kalendar; the ancient baptismal formula is preserved; and the catechism provides an extremely simple and memorable Trinitarian doctrine for young children. Catholic ecclesiology, soteriology and orthodox Christology are interdependent. All three were handed down from patristic and medieval times to the post-Reformation and post-Restoration Church of England in its "popish liturgy."

3.2 Christian organicism in Church and State

The *Book of Common Prayer* transmits Catholic ecclesiology. Catholic ecclesiology presents an organicist conception of the *Church* as (mystical) *body* (of Christ). Whether there be any necessary connection between the two, it was inevitable in the circumstances of seventeenth-century Britain that an organicist ecclesiology should be inextricably entangled with an organicist conception of the state. An organicist conception of ecclesiastical and civil society was then thought to imply an ecclesiastical and civil polity based upon the principle of subordination. The Dissenters' democratic principles were thus at variance with the Act of Uniformity which required "unfeigned assent" to the Prayer Book.

The conception of civil society as a "body," and the use of anatomical conceptions in constructing theories of social processes, was commonplace in the seventeenth and eighteenth centuries. William Petty began his *Political Anatomy of Ireland* (1691) by noting that:

> Sir Francis Bacon, in his *Advancement of Learning,* hath made a judicious Parallel in many particulars, between the *Body Natural,* and *Body Politick,* and between the Arts of preserving both in Health and Strength.[27]

Though there is nothing in either Petty's or Bacon's usage which particularly implies subordination, it would seem that this was understood in the earliest (Tudor) occurrences of the term:

> This realm of England is an Empire ... governed by one supreme *Head* and King ... unto whom the *Body Politick,* compact of all Sorts and Degrees of People ... been bounden and owen to bear a natural and humble obedience.[28]

In such cases the anatomical metaphor of society as "body" gave expression to the belief that order must be predetermined and consciously maintained by the obedience of the many to the few. Archbishop Thomas Secker, an Anglican convert from Dissent, gave classic utterance to this way of thinking in an address to his ordination candidates in 1769, reprinted in 1785 for the edification of Cambridge undergraduates by the reputedly "Latitudinarian," Whiggish Bishop, Richard Watson of Llandaff:

> Without union there cannot be a sufficient degree either of strength or beauty: and *without subordination there cannot long be*

> *union*. Therefore obey, as the Apostle directs, them that have the rule over you.[29]

It is no surprise that Burke should have exploited the metaphor in his answer to Richard Price: "He who gave our nature to be perfected by our virtue, willed also the necessary means of its perfection – He willed therefore the state – He willed its connection with the source and archetype of all perfection." Wherefore "our political system is placed in a just correspondence and symmetry with the order of the world, and with the mode of existence decreed to a permanent body composed of transitory parts ... by ... a stupendous wisdom, moulding together the great mysterious incorporation of the human race."[30] It is this tradition of discourse, I believe, that informed establishment doctrine in the seventeenth and eighteenth centuries and that was generally supposed to authenticate patriarchy in the family, hierarchy in the Church and monarchy in the state. These putative implications of the metaphor are clearly to be seen in the *Book of Common Prayer* and other contemporaneous Anglican sources.

A single organism has only one control centre, or "head," which must be obeyed if its life is to be sustained. For example, if man and woman become "one flesh" in marriage, wives must obey their husbands for "the husband is the head of the wife, even as Christ is the head of the Church."[31] All "members" of a body – hands, feet, ears, eyes – have interdependent functions, some "honorable" and "comely," others "uncomely." St Paul's converts "are the body of Christ, and members in particular:" hence he admonishes them to recognise and respect a differentiation of function in the Church (I Cor. 12: 13–30). Christ is head of the body: but has "poured down his gifts abundantly upon men, making some Apostles, some Prophets, some Evangelists, some Pastors and Doctors, to the edifying and making perfect his Church." Bishops are appointed for the "edifying and well-governing" of the Church, to "teach and exhort," to "banish and drive away all erroneous and strange doctrine," to "correct and punish," and "to be to the flock of Christ a shepherd."[32]

The bishop has a vital part in a structure of authority. His consecration begins with a prayer which includes the people, "that they may obediently follow the same."[33] Priests and deacons promise at their ordination that they will "reverently obey [their] Ordinary, and other chief minister, unto whom is committed the charge and governance over [them]."[34] As priests are subject to bishops, so are deacons to priests. Their function is "to assist the Priest in Divine Service;" prayer is offered that they may "so well behave themselves in this inferior Office, that they may be found worthy to be called unto the higher Ministries in thy church."[35] Even the bishop himself is under authority: he must "profess and promise all due reverence and obedience to the Archbishop and to the Metropolitan Church."[36] It is clear that the ecclesiology of the *Book of Common Prayer* is unequivocally hierarchical. The ecclesiastical polity of the Church of England, re-affirmed at the Restoration in the Caroline Act of Uniformity, is wholly consistent with Catholic ecclesiology transmitted by its "popish liturgy."

It is equally clear, however, that the interdependence and obedience necessary in the body of Christ are seen, in the Prayer Book, as *extending into all parts of human life, whether "ecclesiastical" or "civil."* Wherefore every child must learn:

> To honour and obey the King, and all that are put in authority under him:
> To submit myself to all my governors, teachers, spiritual pastors and masters:
> To order myself lowly and reverently to all my betters.[37]

Whether by accident or design, the organicism of the divine society so obviously entailed by Catholic ecclesiology, has come to be applied to "public society" as a whole. It is evident from the *Book of Common Prayer* and other sources that this extension has been achieved by subsuming the "state" under the "Church."

A perfect illustration occurs in the Litany. Petition is made first "that it may please [God] to rule and govern [his] Holy Church universal in the right way." This is immediately followed by a series of other petitions which unpack "Church" into the contents intended therein: "thy Servant CHARLES, our most gracious King and Governor;" "all the Royal Family;" "all Bishops Priests and Deacons;" "the Lords of the Council and all the Nobility;" "the Magistrates;" and finally, "all thy people." A similar but wider view is afforded by the prayer for the Church in the Eucharist: first "the universal Church;" then "all Christian Kings, Princes and Governors;" and "specially thy servant CHARLES our King;" followed by "his whole Council;" "all Bishops and Curates" (an inversion of the order in the Litany, which ranks the clergy ahead of the Council); "all thy people;" and finishing with a commemoration of (but not a petition for) "all thy servants departed this life in thy faith and fear" (Communion Office). In a Christian country, these prayers imply, the "Church" on earth is one and the same thing as the "state:" not because of an Erastian subordination of the former, but simply because all members of the public society" are united in the body of Christ.

The original version of the English litany (1544) contained a petition for deliverance "from the tyrannye of the bishop of Rome and all his detestable enormities" but this clause was suppressed during the reign of Mary and not restored in the Elizabethan Prayer Book (1559). In principle therefore the prayers for "all Bishops, Priests and Deacons" included the Pope. For the Church of England has always recognized the validity of Roman orders, never re-ordaining converts from that Church though insisting on the episcopal ordination of all Protestant ministers. But it is perfectly clear from the examples cited, as from many other parts of the Prayer Book, that the highest authority in Church and state is exercised not by the Bishop of Rome but by "all Christian Kings, Princes and Governors." The Henrician Act of Supremacy (1534) had declared the King to be "only supreme head in earth of the Church of England;" the corresponding Elizabethan Act (1559) made the Sovereign "the only supreme governor of this realm ... as well in all spiritual or ecclesiastical things or causes as temporal." The legislation claimed with some plausibility to be "restoring to the crown the ancient jurisdiction over the state ecclesiastical and spiritual."[38] Whatever the status of the claims and counterclaims respecting the continually shifting limits of papal and royal jurisdiction in England during the five hundred years from the Norman Conquest to the accession of Elizabeth I (1558), the relatively stable Coronation liturgy throws some light on the traditional – Western – view of Christian kingship.

The English Coronation service descends from the *Pontifical of Egbert* (c. 750), revised for the coronation of Edgar in 973; and the twelfth-century *Pontifical of Magdalen College*. A fourteenth-century order used for Edward II (1307) draws upon each, and became the *Liber Regalis*, used at all subsequent coronations down to and including that of Charles II (1661). The service was translated into English in 1603 when the communion office from the *Book of Common Prayer* replaced the Latin mass. The most important features of *Liber Regalis* are the oath (translated in 1485) and the anointing. The King was required to swear that he will "keepe to the people of England the Lawes and customes and Liberties graunted to the Clergie, and people by [his] Predecessors and glorious King Saynct Edward;" to "keepe after [his] strength and power the church of God to the Clergie. And the people hoole peace and godly concord;" to make to be done "rightfull Justice"; and to defend "such lawes as to the worship of God shalbe chosen by your people." At the heart of the service is the anointing, a quasi-sacramental rite of consecration: the prayers of the celebrant refer to the King's divine calling "to rule and govern;" the choir meanwhile sings an anthem which first appears in the *Pontifical of Egbert:*

> Zadok the Priest and Nathan the prophet anointed Solomon king; and all the people rejoiced and said: God save the king.[39]

The annual services appointed at the Restoration to commemorate the martyrdom of Charles I (30 January) and to celebrate the end of the Great Rebellion (29 May) make frequent mention of God's "Anointed," and on one occasion (30 January, Evening Prayer), refer to his "sacred person."

It would seem from this that the unity of Church and state under a Christian "governor" is neither a Tudor nor a Protestant invention. The King of England is subject to no temporal sovereign but receives his authority from God, the "King of kings, Lord of lords, the only Ruler of Princes" (Morning Prayer). He is bound by law, both divine and positive, the latter requiring the consent of his people. Subject only to that, he may properly be thought of as "head" of a particular "body:" the whole people of God within his realm. There can be no denying, however, that the special circumstances of the Reformation in England gave political urgency to the doctrine, and it is no surprise to discover it stated strongly in one of the Edwardine *homilies* (1547):

> That is Gods ordinance, Gods commandement and Gods holy will, *that the whole body of every realme and all the membres and partes of the same shalbe subject to their hed*, their kyng.[40]

It is important to understand, nevertheless, that the identity of Church and state is contingent, not essential. "A commonwealth is one way, and a church another way, defined," as Hooker acknowledged. But a "schoolmaster" and a "physician" are defined differently, yet "both may be one man." Therefore "in such a politic society as consisteth of none but Christians ... one and the selfsame multitude may in such sort be both:" "For if all that believe be contained in the name of the Church, how should the Church remain ... divided from the commonwealth, when the whole commonwealth doth believe?" This is not the case "under dominions of infidels;" nor is it "in those commonwealths where the Bishop of Rome beareth sway." For in the latter case "the Bishop of Rome doth divide the body into two diverse bodies, and doth not suffer the Church to depend upon the power of any civil prince or potentate." But "in this realm of England ... with us one society is both Church and commonwealth;" and "our Church hath dependency upon the chief in our commonwealth.. . according to the pattern of God's own ancient elect people."[41]

The argument of Section 2 is now complete. I have attempted to show first, that the ecclesiology of the *Book of Common Prayer*, deriving as it does from the Pauline, soteriological conception of the Church as the body of Christ, was then believed to imply an ecclesiastical polity based upon the principle of subordination. The restoration of the hierarchy in 1660 and the imposition of the Act of Uniformity two years later were more than simply political expedients: they were the logical corollary of a deeply held and coherent set of beliefs about Christ and his Church transmitted from Christian antiquity in Scripture and the liturgy. But in a Christian society the state is subsumed under the Church: "Kings, Princes and Governors," "Lords of the Council" and "Magistrates" are functionally differentiated members of the Church in exactly the same way as "Bishops, Priests and Deacons." Only one head can rule the body, exercising supreme authority "as well in all spiritual or ecclesiastical causes as temporal." Civil polity therefore, like ecclesiastical polity, requires a graded distinction of ranks and function and depends upon the principle of subordination. In such a Church/commonwealth indeed, ecclesiastical polity is merely a subset of civil polity and the distinction between the two is uncertain. John Jewel defended parliamentary control of the Tudor Church against papist propaganda *(Apology,* 1562) on the grounds that Parliament was "a body of Christians, both lay and clerical, representing the whole Christian commonwealth."[42]

Dissenters from the Restoration settlement of 1662 dissented first and foremost from episcopalian *ecclesiastical* polity and from the "popish liturgy" which presupposed and ratified it. But in so doing they were obliged to dissent from the wider *civil* polity in which the Church establishment was embedded. Though the Presbyterians among them, at least, might have settled for the principle of subordination within a polity of their own choosing, their failure to hijack the establishment during the Great Rebellion compelled them eventually to look for an ideological alternative in democratical principles.

3.3 Dissent

In Section 1 of this chapter have shown that the *Book of Common Prayer* transmits a Catholic ecclesiology: a biblical and patristic understanding of the Church as a divine society, appropriately represented by the metaphor of the body of Christ; in and through which – *and only in and through which* – each human individual may be reborn in the image of God and live the new life of the Risen Christ. I have also shown that because, on this view, Christ is the "only mediator" between God and man, it is necessary that he should be both God and man. Orthodox (Nicaean and Chalcedonian) Christology implies and is implied by Catholic ecclesiology.

In Section 2, I have shown that Catholic ecclesiology seems to imply, and at any rate is fully congruent with, an organicist conception of the "church militant" here on earth; and therefore with an hierarchical ecclesiastical polity dependent upon the principle of subordination. I have also shown that in a Christian society the Church subsumes the state; that kings and "all that are put in authority" under them exercise a differentiated function within the body; and hence that ecclesiastical polity is a subset of civil polity. The principle of subordination extends to all aspects of human social life, wherein temporal society imitates the eternal. For "God ... appointed his angels and heavenly creatures in all obedience to serve and honour his majestie." In the words of the Elizabethan *homily against Disobedience and Wilful Rebellion*, ordered at the Restoration to be read at the annual celebration of King Charles the Martyr: "it is evident, that obedience is the principal vertue of all vertues, and in deede the very roote of all vertues, and the cause of all felicitie."[43]

Enough should now have been said to make it perfectly clear just what the English Protestant Dissenters were dissenting from. It remains to consider how this may bear upon the putative nexus between the theology and the political theory of Dissent.

According to Joseph Priestley, "as Dissenters we have no peculiar principles of civil government at all." Both he and Richard Price were at pains to disavow republicanism or "a pure Democracy," and affirm their attachment to limited monarchy and "the constitution of England."[44] The Dissenters' apprehension "of any approach towards arbitrary power" is explained as a function of their class location in "the middling and lower ranks of the people." There is no connection between the Dissenters' rejection of a "visible head in religious matters" and "a fondness for equality, and republican maxims in the state." Yet all Dissenters are "friends of liberty;" and Priestley acknowledged that "the religious system of the Dissenters is unfriendly to arbitrary government in any form, but favourable to liberty in general."[45] This began to be politically important during the American War, when "a consistent Americanism was publicized by the Dissenting pulpit thundering forth the rhetoric of liberty," accompanied in many cases by more active opposition to the government.[46] By the end of the century a former Dissenting minister could write that:

> as a collective body of men (for there are some truly loyal and excellent individuals among them) the protestant dissenters have been aiming, to adopt Dr Priestley's elegant term, "to undermine and blow up the constitution.[47]

The revolution in France was viewed by the Dissenters "with an enthusiasm bordering upon frenzy."[48] Despite Priestley's disclaimer therefore, it should be and actually is possible to distil out of his writings and those of Richard Price a coherent set of political principles which stand sharply opposed to the "principle of subordination;" and which may be regarded as characteristic of the most articulate, as well as the most extreme of the "third and last generation" of the Dissenting "Commonwealth men."[49] I shall label these principles "Dissent" to acknowledge that they may not have been held in their entirety by all, even among the tiny intellectual elite known as "Rational Dissent."

"Without Religious and Civil liberty," wrote Price in 1776, man is "a poor and abject animal ... bending his neck to the yoke of every silly creature who has the insolence to pretend to authority over him."[50] "Dissent" begins with a total repudiation of the doctrine that "obedience is the principal vertue of all vertues ... the cause of all felicitie." Obedience of some kind there must be, of course, else society "cannot avoid falling into a state of anarchy," as Price made clear in his famous sermon "On the Love of our Country." But such obedience is due, not so much to one's superiors in Church and state as to those "regulations agreed upon by the community" without which "the ends of government cannot be obtained." The principle of subordination is to be abandoned and a merely instrumental obedience put in its place: "the dominion of Kings changed for the dominion of laws, and the dominion of priests giving way to the dominion of reason and conscience."[51]

By *civil liberty* Price generally means what Priestley more carefully defined as *political liberty*.[52] Now such "liberty ... is too imperfectly defined when it is said simply to be 'a Government by LAWS and not by MEN'." It is essential that "all the members of a state" should participate in the making of such laws;[53] and also that they have the power "of arriving at the public offices, or, at least, of having votes in the nomination of those who fill them."[54]

Priestley had shrewdly noted that Dissenters "agree in nothing but in dissenting from the doctrine and discipline of the Established Church:" as a body they are so scissile that "it is almost impossible for them to act in consent on anything."[55] Price argued that "to be free is to be governed by one's own will, and to be governed by the will of another is characteristic of *Servitude*."[56] It is this view of the intransigent individualism of their co-religionists that informs the "Dissenting" conception of society inspired by John Locke, as "a number of persons united by their common interest, and by the use of the same measures to promote that interest." The state is not a divinely ordained *remedium peccatorum*: it is a man-made instrument for achieving the – strictly limited – common goals of a number of radically isolated individuals. It follows immediately from this that *"all civil power is ultimately derived from the people"* and not *"from God;"* that "kings, senators, or nobles ... are the *servants of the public;"* and that "an injured and insulted people" may "assert their natural rights" and punish a delinquent executive. In the seventeenth century most "friends of liberty" still believed in the *"sacredness of kingly power;"* but "whenever that superstitious notion shall be obliterated" then even regicide "will appear an immortal honour to this country." What also follows is what, with the aid of hindsight, we recognize as the most literally "radical" doctrine of "Dissent:" which is that "the good and happiness of the members, that is *the majority of the members of any state*, is the great standard by which every thing relating to that state must finally be determined."[57] "It was this passage," according to one commentator, "that inspired Jeremy Bentham."[58]

What is the connection between these "democratical principles" so vigorously maintained by "Dissent," and the heterodox Christology of the "*Rational* or *Unitarian* Dissenters" who held that religion ought to be "put upon the same footing with other branches of knowledge"?[59] It has already been noted that no *unique* link exists between Catholic ecclesiology (the Church as body of Christ) and establishment ecclesiastical polity (hierarchy). For as Hooker had conceded, plenary authority was left to "the whole body of the Church," wherefore though bishops may "vouch with conformity of truth that their authority is descended from ... the ... apostles" yet they "must acknowledge that the Church hath power ... to take it away."[60] It is possible in principle for a Presbyterian or Congregational polity to be determined by the authority of the whole Church. It is also possible, moreover, that such a polity might be consistent with a high degree of subordination in civil polity. During the interregnum, as Priestley observed, "The *Presbyterians* ... would have made a most intolerant [establishment], similar to that of the Church of England in those days."[61]

These possibilities are displayed in Figure 3.2, in which the dotted arrows show the trains of reasoning by which an orthodox Dissenter could adopt democratic principles on the one hand (a), and a Rational Dissenter maintain the principle of subordination on the other (b). But in fact these lines of argument seldom appear in the Dissenting literature of the period. Though Hooker had admitted – to use modern terminology – that episcopacy is of the *bene esse*, rather than the *esse* of the Church, the conditions for a change of polity ("universal consent upon urgent cause") are too stringent to have been met either then or previously. As for the second link (b), it has been strongly argued by James E. Bradley that:

> The distinctive contribution of nonconformity to political radicalism is found neither in its orthodoxy nor its heterodoxy, but in its

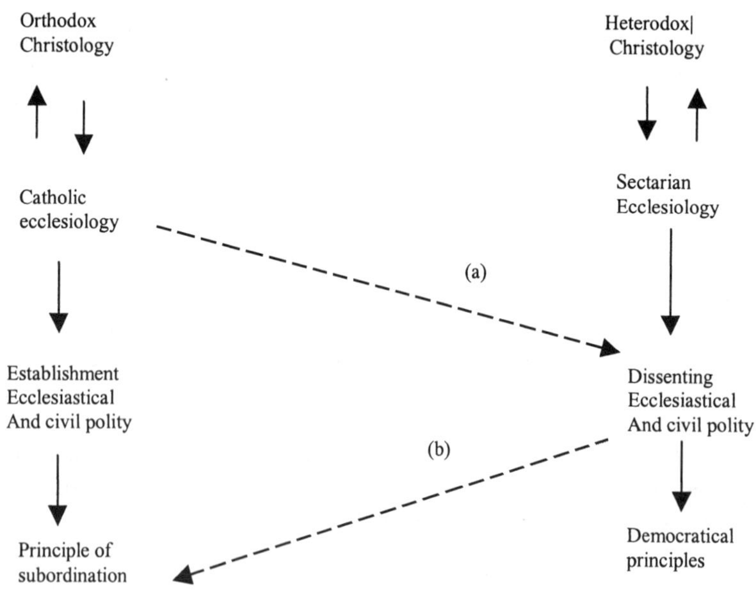

Figure 3.2 Possible connections between Christology and political principles

> interpretation of human autonomy and *ecclesiastical polity*, and the application of these doctrines in a revolutionary setting.[62]

Though many or most Dissenters, especially among the wealthy, may have insisted upon subordination in their households and businesses,[63] their political principles are adequately described by the "Dissent" of Priestley and Price. Though it cannot be rigorously demonstrated, therefore, that "radical politics" are actually "entailed by Socinianism" there is much to be said for Priestley's view that "the religious system of the Dissenters is unfriendly to arbitrary government in any form." Above all this is the case with the "religious system" of the Rational Dissenters.

For – to paraphrase the quotation from Jonathan Clark at the beginning of this article – a Christology that denies the two natures in the one Person of Christ (and that therefore rejects the (doctrine of the Holy Trinity) destroys the conception of Christ as a mediator between God and man. It is not possible for humans to restore the *imago dei* defaced by the Fall – to achieve union with God by union with Christ in his mystical body, the Church. The Pauline conception of the church, and its patristic and medieval elaboration in Catholic ecclesiology, can be dispensed with. The Church can be regarded not as a divine society outside of which there can be no salvation, but simply as a voluntary association: an option for "those that dislike that mode of worship which is prescribed by public authority."[64] The link between ecclesiology and ecclesiastical polity is broken; and so is that between ecclesiastical and civil polity. Church order becomes a private matter to be decided by each association of like-minded believers choosing to describe itself as a "Church." As to public order, "liberty" can take pride of place and "obedience" be radically deflated. A substantial portion of Priestley's published works – certainly more than 50 percent – can be regarded as a sustained attempt to establish these propositions.

Priestley's fundamental assumption is that Christianity is and ought to be capable of being "properly *understood*." His "great outline" takes exactly one hundred words and contains "nothing that any person could imagine would lead to much subtle speculation at least such as could excite animosity." Any appearance of "mystery" in Christianity is evidence of some "corruption" of its pristine intelligibility. The use of ritual, symbol, artistic or poetic imagery is always "superstitious" and must be ruthlessly eliminated by all the "rational." An example of the Corpus Christi procession in Paraguay "fully confirms ... the boundless exuberance of the human imagination in things of this nature, and thereby supplies a strong argument for an early and vigorous opposition to them." "The friends of genuine ... and rational Christianity" are therefore grateful to infidels such as Gibbon, for by their attacks "whatever has been found to be untenable has been gradually abandoned."[65]

Upon this basis Priestley constructed his *Institutes of Natural and Revealed Religion*, the *History of the Corruptions of Christianity*, *An History of Early Opinions Concerning Jesus Christ*, *A General History of the Christian Church*, *Discourses on the Evidences of Revealed Religion*, and *Notes on all the Books of Scripture*.[66] Texts of scripture which seem to support the doctrine of the Incarnation (e.g. John 1: 14, 8: 58, 9: 38, 20: 28; Matt. 28: 18, etc.) are dismissed as "figurative," or given a naturalistic explanation, or simply ignored. Passages in the gospels (e.g. John 14: 6) or Epistles (e.g. I Cor.15: 22) which imply that Christ mediates between God and man and that the latter have access to God, "in" "through," "by," or "with" Christ are similarly treated. All evidence that Christ made "Atonement" for the human race is merely "figurative." It is to the Pelagian controversy that "we owe the doctrines of *original sin, predestination* ... and ultimately that of *atonement*:" because of St Augustine's "superstitious and absurd opinion" that baptism is a "washing away sin." It "immediately follows" from the recovery by Faustus Socinus of "the original doctrine of the proper *humanity of Christ*" that "his death could not in any proper sense of the word, atone for the sins of other men."[67]

Because of Priestley's denial of, or inability to comprehend, the Pauline doctrine of the Church as the "mystical body of Christ" there can be "no *human authority* in matters of religion."[68] Moreover, even were Christ incarnated in the Church, a denial of the Atonement destroys any idea of the Church as "the ark of salvation." There can be no reason to insist on unity, hence no meaning to be attached to Hooker's conception of the identity of Church and state in "this realm of England." Priestley derided Burke for his view of "church establishment ... as ... an indissoluble union;" that for the English, "church and state are ideas inseparable in their minds."[69] His polemical strategy, repeated in many of his writings, is to ignore the theologically defensible conception of a *union* of Church and state; attribute to his opponents the entirely different, Warburtonian theory of an *alliance* between the two; and then to dispose of the latter. Now Warburton's "alliance" is fatally vulnerable to the necessity of regarding "Church" and "State" as separate entities. By ignoring the conception of union and concentrating on that of alliance, therefore, the way was opened for Priestley's view of the state as an instrumental association of individuals, and to his claim that "a multiplicity of sects" is "beneficial to the state."[70]

I hope it is clear from the foregoing that the correspondence between the Arian or Socinian Christology of the Rational Dissenters and the democratic principles of "Dissent" is sufficiently close for the former to be justly regarded as "conceptually basic" to the latter. Whether the orthodox Dissenters should therefore be regarded as "irrational," and whether all the Rational Dissenters perceived the congruence between their theology and their politics as acutely as Priestley did, are matters for a larger inquiry. One thing is certain: the Rational Dissenters as a class had the intellectual ferocity to go for the jugular: the central Christian conviction that Jesus Christ is mediator between God and man. The destruction of this one proposition, as I have shown in Sections 1 and 2, is sufficient to bring down the entire structure of establishment social theory.

Professor Horton Davies[71] has described the evolution of Unitarian worship during the eighteenth century. Beginning with proposals by heterodox Churchmen such as Clarke and Whiston for revision of the Prayer Book, and ending with William Wood's Mill Hill liturgy of 1801, there are certain common themes. In the first place a pre-composed liturgy was desired by the Rational, partly as a safeguard against enthusiasm, partly to distinguish their meetings from the uncouth spontaneity of their social inferiors, the orthodox Dissenters. Secondly, the three creeds, the *Gloria Patri* and all other references to the Holy Trinity were to be expunged. Thirdly, the traditional endings of prayers, "through Jesus Christ ...," all prayers addressed to Christ, all biblical material relating to the Incarnation of God in Christ (including the Virgin Birth), and any hint that baptism is incorporation into the body of Christ were gradually abandoned; for these imply or suggest in some way the fundamental doctrine to be denied: that Christ is mediator. Finally, as the inevitable (and sought for) consequence of that denial, hierarchy and the sacerdotal function of ministers were eliminated: absolutions and consecrations had to go and even blessings were suspect.

If Christ be not the mediator, the question must arise as to his function, if any, in a theology of true religion. In general, Rational Dissenters were silent on this point though Priestley at any rate believed in the Resurrection and the Second Advent.[72] Whether from habit or deference to scriptural phraseology both Price and Priestley made occasional, infrequent allusion in their sermons to Christ as "saviour."[73] The question then becomes, from what are humans believed to be "saved" by Christ?

Any recognition of the traditional Pauline, Augustinian doctrine of original sin and the Fall was deeply repugnant to Rational minds, partly because it mocked their religion of benevolence and enlightenment, partly because any useful theory of the Atonement seemed bound to lead back to the "gloomy and cruel superstition" that God and man are estranged, and that Christ can reunite the two.

It was left to the last generation of Rational Dissent to accept the full implications of this difficulty; and also to push to its furthest extent the repudiation of the principle of subordination. In that he weakened somewhat in his third edition, the second edition of William Godwin's *Political Justice* (1796) represents the intellectual high-water mark of Rational Dissent. Godwin acknowledges an "author of the universe" who allows no miracles and whose functions seem to extend no further than creation. There is no mention of Christ. It is impossible that there should be any "innate perverseness" in humans, "for man is thought, and, till thought began, he had no propensities either to good or evil."[74] Hence the "original sin of the worst men" is the perverseness of social institutions such as private property, marriage, wage labor and the like which cause men to become evil. In principle all of these can be reduced to inequality of property, hence "salvation" is a strictly temporal objective to be achieved by the equalization of property.[75] This is not to be performed within political society however, but rather by the dissolution of political society. For even the instrumental "obedience" admitted by Price and Priestley is objectionable; moreover, all society requires interdependence, and it is interdependence that underlies the principle of subordination, as Jonathan Swift explained so lucidly in his sermon *On Mutual Subjection*:

> God Almighty hath been pleased to put us into an imperfect state, where we have perpetual occasion of each other's assistance. There is none so low, as not to be in a capacity of assisting the highest; nor so high, as not to want the assistance of the lowest.[76]

Now "where there is a mutual dependence, there must he a mutual duty, and consequently a mutual subjection;" and "the practice of this duty ... would make us rest contented in the several stations of life, wherein God hath thought fit to place us."[77] Godwin faced the problem squarely and frankly "Everything," he declared with characteristic rigour, "that is usually understood by the term cooperation, is in some degree, an evil." All human society, even that of marriage and the family, must go: "we ought to be able to do without one another."[78] Neither "rationality" nor "dissent" could further go.

4

Intellectual Foundations of Tory Doctrine

The meaning of a text is known as much by what it omits as by what it relates. So we are told by the most fashionable authorities, who omit to tell us their own political agendas. According to this bizarre and truly academic way of using English, the whole of human experience is "text." We may therefore discover the political culture of any of the possible "Australias" which made Eugene Kamenka by examining those elements which none of them contain.

In John Buchan's most famous novel a colonial adventurer harangues a Labour Party meeting in Britain and wins tumultuous applause by announcing that he comes from a land where there are no *Tories* whatsoever, "Only Labour and Liberal." There could be no sharper contrast between Australia and her sister-Dominion, where Buchan became Governor-General and in which I have lived most of my adult life.

English Canada was formed as an act of Tory defiance against the Whiggish ideology of the American Revolution. In French Canada, for more than a century after Bastille and the Reign of Terror, the Roman hierarchy taught the faithful that a Protestant monarch is better than a godless republic. The potent symbols of monarchy have always been Canada's first line of defense against the cultural imperialism of her powerful neighbor to the South. Though affection for the "antient constitution" has waned in my lifetime, there is no trace of that godly fervor for a godless republic which now possesses so many otherwise sane Australians.

In a conference assembled to honor Eugene Kamenka for his many contributions to the scholarship of politics, it seems appropriate to consider the cultural context of his work. As we may only define "dissent" by understanding that from which "dissenters" differ, so we may only fully understand Australian political culture by an informed awareness of that vast *lacuna* which distinguishes it from that of Canada and Britain. I therefore propose to explain the pure theory of Tory politics which gradually took shape in England between the Restoration of Church and King after the Great Rebellion, and the American Revolution of the 1770s. And in so doing I shall identify a distinctive ingredient in the political culture of the South of England: which is the only place on earth, Eugene told us, where he could say "We Australians ...".

4.1 Subordination

"England from 1660 to the 1770s," said Jacob Viner in his Jayne Lectures for 1966:

> strikes me as presenting a unique case of a spontaneous union of theologians, philosophers, economists and intellectuals in general, maintained without serious breach of continuity for over a century and without a single outstanding heretic or dissenter, and dedicated to the justification of an existing social structure and especially of its economic and social inequalities.[1]

What seemed to Viner to unite Whig magnates, bourgeois Dissenters and Scotch political economists with Cambridge wranglers, Tory pamphleteers and non-juring divines was "the belief in the providential origin of the stratified social order."[2] Though Viner is correct to say that everyone in this period, even the Rational Dissenters, thought that the lower orders should know their place, their reasons for thinking so differed widely. What I shall understand in this paper as the *grand scheme of subordination*[3] is both more and less than a belief in the providential origin of the stratified social order. And it is clearly separable, as I shall attempt to show, from the idea of the Great Chain of Being made famous by the work of A. O. Lovejoy.[4] In its most articulated form it is nothing less than the intellectual basis of eighteenth-century Tory doctrine, sunk without a trace by the Whig historiography of present-day intellectual historians.[5]

A brief statement of the principle is contained in an address to his Ordination candidates by Archbishop Thomas Secker in 1769, conveniently re-printed for Cambridge undergraduates in 1785 by the reputedly "latitudinarian," certainly Whiggish, Bishop Richard Watson of Llandaff:

> Without union there cannot be a sufficient degree either of strength or beauty; *and without subordination there cannot long be*

> *union*. Therefore obey, as the Apostle directs, them that have the rule over you ...[6]

Subordination is a necessary condition of "union." Moreover, it is a religious duty. However, in order to identify the principle precisely, so as to distinguish it from the redundant metaphysical trappings of the Great Chain of Being and to locate if without ambiguity in the Tory tradition of political discourse, more detail is required. I therefore propose a "rational reconstruction" of the scheme of subordination in eight propositions, some of which were acknowledged by all; but all of which were acknowledged only by Tories:

(1) A society in which all are politically equal is unfeasible: some must be inferior to others in the sense that certain decisions that affect them are made only by their superiors.
(2) This implies no *natural* inequality. In all save the politically relevant respects, inferiors are upon the same footing as superiors (and may actually exceed them in such matters as natural ability, education and virtue).
(3) The obedience of inferior to superior is a necessary condition of peace and social order.
(4) Superiors have obligations to their inferiors and duties to them which they must perform. There must be, in the phrase made famous by Swift, a *mutual subjection* of each to all.
(5) The conditions in (3) and (4) arise because human society, the *body politick*, resembles a single organism in this respect: that without mutual subordination of the several parts there must be disease, death and dissolution.
(6) In a Christian society the body politick is *ipso facto* the church, the body of Christ.
(7) Thus in a Christian society – in post-Restoration England, at any rate – obedience and mutual subjection are Christian duties. The state, like the church, is held together by love.
(8) All must obey someone, if only the king: and the king must obey God. Thus a Christian society, in which obedience is voluntary and an act of love, imitates the divine society, in which angels and archangels rejoice to worship and obey the King of Heaven.

Propositions (1), (2) and (3) could be and were assented to by those regarded as Whigs. Propositions (1), (3) and (4) but not (2) seem to be consistent with the position of those few eighteenth-century authors who defended subordination in terms of the Great Chain of Being. Tories accept all of the first four propositions and explain then in terms of (5). But as I shall show, (5) leads to (6), the characteristically Tory conception of the unity of church and state, of which (7) and (8) are corollaries. It is well-known that "Whig" and "Tory" are large categories, each of which lumps together very considerable variety of belief and practice. It is evident, moreover, that the content of neither was stable during the century between the Glorious Revolution and Burke's Reflections. Like all rational reconstructions my model is a deliberate caricature: not history but a tool of history. In this respect it closely follows Adam Smith's treatment reported below.

In part 2 of this paper, I shall explain how Whigs could accept the first three propositions. In part 3, I shall discuss the relation of subordination to the Great Chain of Being. And in part 4, I shall present the alternative, genuinely Tory account in terms of theological organicism.

4.2 Instrumental justification of subordination

There can be few more Whiggish documents than Richard Price's sermon *On the Love of our Country*, delivered to the Revolution Society a few months after the fall of the Bastille and the occasion of Burke's famous rebuttal in his *Reflections*. Price welcomed the American and French Revolutions in high-flown, even apocalyptic language and celebrated the principles of the English Revolution: the right to liberty conscience in religious matters; the right to resist power when abused; and the right "to chuse our own governors, to cashier them for misconduct, and to frame a government for ourselves."[7] However, whilst denouncing "the odious doctrines of passive obedience, non-resistance and the divine right of kings," and noting the "dangerous tendency" of "adulation and servility" to those with power, Price was careful to warn against the opposite tendency: that of "spurning at all public authority, and throwing off that respectful demeanor to persons invested with it, *which the order of society require*."[8] For though civil government is only "an instrument of human prudence" for protection of persons and property, magistrates and officers must be obeyed: for "without this obedience the ends of government cannot be obtained, or a community avoid falling into a state of anarchy ...".[9] It is clear that Price accepts propositions (1 (2) and (3). It is quite evident from the writings of both Price and Priestley moreover, that Rational Dissenters, like less extreme Whigs, insisted on subordination both in their own households and in their business establishments.[10]

What distinguishes what I shall call the "pure Whig" understanding of subordination from that of either Soame Jenyns or Samuel Johnson is that for the "pure Whig" subordination is merely instrumental and has no necessary metaphysical or theological justification; though as an observable feature of actual human societies it might have naturalistic explanation in terms of human psychology. The clearest account of both the positive and the normative understanding of subordination in Whig discourse is to be found, as one might expect, in Adam Smith.[11]

Samuel Johnson had defended subordination as "conducive to the happiness of society" on the grounds that there is "a reciprocal pleasure in governing and being governed."[12] Adam Smith devoted many pages, in *The Theory of Moral Sentiments* to explaining the "pleasure in being governed" as a "natural" propensity to deference closely related to the tendency to overvalue "the pleasures of wealth and greatness." When transmuted into emulation the latter affords the motive power of economic activity. Individuals are "led by an invisible hand" to bring about an optimum of production and distribution, "and thus, without intending it and without knowing it, to advance the interest of the society ...".[13] In the *Lectures on Jurisprudence* this positive notion "that everyone naturally has a disposition to respect an established authority and superiority in others" is labeled the *principle of authority* and contrasted with the normative "principle of common or general interest" subsequently called the *principle of utility*:

> Every one is sensible of the necessity of this principle to preserve justice and peace in society. By civil institutions the poorest may get redress of injuries from the wealthiest and most powerful, and tho' there may be some irregularities in particular cases ... yet we submit to them to avoid greater evils. It is the sense of public utility, more than private, which influences men to obedience.[14]

It is evident that this analysis both justifies the Whig agreement with the first three propositions of the grand scheme of subordination, and makes it clear that there is no conflict between these propositions and the new paradigm of "spontaneous order"[15] by which authors of the Scottish Enlightenment are supposed to have inaugurated a "fully secular discussion of their society and its destinies." The normative principle allows the former; the positive principle is consistent with the latter. The principle of authority, moreover, suggests an explanation of why Tories might feel it necessary to add more to the picture:

> In all governments both of these principles take place in some degree, but in a monarchy the principle of authority prevails, and in a democracy that of utility. In Britain, which is a mixed government, the faction's [sic] formed sometime ago under the names of Whig and Tory were influenced by these principles; the former submitted to government on account of its utility and the advantages which they derived from it, while the latter pretended it was of divine institution, and to offend against it was equally criminal for a child to rebel against it's [sic] parent. Men in general follow these principles according to their natural dispositions. In a man of a bold, daring and bustling turn the principle of utility is predominant, and a peaceable, easy turn of mind is usually pleased with a tame submission to superiority.[16]

Before turning to the Tory pretence of "divine institution" so despised by the "bold, daring and bustling" Whigs, we must dispose of a metaphysical conception which, though widely held in Britain during the eighteenth century, was neither necessary for, nor congenial to, either way of talking about politics.

4.3 The Great Chain of Being

In his famous study of *The Great Chain of Being*, A. O. Lovejoy traced the transmission of this "unit-idea" from its origins in Plato through to the beginning of the nineteenth century. The idea was shown to inform Renaissance and early modern cosmology; the philosophy of Locke, Leibniz, Spinoza and Kant; and eighteenth-century biology. It is found in the theology of Hooker, the politics of Bodin,[17] and in the poetry of Alexander Pope. And as all the world knows, it found its way through the last of these into Soame Jenyns's notorious *Free Inquiry into the Nature and Origin of Evil*,[18] there to form the occasion and the target of Samuel Johnson's "most exquisite critical essay."[19]

According to Lovejoy, who bases this judgment chiefly on Pope and Jenyns, the Great Chain of Being could be, and was in eighteenth-century Britain, used in two different ways to inform the ideological defense of economic and social inequality. In the first place, it is an implication of this idea that humans occupy an intermediate position in the universe:

> Plac'd in this isthmus of a middle state,
> A being darkly wise and rudely great ...

> Created half to rise and half to fall,
> Great lord of all things, yet a prey to all.

"A creature so limited," Lovejoy pointed out, "must necessarily be incapable of attaining any very high level of political wisdom or virtue, and ... consequently no great improvement in men's political behavior or the organization of society would be hoped for."[20] In the second place, human society is, or ought to be, an example of the Great Chain of Being, and – as with the universe as a whole – to reveal a finely graded ordering from highest or greatest, to lowest or least. In Jenyns's words "the universe is a system whose very essence consists in subordination; a scale of beings descending by insensible degrees from infinite perfection to absolute nothing."[21] Tories like Johnson shared Jenyns's political pessimism, though for quite different reasons as we shall see. But neither Whig nor Tory could swallow the metaphysical justification of subordination because neither (though here again for very different reasons) could accept the "natural" or divinely ordained human inequality that it entails.

The difficulty thus presented to Tories at any rate is illustrated in Lovejoy's treatment of the theme. He began by quoting Pope:

> Order is Heav'n's first law; and thus confest,
> Some are, and must be, greater than the next,
> More rich, more wise; ...[22]

According to Lovejoy, a "fundamental premise" of the *Essay on Man* is that "hierarchic gradation is everywhere required by the divine Reason" hence:

> The doctrine of the Chain of Being thus gave a metaphysical sanction to the injunction of the Anglican Catechism: each should labor truly "to do his duty in that state of life" – whether in the cosmical or the social scale – "to which it hath pleased God to call him."[23]

Now it is obvious that Lovejoy could only make his point by an egregious misquotation of the *Book of Common Prayer*. The actual words of the Prayer Book Catechism in every version from 1549 to its final form in 1662 are a promise by the catechumen "to do my duty in that state of life unto which it *shall* please God to call me" (my italics). The future tense leaves everything open and undetermined. Who shall say whether the child in the pew will follow the humble trade of his forefathers, or rise by his own efforts and abilities to the highest offices in church and state? Laud was the son of a small-town clothier; Secker of a Dissenting yeoman. Whatever may be the case in the *Essay on Man*, there is no textual warrant for interpreting the Anglican formularies as though they sanctioned the supposition that "to leave one's place in society is also 'to invert the laws of Order.'"[24]

Few Englishmen of the eighteenth century had a clearer grasp of this than Samuel Johnson. There can be no reason – save a desire to cloud the issue for polemical purposes – for treating his political theory with the disdainful contempt which has become common since Sir Leslie Stephen.[25] Johnson's review of Jenyns takes for granted both the fact and the necessity of subordination. But Johnson rejected the "metaphysical sanction" partly because the metaphysics were seriously defective; partly because he denied that subordination implied "imperfection" (and therefore natural inequality); and partly because he detested the political inferences that Jenyns drew.

"This doctrine of the regular subordination of beings, the scale of existence, and the chain of nature" Johnson had "often considered, but always left the inquiry in doubt and uncertainty." This is for the simply logical reason that "the scale of existence from infinity to nothing cannot possibly have being. The highest being not infinite must be ... at an infinite distance below infinity ... Between the lowest existence and nothing, is another chasm infinitely deep ... Nor is this all. In the scale, whenever it begins or ends, are infinite vacuities." The idea of a "scale of being," Johnson concludes, is "raised by presumptuous imagination," and we are therefore "little enlightened by a writer who tells us, that any being in the state of man must suffer what man suffers, when the only question is, Why any being is in this state?"[26]

Johnson's second line of argument was to question "whether subordination implies imperfection." For if it does, then some humans are lesser creatures than others, their inferiority is "natural" and their condition unalterable. But the weed as a weed is no less perfect than an oak. In his *Dictionary* (1755) Johnson illustrated the adjective "subordinate" from a passage in one of South's sermons: "It was *subordinate*, not enslaved to the understanding; not as a servant to a master, but as a queen to her king, who acknowledges a subjection, and yet retains a majesty."[27]

It was important for Johnson to detach subordination from any idea of natural inferiority because of the morally obnoxious inferences Jenyns and those like him drew from that comfortable doctrine.

"It is utterly impracticable," Jenyns had asserted, "for infinite power, to exclude from creation this necessary inferiority of some to others.

All that it can do is to make each as happy as their respective situations will permit ...." Thus "poverty, or the want of riches is generally compensated by having more hopes and fewer fears, by a greater share of health, and a more exquisite relish of the smallest enjoyments ...." The poor are to be maintained in this happy condition by lack of education. "Ignorance, ... the appointed lot of all born to poverty, and the drudgeries of life, is the only opiate capable of infusing that insensibility which can enable them to endure the miseries of the one, and the fatigues of the other."[28]

Johnson "loved the poor," said Hester Thrale, "as I never yet saw anyone do, with an earnest desire to make them happy."[29] "Sir!" cried he to Boswell in 1763, "all the arguments which are brought to represent poverty as no evil, shew it to be evidently a great evil. You never find people labouring to convince you that you may live very happily upon a plentiful fortune."[30] Johnson dismissed Jenyns's "gentle paraphrase" of poverty as "want of riches." Poverty means "want of necessaries," which is impossible to justify. "To entail irreversible poverty upon generation after generation" by denying them education is "cruel, if not unjust." Granted that the poor do not feel the "little vexations which sometimes embitter the possessions and pollute the enjoyments of the rich," But "this happiness is like that of a malefactor, who ceases to feel the cords that bind him when the pincers are tearing his flesh."[31]

The Chain of Being is nonsense. Even were it not, the differential imperfection of humans is untenable. And even if it were tenable, the political implications are ethically outrageous. "Subordination in human affairs is well understood," Johnson pointed out, but "when it is attributed to the universal system, its meaning grows less certain."[32] It is indeed the case that the necessity of subordination arises out of human imperfection, but not at all in the way Jenyns pretended. Man was created "perfect in a sense consistent enough with subordination, perfect, not as compared with different beings, *but with himself in his present degeneracy* ...."[33] The very real evils of poverty and inequality are not to be hidden by metaphysical whitewash. They are a consequence of Original Sin. For Johnson, like many Christian thinkers after St Augustine, the state with its necessary apparatus of subordination, inequality and injustice, bad as it is, is better than anything else. It is both God's punishment of human sin, and his remedy for it.[34]

After Johnson's review, continually cited by other authors for the next eighty years or more, we hear very little of the Great Chain of Being. Viner reported that the doctrine "was unquestionably used by some eighteenth-century writers as a kind of support for the legitimacy of social inequality, but rarely by important writers... . Most writers on social issues paid no attention whatsoever to the doctrine."[35]

4.4 Tory theory of subordination

Whigs and Tories agreed[36] on the first three propositions and differed from Soame Jenyns on the second of these. Whigs left it at that, but Tories went on to attempt an explanation of subordination in theological terms: not because they were "pleased with a tame submission to authority" as Adam Smith had affected to believe, but because to be a Tory[37] meant to be a high-church man; and to be a high-church man meant to inherit and take seriously a pre-Reformation tradition of catholic ecclesiology.

According to the elder Pitt the Church of England had "a Calvinist Creed, a Popish liturgy and Arminian clergy."[38] What Pitt meant by a Popish liturgy is that the *Book of Common Prayer* consists almost entirely of translations of pre-Reformation service books. When in 1661 the Prayer Book was restored along with the Crown and the episcopate it became the shibboleth which ever after divided churchmen (and the church party, the Tories) from Dissenters, latitudinarians, Socinians and infidels, all of whom were normally allied as Whigs. Its theology and its idiom saturated the consciousness of ten generations of English men and women and determined the largely unspoken presuppositions of Tory political thinking.[39]

I have elsewhere shown[40] that the *Book of Common Prayer* both assumes and inculcates a conception of the church not as a voluntary, human society of like-minded believers, but as a divine society, transcending space and time, uniting the living and the dead, joined with "Angels and Archangels, and with all the company of heaven" in the everlasting praise of God. Entering the church by faith and baptism the individual is "grafted into the body of Christ's church," made "a member of Christ;" in the eucharist worshippers believe that they are "very members incorporate in the mystical body of Christ;" in matrimony the congregation is reminded that its individuals are become "members of his body, of his flesh, of his bones."[41] It is precisely by incorporation in the Body of Christ that the individual may share in the resurrection, for only the body of Christ rises from the dead.[42] In summary, humans find "salvation" through union with Christ; union with Christ means union with his church, his (mystical) Body; hence follows the patristic slogan: *Salus extra ecclesiam non est.*[43]

A theological organicism of this kind informed the ecclesiastical polity of the Church of England. A single organism has only one control centre, or "head," which must be obeyed if its life is to be sustained. Christ is the head of the body, but has left plenary power to his church. The episcopate transmits the power first committed to the Apostle, and each bishop plays a strategic part in the structure of authority: priests, deacons and the laiety are bound to obedience, and the bishop himself must "profess and promise all due reverence to the Archbishop and to the Metropolitan church."[44] The question is, of course, whom should the Archbishops obey?

Whether Pope or Emperor should be supreme head of Christ's people on earth was never agreed in the centuries between the accession of Constantine the Great and the Reformation, nor were the scope and limits of royal as against papal jurisdiction in England.[45] The Henrician *Act of Supremacy* (1554) declared the king to be "the only supreme head in earth of the Church of England;" the corresponding Elizabethan Act (1559) made the sovereign "the only supreme governor of this realm ... as well in all spiritual or ecclesiastical things or causes as temporal."[46] Behind that claim lay the English Coronation service, originating in the *Pontifical of Egbert* (c. 750), revised for the coronation of Edgar in 973, and the twelfth-century *Pontifical of Magdelen College*. The *Liber Regalis*, used for Edward II(1307) and at all subsequent coronations down to that of Charles II (1661) draws upon each. In the Coronation oath (translated in 1498), the king declares that he will:

> keepe to the people of England the Lawes and customes and Liberties graunted to the Clergie, and the people by [his] Predecessors and glorious King Saynct Edward; and to keepe after[his] strength and power the church of God to the Clergie. And the people hoole peace and godly concord ... [and to defend] such lawes as to the worship of God shalbe chosen by your people.[47]

At the heart of the ritual is the anointing, a quasi-sacramental rite of consecration. The annual services appointed at the Restoration to commemorate the martyrdom of Charles I (30January) and to celebrate the end of the Great Rebellion (29 May) make frequent mention of God's "anointed;" and on one occasion (30 January, Evening Prayer) refer to the "sacred person" of the monarch.

It would seem, both from the development of the English coronation ritual and from the more general evidence presented by Ullman,[48] that the unity of church and state under a Christian "governor" is neither a

Tudor nor a Protestant invention – though the circumstances of the Reformation gave peculiar urgency to the doctrine and it is no surprise to discover it stated strongly in one of the Edwardine *Homilies* (1547) ordered to be read in churches:

> That is Gods ordinance, Gods commandment and Gods holy will, that *the whole body of every realme* and all the membres and partes of the same shalbe subject to their hed, their kyng ...[49]

Nevertheless, as Richard Hooker was careful to point out, the identity of church and state is contingent, not essential. "A commonwealth is one way, and a church another way, defined." But a "schoolmaster" and a "physician" are defined differently, yet "both may be one man." Therefore, "in such a politic society as consisteth of none but Christians ... one and the selfsame multitude may in such sort be both." Whatever may be the case "where the Bishop of Rome beareth sway," it is evident that "in this realm of England ... with us one society is both Church and commonwealth;" and "our Church hath dependency upon the chief in our commonwealth ... according to the pattern of God"s own ancient elect people... ."[50]

It was precisely this understanding of the nature of church and state that was contested before and during the Great Rebellion and restored in England in 1661. And it is precisely this which soon divided those who came to be known as Whigs and Tories. By 1755 Samuel Johnson could define *Tory* in his Dictionary as "one who adheres to the antient constitution of the state, and the apostolic hierarchy of the church of England, opposed to a Whig." But the "antient constitution" and the "apostolic hierarchy" are rooted in theological organicism. In Hooker's words, "... the very Deity itself both keepeth and requireth this for ever to be kept as a law, that whensoever there is a coagmentation of many, the lowest be knit to the highest by that which being interjacent may cause each to cleave to the other, and so all to continue one." Subordination is therefore a function of the organic nature of society. "Without order there is no living in public society, because the want thereof is the mother of confusion, whereupon division of necessity follows, and out of division inevitable destruction... . And if things or persons be ordered this doth imply that they are distinguished by degrees. For order is a gradual disposition."[51]

It has often been pointed out[52] that the metaphor of the *body politick* could be and was used by many who were not Tory. Sir William Petty, for example, began his *Political Anatomy of Ireland* (1672) by noting Bacon's "judicious parallel ... between the *Body Natural* and the *Body Politick*;"[53] and went on to employ the metaphor extensively himself. Sir Robert Filmer[54] and Jean Bodin each made the body politick resemble the family and thus extended patriarchy from the latter to the former.[55] What characterizes Tory organicism, I believe, is its dependence upon the primary theological image of the Body of Christ. It is because of the belief that "in this realm of England ... one society is both Church and commonwealth" that the biblical – specifically Pauline – ideas of organic union, interdependence and subordination of members could be transferred without modification from "church" to "commonwealth."

It has been seen that the ecclesiology of the *Book of Common Prayer* is unequivocally hierarchical. The ecclesiastical polity of the Church of England, reaffirmed at the Restoration in the Caroline *Act of Uniformity*, is wholly consistent with catholic ecclesiology transmitted by it "popish liturgy." It is equally clear however that the obedience necessary in the Body of Christ is seen, in the Prayer Book, as extending into all parts of human life, whether "ecclesiastical" or "civil." Wherefore each child must learn, for example:

> To honour and obey the King and all that are put in authority under him:
> To submit myself to all my governors, teachers, spiritual pastors and masters:
> To order myself lowly and reverently to all my betters[56]

The subordination required of members of the Body of Christ is completely fused and identified with the subordination required of members of the body politick.

For that very reason, however, Tory writers tend to stress not only obedience but *interdependence:* a notion which was far less congenial to the "bold, daring and bustling" Whigs.

Swift's sermon on *Mutual Subjection* presents the familiar "comparison which St. Paul maketh between the Church of Christ and the natural Body of Man: for the same Resemblance will hold, not only to Families and Kingdoms, but to the whole Corporation of Mankind." We learn from this, "the Nature of that Subjection which we all owe to one another. God Almighty hath been pleased to put us in a state, where we have perpetual Occasion of each other's Assistance." It therefore "plainly appears that no one human Creature is more worthy than another in the sight of God." For in all the Relations between Man and Man, there is a mutual Dependence, whereby the one cannot subsist without the other." And "where there is mutual Dependence, there must be a mutual Duty, and consequently a mutual Subjection."[57] For Christians mutual subjection is an act of love. There is negligible difference between Swift's doctrine and that of Thomas Starkey's *Dialogue between Pole and Lupsett* (c.1530) two centuries earlier, an important example of Henrician political thought that made love the cement that holds society together:

> ... yf al the partys of the cyty wyth love be not knyt togyddur in unyte as membres of one body, ther can be no cyvylyte ... [but] ... there ys perfayt cyvylyte ... where ... al the partys ... be knyt togyddur in perfayt love & unyte, evey one dowying hys offyce & duty ... & wythout envy or malyce to other accomplysh the same ...[58]

Swift used the strongest possible illustration to make the same point: mutual subjection "our Saviour himself confirmed by his own Example; for he appeared in the Form of a servant, and washed his Disciples Feet."[59]

According to Samuel Johnson, deliberately parodying himself as a corrective to Boswell's romantic conservatism,[60] "A high Tory makes government unintelligible: it is lost in the clouds."[61] What Johnson meant by that, perhaps, is the doctrine of the Elizabethan *Homily against Disobedience and Wilful Rebellion* (1570), ordered at the Restoration to be read as an alternative to the sermon at the annual fast day in commemoration of King Charles the Martyr. "The piety of Dr. Johnson was exemplary and edifying," Mrs Thrale tells us: "he was punctiliously exact to perform every public duty enjoined by the church."[62] He and Boswell must often have heard the Homily read in church on 30 January. Developing the idea that a Christian society is united by love in the divine society of Christ, the Homily goes on to draw an explicit parallel between the temporal and the eternal kingdom. For in the latter, "God ... appointed his angels and heavenly creatures in all obedience to serve and honour his majestie." Subordination is characteristic of the heavenly kingdom, thus in practicing it here on earth we imitate and prepare for the life to come. Wherefore "... it is evident, that:

> obedience is the principal vertue of all vertues, and in deede the very roote of all vertues, and the cause of all felicitie."[63]

I have attempted, in this part of the paper, to show that although, as Johnson stated, "A wise Tory and a wise Whig ... will agree" – on the necessity of subordination at any rate – "their modes of thinking are different."[64] Whigs accepted subordination, as Adam Smith correctly noted, because of its utility. The Whig conception of society could be individualistic, contractual and fully compatible with the emerging paradigm of "spontaneous order." But he missed the point in saying that Tories accepted subordination because they pretended that government "was of divine institution;" even though many, notably Burke in his final, quasi-Tory phase, made use of that idea.[65] Tories were steeped in the language and idiom of the *Book of Common Prayer* and other Anglican formularies such as the Coronation ritual, the books of Homilies, and the post-Restoration liturgies for 30 January and 29 May. By transmitting and inculcating the theological organicism of pre-Reformation Christianity this literature created minds which perceived social and political relations in terms of a sacramental unity in the Body of Christ and an earthly foreshadowing of the eternal city.

This would have required no explanation in the eighteenth century, which is why there is no systematic exposition of Tory doctrine, only occasional and belated sketches such as that by Daubeny,[66] and Keble's magnificently anachronistic sermon on *National Apostasy* (1833) which launched the Oxford Movement on its reactionary course. But the world had changed by 1829. The two most substantial theoretical contributions to arise from the crisis over Catholic Emancipation in the early nineteenth century, those by Coleridge and Gladstone, abandoned the assumption of the *unity* of church and state and accepted, more or less willingly, Warburton's very different – indeed, diametrically opposed – conception of an *alliance* between the two.[67] For it was in the years between 1827 and 1832, as Jonathan Clark has so clearly shown, that the *ancien régime* finally came to an end in Britain.[68] By the second quarter of the new century the idea of the unity of church and state was politically dead: and so, therefore, was Tory doctrine and the Grand Scheme of Subordination.

5
A Cambridge *"Via Media"*

Significant exceptions to the pervasive latitudinarianism of the Georgian Church have lately been found in the domain of public worship.[1] With respect to theology however the traditional view canonized by Sir Leslie Stephen ("Oxford was then at the very nadir of intellectual activity:" at Cambridge "the intellectual party of the Church was Socinian in all but name") remains undisturbed.[2] It is my object in this chapter to reappraise the performance of "the intellectual party" in the latter part of the eighteenth century.

Roughly speaking "the intellectual party" seems to comprise those wranglers of the mid-century or later who achieved note as theologians without quitting the Church: say from Thomas Balguy (1716–95, BA 1737) to George Pretyman (1750–1827, BA 1772). More narrowly it means what Stephen misleadingly labelled "Paley and his School;" more narrowly still, John Hey (1734–1815, BA 1755), Richard Watson (1737–1816, BA 1759) and William Paley (1743–1805, BA 1763).[3] It is evident that a marked variation exists between the intransigent orthodoxy of the younger members of this "school" such as Pitt's tutor, Pretyman, and the bland toleration – not to say laxity – evinced by its putative father,[4] Edmund Law (1703–87, BA 1723). I wish to suggest that a change of heart and mind may gradually have occurred during the last quarter of the eighteenth century; that the increasing orthodoxy of "Paley and his School" was more than merely a response to political changes; and that it proceeded rather from the inner logic of a common set of deeply held theological convictions. In particular I shall argue that "the intellectual party" was united by a non-Laudian, characteristically Cambridge vision of an Anglican *via media* between "blind attachment to system" on the one hand and the "contagion of Infidelity" on the other.

No completely new evidence will appear: only a re-examination of Hey's Norrisian lectures which seem not to have been read since 1876; and a consideration of certain other literature – such as Watson's *Collection of Theological Tracts* and some of Paley's sermons – which though readily accessible have never before, or only recently,[5] been brought to bear on the question. Most of what I have to say will relate to this evidence and is found in part 2. But since "externalist" and "internalist" explanations are complementary in the history of ideas, part 1 will review the connection between theology and politics in the latter part of the century. And since part of the reason for questioning Stephen's judgment lies in the fact that "Paley and his School" were treated with the utmost deference by the next two generations of High Churchmen and (non Calvinistic) evangelicals, part 3 relates some interesting pre-Tractarian responses.

5.1 The political and theological climate of opinion

According to what might be characterized as the "naive" model of Cambridge theology in the eighteenth century, "unitarianism" flourished under the benign patronage of Edmund Law until the subscription controversy, whereupon "men of more tender conscience" (Lindsey, Wakefield, Jebb, Disney and Evanson) seceded. "Paley and his School" held on to their benefices because "they attached too little importance to their dogmas to care for a collision with the Thirty-nine Articles." Whether or not they were sincere in their professions of orthodoxy, their "methods of reasoning led naturally to the Unitarianism which presents the nearest approach to a systematic evolution of opinion in the latter half of the eighteenth century." And the result of their attempts at self-justification was "to produce a literature more meager and superficial than any which had flourished since the days of Hooker."[6] Such was the judgment of Sir Leslie Stephen, concerned to represent eighteenth-century thought as a triumphant progress towards the *telos* of late-Victorian agnosticism. A reappraisal of the naive view must begin with intellectual life in Cambridge before 1771.

If there ever was an identifiable "intellectual party" its locus must have been the Hyson club, a society of resident former wranglers founded in 1757 or thereabouts to drink china tea and engage in rational conversation: Paley had become a member by 1766. There has been a tendency to view the club as a latitudinarian conventicle, possessing "a coherence and unity of purpose that indicate strong ties of sympathy and collaboration."[7] Yet its membership "usually consisted of some of the most respectable members of the university," several of whom were noted rather for rigor than for latitude. Isaac Milner (1750–1820, BA 1774) for example, refused to sign the petition against subscription in 1771–2; in later years he was powerful in stamping out heterodoxy, as were Beadon (1737–1824, BA 1758) and Pretyman.[8] It seems more probable, therefore, that "the intellectual party," though rejoicing in "free disquisition" as such men normally do, was wedded to no particular theological or political program. For the third quarter of the eighteenth century was the high summer of the English Enlightenment, a time when all men of taste and letters however disparate in faith or morals could meet in urbane society – Hume and Rousseau at the home of Daniel Malthus, Jack Wilkes and Samuel Johnson at the table of Mr Dilley – not yet knowing that the day must soon come when such amenities would cease.[9]

During the 1770s the world changed, and those who had come to maturity during the past twenty years found themselves gradually compelled to choose sides. In 1772 the move within the Established Church to abolish subscription was defeated. The failure in 1773 of the first of many petitions to parliament for relief forced the Dissenting Interest into detachment and disaffection: "an alienation from the House of Hanover and their ancient loyalties which grew progressively more complete."[10] The American Declaration of Independence gave heart to "patriots" and "friends of liberty" at home and abroad, being followed by the Gordon riots four years later and the storming of the Bastille after another nine. *Wealth of Nations* and Bentham's *Fragment on Government*, both of which were published in 1776, symbolically inaugurate a fundamentally new way of looking at human society and its ills. The first two volumes of *Decline and Fall* which also appeared in that year marked the beginning of a frontal assault on Christianity; Hume's posthumous *Dialogues concerning Natural Religion* was first printed three years later. Modern labels such as "conservative," "radical" or "progressive," hopelessly anachronistic before the 1770s, gradually became more applicable (from our own standpoint only, of course) to a generation living through an increasing polarization of belief and opinion. And, as J. C. D. Clark has lately argued, there was a high correlation between theological and political "radicalism" in the last three decades of the century: for Socinian theology "*entailed*" a credo of political revolution.[11]

The collapse of the Feathers Tavern petition of 1772 obliged the Anglican elite to make up its collective mind about subscription and even, to a considerable extent, about Trinitarian orthodoxy.[12] When John Jebb (1736–86, BA 1757) resigned his preferments in 1775 he was followed by others into Dissent or infidelity. Two younger members of the Hyson Club, Gilbert Wakefield (1756–1801, BA 1776) and Paley's former pupil William Frend (1757–1841, BA 1780), eventually gave up the struggle to reconcile mind and conscience to orthodoxy. Wakefield, who had at one time been an intimate friend of the conservative Pretyman,[13] went quietly in 1779. At Milner's instigation Frend was ejected with much screaming and kicking in 1793, having been dismissed as college tutor in 1788 by Beadon, who was at that time Master of Jesus.[14]

The men in the middle – Hey, Watson and Paley – found themselves inexorably drawn towards theological orthodoxy and political conservatism during the last quarter of the century. In part, no doubt, this was the mere consequence of advancing years and respectability. In 1780 Hey, who was then forty-six, became the first Norrisian professor. Two years later Watson, aged forty-five, was made bishop of Llandaff; and Paley, aged thirty-nine, archdeacon of Carlisle. But in larger part their seeming change of position was a response of clear, honest and critical minds to the pharisaical scruples of the theological objectors and the artless delusions of political Jacobinism.[15] It is perfectly true, as Le Mahieu asserts, that Paley and his friends believed in "tolerance, in promoting rational inquiry by creating an open market-place of ideas" and that they "demanded a climate of understanding if they were to survive and advance in the ecclesiastical hierarchy,"[16] but they also believed with Watson that "free disquisition is the best mean of illustrating the doctrine and establishing the truth of Christianity."[17] It is a fact that eluded Sir Leslie Stephen and those who have followed him that "rational inquiry" and "free disquisition" in the two decades after 1780 could induce Hey, Watson and Paley to part company with Jebb and his followers and ally themselves with the unashamedly conservative Pretyman and Isaac Milner.

The beginning of this process is well illustrated in Paley's first book, *The Principles of Moral and Political Philosophy*, which became a textbook at Cambridge within a year of its publication in 1785. To George III and other timorous folk the *Principles* might well have seemed alarming. As late as 1802 the *Anti-Jacobin Review* "hesitated not to affirm" that in the *Moral and Political Philosophy* "the most determined Jacobin might find a justification of his principles, and a sanction for his conduct."[18] The famous parable of the pigeons satirizes the "paradoxical and unnatural" distribution of property, ninety-nine out of a hundred "gathering all they got into a heap" and keeping it for "one, and the weakest, perhaps worst pigeon of the flock." The oath of allegiance "permits resistance to the king, when his ill behavior or imbecility is such, as to make resistance beneficial to the community." Slavery is an "abominable tyranny;" the poor have a claim on the rich "founded in the law of nature;" inequality of property is an evil; and "it is a mistake to suppose that the rich man maintains his servants, tradesmen, tenants and labourers: the truth is, they maintain him."[19] In chapter xi of book vi of *Political Philosophy* Paley made proposals for a progressively graduated income tax with exemptions for marriage and child-support which were later advanced by Tom Paine: as a result of which their permanent implementation was delayed until the twentieth century.[20] In theological and religious matters there was the same deceptive appearance of adventure, not to say "radicalism." "Whoever expects to find in the Scriptures a specific direction for every moral doubt that arises, looks for more than he will meet with." The Thirty-Nine Articles were "articles of peace" intended to comprehend in the National Church all save those, such as Papists and Puritans, who would have destroyed it by their intolerance. Yet there should be with respect to public office in the state a "complete toleration of all dissenters from the established church."[21]

It is not in the least to disparage either Paley's integrity or the coherence of his arguments to observe that beneath this Whiggish surface there subsisted a devotion to accustomed ways as secure as that of Burke or Samuel Johnson. A distinguished author has lately remarked indeed that the *Principles* afforded "a more systematic exposition of the intellectual tradition upon which Burke and other conservative writers of the 1790's drew" than Burke's own work. "It is evident that it provided a quarry for arguments paraded by anti-revolutionary publicists after 1789."[22] The parable of the pigeons is immediately followed by an exposition of the expediency of private property which might have been (but probably was not) taken from the *Summa Theologiae*.[23] Abominable though it be, "no passage is to be found in the Christian Scriptures, by which [slavery] is condemned or prohibited."[24] Whether he realized it or not, Paley's defense of the claims of the poor follows that of St Ambrose, and its corollary – that "the real foundation" of property rights in real estate "is the law of the land" – is Augustinian.[25] Moreover, though "a poor neighbor has a right to relief, yet if it be refused him, he must not extort it."[26] Though inequality be an evil "abstractedly considered," it "flows from those rules concerning the acquisition and disposal of property, by which men are incited to industry, and by which the object of their industry is rendered secure and beneficial."[27] The rich man is maintained by his servants but is no happier than they and probably less so.[28] Toleration of religious dissenters must be qualified by the exception of "what arises from the conjunction of dangerous political dispositions with certain religious tenets." It was usual to suppose that this refers to the connection between popery and Jacobitism but John Ehrman has recently pointed out that Paley's qualification was "a significant statement in the middle eighties, for it was just then that doubts were being voiced about the "dispositions" of Dissent."[29]

Paley's *Principles*, like Watson's *Apology for Christianity* which was written in 1776 in answer to Gibbon, is only the beginning of that emergence among the Cambridge moderates of increasingly conservative convictions. And in the case of the *Principles* it is confined to political theory. Not until *Horae Paulinae* (1795) and the *Evidences* (1794) did Paley expound what he took to be Christian orthodoxy in response to the objections of Hume and lesser men.[30] It has been suggested by M. L. Clarke that some of this was inspired by Paley's "discussions in the Hyson Club and his defense of the *status quo* against reformers such as Jebb."[31] What does seem to have been the case is that Paley, like all intellectuals of good will, tried very hard during his early years to maintain communication with and to do justice to the deeply held convictions of all his friends and colleagues;[32] that this was at least partly feasible in the eirenical climate of the 1760s and early 1770s; but that from the late 1770s he was increasingly forced by the logic of his own spiritual and intellectual commitments into criticism of and confrontation with the "reformers." Something of the same kind may well have happened in the case of John Hey and even to a lesser extent, notwithstanding Stephen's cynical presuppositions, in that of Richard Watson.[33]

5.2 Watson, Paley and Hey

The clearest statement of the convictions which united "Paley and his School" was supplied by Richard Watson. In 1785 he published a *Collection of Theological Tracts* in six volumes. What I have called a "Cambridge *via media*" is contained in the "Preface."[34]

As Paley laid himself open to misrepresentation by his jocular remark about being unable to afford a conscience, so to a far greater extent did Watson in the belated publication of his *Anecdotes*.[35] ("I determined to study nothing but my Bible, being unconcerned about the opinions of councils, fathers, churches, bishops, and other men as little inspired as myself.")[36] Even Norman Sykes was deceived by Watson's boastful, octogenarian ramblings about the events of forty-six years before. "From ... a robust application of the principles of scientific enquiry to the doctrines of Christianity, and ... a contumelious disregard for tradition as representing but the speculations of the dark ages, Watson evolved an individual creed, strikingly different from the official articles of the Established Church."[37] A somewhat different picture appears in the *Theological Tracts*, a set of readings for ordinands from "books of such acknowledged worth, that no Clergyman ought to be unacquainted with their contents." To be sure, "Their Bible is the only sure foundation upon which they ought to build every article of faith which they profess," but the "decisions of councils, the confessions of churches, the prescripts of popes, or the expositions of private men" ought not "to be fastidiously rejected as of no use:"

> for although the Bible be the one infallible rule by which we must measure the truth or falsehood of every religious opinion, yet all men are not equally fitted to apply this rule, and the wisest men want on many occasions all the helps of human learning to enable them to understand its precise nature, and to define its certain extent. These helps are great and numerous, they have been supplied in every age, since the death of Christ.[38]

Watson printed twenty-three excerpts of eighteenth-century divinity, some of them of book-length, with annotations and suggestions for further reading which leave no doubt that he at any rate had read and was reading very widely.[39] Notwithstanding his declared hostility to "system" in theology, he would himself have composed a "Systematic Treatise" but was "seized with a disorder three years ago." For those so inclined he recommends numerous authorities including several of "our English Divines," and the *Summa Theologica* [sic] of *Thomas Aquinas:*

> for notwithstanding the ridicule which usually, in these days, attends the mere mention of the *Angelic* Doctor, I will venture to affirm, That in that work there are, mixed indeed with many difficult subtleties and perverse interpretations of Scripture, not a few Theological questions of great moment stated with clearness and judgment.

Moreover, in divinity, "new books are not so much wanted ... as inclination in the younger clergy to explore the treasures of the old ones."[40]

Watson claimed to have "no regard for latitudinarian principles"[41] and it is by no means clear from the *Tracts* that his "individual creed" was "strikingly different from the official articles of the Established Church." His position on the final authority of scripture is entirely consistent with Article VI and not contrary to Article XX, "Of the Authority of the Church." His repeated insistence on the fallibility of each individual and of all churches is wholly in the spirit of Article XIX. The Anglican biblicism he took for granted was sufficiently protected by theological learning against "enthusiasm," and clearly separated, by its rejection of "system," from the related errors both of popery and of Calvinism.

It is at this point that the unique character of the Cambridge *via media* appears. Far from seeking a middle way between Rome and Geneva, Watson and his colleagues repudiated both alike for their "blind attachment to system:"

> The effect of established systems in obstructing truth is to the last degree deplorable ... *Calvin*, I question not, thought it almost impossible that the Scriptures could ever have been so far perverted as to afford the Romanists any handle for their doctrine of Transubstantiation, ... yet this same *Calvin* followed St. *Augustine* in the doctrine of absolute reprobation and election, inculcating it as a fundamental article of faith, with nearly the same un-christian zeal which infatuated him when he fastened *Servetus* to the stake.[42]

"Established systems" obstruct truth because they attempt to elicit more information from Scripture than the data are able to afford. The equal and opposite error is that "which spurns with contempt the illusions of fanaticism and the tyranny of superstition ... magnifying every little difficulty attending the proof of the truth of Christianity, into an irrefragible argument of its falsehood." Those who are "unlearned in Scripture knowledge" are easy victims in "this dissolute but enlightened age" to the infidelity of "Bolingbroke, Voltaire, Helvetius, Hume," the "*Esprit forts* of France, and the *Frey-Geisters* of Germany."[43]

At first glance all this seems little more than yet another iteration of Chillingworth's "bible the religion of protestants," enlivened by Warburton's double-barreled polemic against "bigots" on the one hand and "libertines" on the other.[44] What distinguishes "Paley and his School" from such exemplars is the attempt to locate and deal with the cause of those equal and opposite errors between which theological truth is to be sought. As Watson put it with precision, difficulty arises:

> in many cases, from our not understanding the full meaning of the language, be it common or figurative, in which a doctrine is revealed; in some it proceeds from our attempting to apprehend definitely, what is expressed indeterminately, or clearly what God hath not thought proper clearly to reveal.[45]

Men of "system" succumb to the temptation to tidy up God's language. "Sophists" and "profane men" yield to the opposite temptation: to give up trying to extract *any* meaning from scripture and so to ignore altogether the data of divine revelation. The proper course is to affirm all that scripture clearly affirms, and where scripture is doubtful or ambiguous to go on using the actual works and phrases without insisting on any exclusive interpretation. The *via media* is thus essentially epistemological: a becoming modesty about the possibility of theological knowledge which steers well clear of Hume's defeatist skepticism. There are several examples in Watson's "Preface,"[46] but the technique is best studied in Paley's sermons.

Stephen's tendentious remark about the "Unitarianism" implied by Paley's "method of reasoning" has been noted above. As Le Mahieu has demonstrated convincingly, Stephen missed the point. The "method of reasoning" actually employed by Paley in his books depended upon the ruthless application of Ockham's razor to every dialectically redundant concept. His terse dismissal of the "moral sense" at the very beginning of his first work is a typical example.[47] But Christian theology, purporting to be knowledge of God, is more than a "method of reasoning" and is in part expressed in other discourse such as prayer, liturgy and the rhetoric of preaching. Paley and those like him in the church differed sharply from the "Unitarians" in their unhesitating willingness, when occasion required, to employ such discourse in its most orthodox and traditional forms.[48]

A striking example of this is to be found in a sermon on the "Unity of God." Having expounded the Hebraic tradition of monotheism and shown that "the Christian dispensation entirely confirms and repeats what the Jewish Scripture of the Old Testament had before delivered," Paley declared that:

> We hear, nevertheless, of three divine persons – we speak of the Trinity. We read of the "Father, Son, and Holy Ghost"... . What is

> that union which subsists in the divine nature; of what kind is that relation by which the divine persons of the Trinity are connected, we know little – perhaps it is not possible that we should know more: but this we seem to know, first, that neither man nor angel bears the same relation to God the Father as that which is attributed to his only-begotten Son, our Lord Jesus Christ.[49]

The language, though deliberately rhetorical, is carefully used. We "seem to know" because of the evidence of Scripture, which for Paley, as for Watson, was impossible to ignore. We "speak of the Trinity" because the word is used in the formularies and liturgy of the church to which Paley had given his solemn assent. If the New Testament really "attributes" to "our Lord Jesus Christ" a unique relationship with "God the Father" then we must simply accept this essentially mysterious dogma. Too much subtle theology may be unhelpful: "perhaps it is not possible that we should know more."

In very much the same manner, Paley dealt with the doctrine of the atonement in a sermon for Good Friday. "The full magnitude and operations of those effects which will result from the death of Christ we can only comprehend ... from general expressions used in the Scripture." Certain theories, such as that God is a "harsh and austere character" whose "enmity was to be reconciled by the blood of his Son" can be ruled out: not because they are irrational (which they need not be) but because they are unscriptural." God is never said to be *at enmity with* us ... but we are said to be *at enmity with God.*" It is probable that Christ was sent to teach holiness to the human race "lost in an *almost total depravity,*" but "*we are not to stop at this*:"

> in various declarations of Scripture concerning the death of Christ ... there are other and higher consequences attendant upon this event: the particular nature of which consequences, though of the most real and highest nature, we do not understand, nor perhaps are capable of understanding ... until we be admitted to more knowledge than we at present possess of the order and economy of superior beings, of our own state and destination after death, and of the laws of nature by which the next world will be governed, which are probably very different from the present.

The sermon ends with a scriptural doxology glorifying God in Christ "in whom we have redemption through his blood."[50] It may be noted here that even in his *Natural Theology* Paley was willing to speak, though in a footnote, of the "propitiatory virtue" of Christ's passion.[51]

It is true that by comparison with earlier or later generations of Anglican divinity the use of liturgical and mystical language by Paley and his Cambridge contemporaries was slight. But it was not negligible and there is a difference between their didactic writing on the one hand and their homiletic or pastoral writing on the other. In the former the conceptual apparatus is stripped down to the bare minimum required to sustain the argument. Wherever possible it is naturalistic. The authority of scripture is invoked only where no other argument will do, as in Paley's defense of the killing of animals for food.[52] In the latter sort of writing the traditional dogmas of the Church – incarnation, atonement, Trinity – are presented inasmuch, and only inasmuch, as they seem to be required by the liturgy and are consistent with scripture. Subject to the last criterion "Paley and his School" were also willing to use and recommend the more highly-flavored pastoral and devotional literature of former ages. Watson listed the *Imitation of Christ* and *Holy Living and Dying* for his students; as archdeacon of Carlisle, Paley authorized and caused to be reprinted for his clergy a Caroline compendium of High Church pastoralia, *The Clergyman's Companion in Visiting the Sick*, which his father had used in Giggleswick.[53]

The Cambridge *via media* is clearly visible in the attitude of "Paley and his School" to subscription. General historians such as Halévy have evaded this issue with the astonishing claim that "nobody was obliged to believe the thirty-nine articles or even read them."[54] But John Hey, first Norrisian professor of Revealed Religion (from 1780 to 1794) devoted the greater part of his four volumes of lectures to a minute examination of the Articles and the grounds for assenting to them: for "In treating of religious Societies in the present Times, the great business seems to be, to give a right account of what are called *Articles of Religion*."[55] Hey's lectures, being delivered in English, attracted a large and attentive audience, and before long the bishops required all Cambridge ordinands to present a certificate of their having attended them.[56]

Hey, like Watson and Paley, has had a bad press. According to Stephen, he held "with the deists that talk about the Trinity is little better than unmeaning gibberish."[57] Overton and Relton, desiring to distinguish "Evangelicals and High churchmen" (good) from the "intellectual party" which held no "very definite opinion of a positive sort" (bad), remark of Hey – seemingly without the benefit of any acquaintance with his work – that he was "even more inclined to liberalism in theology than either Watson or Paley."[58] Sykes and all later historians ignore him.

In fact, Hey declared that he was "not the person, who would insinuate, that any of our own Articles stand in need of anything beyond plain interpretation." He believed that "the most improved *comments* on the Scriptures would rather confirm our Articles than overthrow them." As a parish priest:

> I have sometimes told my Congregation, in Sermons, that I speak as a minister and not as a man; that, though I believe the doctrines I preach, I deliver them not as my own, but as the doctrines of the Church: and on this account such doctrines demand greater attention.[59]

Nevertheless the "doctrines of the Church," being based upon God's self-revelation in Scripture, are *for precisely that reason* a matter of some uncertainty. In a brilliant, analytical treatment "Of Assenting to Propositions, which are Unintelligible" Hey laid bare the epistemological assumptions of Cambridge theology and their logical relation to Anglican orthodoxy.

Unintelligible propositions arise both in natural and in revealed religion. "Men may run into two faulty extremes about them: too easily receiving them leads to error, and fruitless controversy; ... too easily rejecting them, tends to ignorance and disorder; and finally to the obstruction of religious authority." The problem is acute in revealed religion. "When we have things communicated to us from above by Language, we have to consider and investigate the precise meaning of expressions." Yet much of Scripture is strictly unintelligible. The attempts to solve many of the problems thus presented by a view of Scripture as "God's word written" (Article XX), even relatively successful attempts to rule out demonstrable error, "have greatly increased the number of unintelligible propositions." Yet in many cases we ought to assent: "not for the sake of *truth* (for that cannot be ascertained in such cases) but for that of *utility*, that is, avoiding some evil, or attaining some good."[60]

Just what is the evil to be avoided, or the good to be attained? The question has nothing to do with social order but concerns rather, the more fundamental issue of theological knowledge:

> It is an evil to neglect or throw aside anything, which it has pleased God to reveal to mankind: if he sends a message, whether it be understood or not, it is to be carefully preserved; it is to be noted

> and registered faithfully and simply; nay, the more exactly, for not being understood.

If we "write what we understand" we may safely correct and edit, but "if we copy a language which we have never learned, we must copy everything, even blots and mistakes." For though at present we may not understand, we do not know how soon we may. Conservation of all potential information is one thing, assent another. Why require the latter? Because:

> it is not conceivable, that we should value Scripture, and not throw the expressions of it into some forms; of doctrine or devotion: into sermons, prayers, hymns &c – these are necessary if we are only to remind men of what has been revealed.

The semi-poetic, seldom univocal language of scripture, liturgy, devotion and homiletic is precious and venerable, for it transmits "notices from heaven." Even creeds and other doctrinal formularies must – if they are not to "*lower* the things revealed to what we fancy is common sense" – employ language in this ritual, incantatory way. Hence a creed "will become a kind of Hymn... . On this principle it may be, perhaps, that Creeds are sometimes *sung*."[61]

Hey buttresses the argument with sensible remarks about the differential capacity of humans to apprehend revealed truth. The Church will include "the young and old, the wise and the unthinking ... it is not conceivable, that such a Society should be carried on, without some members assenting to what they did not understand: ... all ranks must join in creeds, catechisms and Liturgies." The young child is taught the catechism by rote. "At first the whole will be unintelligible to him, and always some part: yet it is right upon the whole, that he should repeat it."[62] But at its heart is the central conviction of "Paley and his School:" the reality of God is beyond the power of *any* human mind to comprehend; it is a misuse of "reason" to articulate putatively univocal propositions about the divine nature. Such things are a *mystery* – where "the True scriptural notion of μυστηρτον is, a design of God not yet executed, or made manifest. Mysteries, according to this notion, may both be "'kept secret since the world began' – and be revealed or made known."[63] Hey quotes with approval Dionysius of Alexandria on the Apocalypse:

> Some ... have utterly rejected and confuted this Book .... But for my part I dare not reject the Book, since many of the Brethren

> have it in high esteem: but, allowing it to be above my understanding, I suppose it to contain throughout some latent and wonderful meaning ... not measuring and judging these things by my own reason, but ascribing more to faith, I esteem them too sublime to be comprehended by me.

"As Dionysius reasons on the mysteries of the Apocalypse," concludes Hey, "we might reason on any other mysteries."[64]

To a mind so resolutely closed against theological knowledge as that of Sir Leslie Stephen it is proper to describe discourse of this kind as "unmeaning gibberish." But it is quite another matter to regard Hey as "a decided rationalist" and to assume that he and his colleagues were "Socinian in all but name." As Stephen himself pointed out, the movement towards Socinianism was accompanied by, and related to, "the attempt to banish mystery from theology."[65] It was precisely this attempt that Hey resisted. There is a world of difference between his humble acknowledgement of the incomprehensibility of God, and the arid rationalism of such as Price and Priestley. Hey's conception of the "latent" meaning contained in Scripture actually has far more in common with Tract 87, *On Reserve in Communicating Religious knowledge.*[66]

5.3 Intellectual legacy of "Paley and his school"

Whether Isaac Williams had actually read Hey, or whether the verbal similarities noted above were coincidental, there is evidence that the writings of the Cambridge school were highly regarded by churchmen of various kinds during the first three or four decades of the nineteenth century. The best complete edition of Paley's *Works* is that produced by his son in 1825, but many other editions appeared and the demand for individual books, especially *Evidences*, remained strong to the end of the century and beyond. Hey's *Lectures* went to a third edition in 1841.[67] Watson's *Tracts* fared less well, but as late as 1828 a complete set of the 1785 edition was presented to the new (exclusively Anglican) University of Upper Canada by the Society for the Propagation of Christian Knowledge.[68]

What is more significant is the way these works were treated by those among the next generation or two who made no secret of their affection for orthodoxy. In 1799 Pretyman brought out what must be regarded as the first modern textbook of theology, in response to "the great deficiency with respect to professional knowledge, which I frequently found in the Candidates for Holy Orders." Most of the second volume consists of a exposition of the Articles, which must be subscribed "in their plain and obvious sense."[69] Pretyman listed 179 volumes which ought to be found in the library of "a respectable and useful Parish Priest." Works by Clarke, Lardner, Macknight and Secker are among those excerpted or reprinted by Watson.[70] There is no discussion of "evidences" in Pretyman's book "as upon that point" he wished "to refer the reader to ... Dr. Paley." *Evidences, Horae Paulinae* and *Natural Theology* are recommended.[71]

Charles Simeon (1759–1836, BA 1783) and his pupil John Bird Sumner (1780–1862, BA 1803) "first evangelical Archbishop of Canterbury," were disqualified – as King's men – from membership of "the intellectual party."[72] But it is clear that they were by no means immune from the theological climate of the place and time; in particular that their evangelicalism, especially that of Sumner, was appreciably modified by "Paley and his School." After Simeon became dean of divinity in 1789, King's College library – having acquired only 102 titles in the previous forty-six years – added 254 in the next decade. Paley's *Moral Philosophy, Horae Paulinae* and *Evidences* were acquired by 1795; "Dr Hey's *Lectures on Divinity*" in 1798. Acquisitions included works by other members of the intellectual party" such as Balguy, Jortin and Milner, along with the evergreen *Pearson on the Creed, Shepherd on the Common Prayer* and the first edition of *Wilberforce on Practical Religion*.[73] The impact of Cambridge theology upon the early evangelicals is best seen in the controversy over Calvinism.

Strong dissonance between evangelicalism and Cambridge theology had appeared as early as 1802 in a work by John Overton, *The True Churchman Ascertained: or An Apology for those of the Regular Clergy of the Establishment who are sometimes called Evangelical Ministers* (1802). Overton sought to establish that "evangelical ministers" read the Articles in "the literal and grammatical sense;" that attacks upon Calvinism both by High Churchmen such as Daubeny and by the Cambridge moderates impugned that sense; and hence that the "evangelical ministers" were "true churchmen," and "Drs. Paley, Hey, Croft; Messrs. Daubeny, Ludlam, Polwhele, Fellowes; the Reviewers, etc., etc." were not. There was a grain of truth in this, but also much paranoia, confusion and serious misrepresentation. The central issue was whether the Articles imply, *and only imply*, a Calvinist doctrine of justification, predestination and election. "Paley and his School" were reviled for asserting, "with almost one voice, like the Papists, the doctrine of *freewill*: our Reformers say, that the Papistical doctrine of freewill is *abominable* in the sight of God, and to *be abhorred* by all

Christian men." Watson's charges and Hey's lectures were selectively quoted in ways that destroyed their careful balance; Paley was smeared by association with his "great model, the late Bishop Law," and condemned for his "Jacobin" parable of the pigeons which puts St Paul on the "same footing" as Tom Paine.[74] Sumner's *Apostolical Preaching* (1815) was intended to distance the more moderate and respectable evangelicals from this kind of thing: John Henry Newman, at any rate, acknowledged that it was Sumner's work which caused him to give up his "remaining Calvinism" in 1824.[75]

Very much in the manner of Watson, Sumner began with the aim of "a complete freedom from all party designs, and without aiming either to defend or confute any man or set of men. He reverences no party ... and appeals to no authority, except that which all profess to acknowledge." In particular he repudiated the authority of Calvin. In the matter of predestination, Sumner preferred "the examples of our own Church ... who, in her Articles, considers this doctrine separately, as a speculation distinct from the essential points of the Christian faith." Calvin's doctrine of "decrees" is at variance with the idea that this world is a "probationary state." The doctrine of election is as objectionable as that of predestination for it implies that Christ's propitiatory sacrifice is not efficacious for all.[76] In the chapter on justification, however, his principle target is the Church of Rome. ("The fatal, though predicted, apostasy in Papal Rome" recalls his early Hulsean essay.[77]) "It is the nature of the Roman Catholic religion silently to undermine the true nature of Christian justification."[78] His final chapter displays precisely that desire to find a middle way between "blind attachment to system" and the "meer light of reason" which, as I have argued above, was characteristic of Watson, Paley and Hey. "There are two characters of preaching ... at opposite extremes equally removed from the spirit and practice of the Apostles." The first presents the Gospel through the distorting lens of the Calvinistic system. Though "Calvinistic tenets" *may* be deduced from Scripture, may be consistent with one possible interpretation of the Articles, they are nothing but conjecture: and "as far as concerns the Christian preacher or teacher, these doctrines are as if they were not true." The other "very different character of preaching," equally unsatisfactory, fails to see that the Gospel unfolds to mankind "certain terms and means of entering into everlasting life, which their natural reason would never have discovered." Sumner had in mind "what Priestley called rational Christianity ... not deduced from Scripture, but the coinage of his own imagination."[79] Sumner's apparent continuity with the Cambridge moderates is confirmed by an examination of his use of their work in his own writing. Paley's *Sermons* and *Horae Paulinae* were cited with approval several times in *Apostolical Preaching* alongside such Anglican authorities as Burnet, Butler, Hall, Hooke, Horsley, Milner, Taylor, Warburton and Wilberforce.[80] Overton was altogether ignored. Paley's *Natural Theology* was cited eight times in his famous *Treatise on the Records of the Creation* (1816), more than any other authority, and *Moral Philosophy* once.[81] In his slightly later work, *The Evidence of Christianity* (1824), Sumner cited Paley's *Moral and Political Philosophy, Evidences* and *Horae Paulinae,* Watson's *Letters to Gibbon,* and Hey's *Lectures*. The last is his chief authority in that work, cited respectfully and often at length in five places.[82] That his use of Paley, Hey and Watson was no mere resort to the authorities he had read as an undergraduate is shown by the complete absence of any reference to Pretyman's highly popular textbook which first appeared in Sumner's second year at King's. It is, however, consistent with the surmise that he was much influenced as a young man by Charles Simeon. For like the Cambridge moderates, the latter based his "sentiments" not on "the dogmas of Calvin or Arminius, but in the Articles and Homilies of the *Church of England,*" being persuaded that these are "manifestly contained in the Sacred Oracles."[83]

The authority of "Paley and his School" was by no means confined to Cambridge circles. There are signs that Oxford had begun to ascend from "the nadir of intellectual activity" before the end of the century, and that this was associated with a gradual penetration of Cambridge theology.

It is well known that Oriel College was the primary meristem of new growth at Oxford. Its *Register of Books taken out of the Library* affords *prima facie* evidence of intellectual life. Throughout the 1770s and 1780s there were many entire years during which no single volume of divinity was removed from the library. Yet virtually every resident fellow had borrowed Robertson's *America* within a year or so of its acquisition in 1777. Paley's *Principles, Evidences* and *Horae Paulinae* were acquired in 1786, 1796 and 1794 respectively, and *Natural Theology* in the year of publication, 1802. All began to be borrowed by fellows, resident bachelors and undergraduates. Watson's *Theological Tracts* were acquired in 1791 and the first edition of Pretyman's *Elements* in 1799. In the early 1820s E. B. Pusey, then a newly-elected fellow, recommended the purchase of "Dr. Hey's Norrisian lectures" (the 1822 edition was acquired) and "Bp. Watson's collection of Theological Tracts," not realizing that the latter had been held in the library since before he was born. Sumner's *Apostolical Preaching* (the same that

Hawkins persuaded Newman to read) was acquired in 1817, *Records of the Creation* in 1818 and *Evidences* in 1824.[84] Edward Copleston, who by then was Provost, had mastered *Records of the Creation* by early 1819.[85] The spirit of "Paley and his School" lived on in Copleston's pupil, Richard Whately, greatest of the pre-Tractarian Noetics.[86]

All this is very fragmentary and inconclusive. But it does suggest that the Cambridge *via media* had an appeal which outlived its first proponents; and that its literature, "more meagre and superficial than any which had flourished since the days of Hooker," remained in demand, and exerted a detectable influence, for more than half a century. "The giants of those days" may have been "but dwarves compared to their predecessors or their successors."[87] But that was not the way it seemed to the first two generations of the successors.

6
Wealth of Nations as Theology

> If a great book such as Smith's *Wealth of Nations* is read repeatedly, on even a fifth or a tenth reading one continues to learn new things. I doubt whether anyone will ever fully apprehend all the things that Smith wished to express, and there is even more to learn from an interesting mind than its owner wished to teach us.[1]

6.1 Introduction

Economics is a scientific enterprise. Well-grounded theory is continually refined. Observations of social phenomena are made in light of that theory. Inferences are constructed by means of the best statistical techniques. Putatively falsifiable predictions are made. This is all as it should be.

Yet what Heyne[2] calls "the economic way of thinking" is more than just a science. It is a way of looking at society that rests on certain assumptions about the human condition. Those assumptions are neither innocent nor uncontroversial, for they stir up baffling moral and theological questions. Is there a higher good than economic welfare? If economics is about scarcity, and scarcity is an evil,[3] why does God allow scarcity? If individuals actually are as rational and self-interested as we assume, ought they to be?

Because economists are human beings, our utterances reveal our preconceptions and values. They also function so as to recommend those preconceptions and values. We have all been made aware in recent years of the "rhetoric of economics."[4] Every text written by every economist is and must be to some extent an "essay in persuasion." It is a short step from this awareness to recognizing the theology latent or implicit in much economic literature. Many have noted that present-day economics offers a new kind of "modernist faith" with its own "Ten Commandments and Golden Rule," its "nuns, bishops and cathedrals," and its "trinity of fact, definition, and holy value."[5] Policy debates between economists and environmentalists sometimes look like "wars of religion" between "two faith communities."[6] Modern economics has been plausibly represented as the last gasp of a liberal-protestant "social gospel" that the American Economic Association was founded to promote: a grand if futile gesture explained in *Reaching for Heaven on Earth: The Theological Meaning of Economics*.[7] The author of that work has now given us a more focused analysis in *Economics as Religion: From Samuelson to Chicago and Beyond*.[8]

It is my purpose in this article to show by means of a classic case that what Nelson calls "economic theology" has been with us for a long time before Samuelson. Not only was Karl Marx "the most successful of all theologians since the Reformation," as Paul Tillich[9] once observed, but Marx's chief exemplar was Adam Smith. If we reread Smith's[10] "great book" with proper attention, we may learn from his "interesting mind" a lot more than "its owner wished to teach us." For *Wealth of Nations* may be read – and conceivably was sometimes read – as a work of "natural theology;" rather as Newton's *Principia* was read by Cambridge undergraduates as natural theology for most of the 18th century.

It is important for me to state what I am not trying to do. I am not trying to discover what Adam Smith actually believed in 1776. A person's religious beliefs or unbeliefs are seldom stable or coherent, seldom completely understood by that person or by others, and never reliably signified by documentary evidence alone. I have therefore deliberately ignored the interesting and important work of other scholars who have discussed Smith's theology in a more orthodox, intellectual–historiographic fashion.[11] I have instead attempted to follow the literary–theoretic suggestions of Vivienne Brown[12] according to which "the richness of a text may be explored independently of the question as to whether the author was aware" of the "textual devices" he employed, such as "style and figurative language," which may now tell us more than he was either willing or able to say in a more straightforward way at the time of writing.

What follows is in four parts: first, an investigation of the theological work done by Smith's pervasive and ambiguous conception of "nature;" second, an account of human "interest" and of the ethical and political problems this appears to create, third, Smith's reconciliation of the two in a theodicy of social life that explains in part how a "divine Being" produces "the greatest quantity of happiness;" and finally, some concluding remarks.

6.2 Nature

"Nature" and its cognates ("natural," "naturally," "unnatural") is one of the most frequently used such families of words in *Wealth of Nations.*[13] (See Table 6.1) Moreover, it is often used as either synonymous with, or as thematically connected with, "necessity" and its cognates. The frequency of "nature" and "necessity" combined is 1529, which exceeds that of the otherwise most important significant term, "price." "Nature" is sometimes used innocently, to denote the characteristic properties of some entity, as in the title of Smith's *Inquiry* itself and in such phrases as "the nature of its laws and institutions," "the nature of its soil and climate," and so on (*WN* 89, 111). But it is also used to denote that which exists, or at any rate the whole created, material universe in which man is located: "the call of nature," "the order of nature," "the nature of things," the "great phenomena of nature" (*WN* 100, 145, 515, 766, 767). And in some cases there is slippage between this sense and a hypostasis of the term that seems to refer to or imply a putative creator of "nature:" "talents with which nature has distinguished its fellows," "nature labours along with man,"

Table 6.1 **Frequency of important terms (and their cognates) in WN**

Price	1388
Labor	1199
Value	907
NECESSSITY	873
Profit	785
Capital	765
NATURE	765
Land	645
Market	600
Goods	581
INTEREST	557
Rent	502
Wage	354
Landlord	199
Cost	118

Source: Glahe (1993).

"the work of nature," the "difference which nature has established between corn and almost every other sort of goods," "the nature of things has stamped upon corn..," "nature does herself the greater part of the work," "Nature does nothing for him" (*WN* 30, 363, 364, 515, 515, 694, 695). The most striking and suggestive example of such usage – an important clue to the theodicy I will suggest in section 4 – is to be found in Smith,[14] in a context that criticizes Quesnay. In the "political body," Smith maintained against the Physiocrats, "the wisdom of *nature* has fortunately made ample provision for remedying many of the bad effects of the folly and injustice of man" (*WN* 674; my italics). In each of these cases and especially the last, "nature" is nearly synonymous with the God referred to in *WN* (770, 772) as "the Deity."

The substantive use of "nature," however, is less frequent than either the adjectival or the adverbial. For whereas "nature" occurs 149 times in *The Wealth of Nations*, "natural" occurs 232 times and "naturally" 272 times.[15] And it is in these that Smith's ambiguous, not to say equivocal, treatment of "nature" – sometimes teleological and/or normative, sometimes merely positive and/or naturalistic, sometimes dubiously either or both – is most evident.

"Natural" is occasionally used adjectivally of {"nature" = the created universe} as in "natural philosophy"(*WN* 766) and in these cases is neither teleological nor normative. The beginning of some trace of the normative is to be seen in such uses as "the natural aristocracy of every country" and "the natural and respectable distinctions of birth and fortune" (*WN* 622, 944). These social arrangements may, indeed must, exist in consequence of "the nature of things," but they do seem to be regarded with approval, and they do not seem to be quite inevitable. "Natural liberty and justice," (*WN*,157, 470, 530, 606) being a state of affairs that ought to exist but that may not, is more obviously both teleological (it is an end, or part of the purpose of "nature") and normative (we must try to see that it is maintained).

The most interesting uses of "natural" occur within an explicit economic-theoretic context. A key example is found at the beginning of Book III, "Of the Natural Progress of Opulence." "According to the natural course of things ... capital of every growing society is, first, directed to agriculture... . This order of things is so very natural" (*WN* 380). The redundant phrase "so very" is a giveaway. In a strictly positive sense, something is either "natural" or it is not. Smith was here using "natural" to mean "good" or "desirable." For "in all the modern states of Europe," this "natural order of things" has been "entirely inverted" to become an "unnatural and retrograde order" (*WN* 380,

422). The "natural progress of improvement" (*WN* 708), the "natural proportion which would ... establish itself between judicious industry and profit" (*WN* 758), a "natural distribution of stock" the "derangement" of which is "necessarily hurtful" (*WN* 632), and Smith's rhetorical, fourfold incantation of the sonorous phrase "natural and free state" (*WN* 608) are all equally normative, with more than a hint of the teleological. The "natural price" (*WN* 77 etc.) and "natural rate" of wages (*WN* 79 etc.) are ambiguous. They may be conceived positively as long-run equilibrium outcomes, but there is always some suggestion of the normative in their use. The corresponding value of the "profits of stock" is actually referred to at one point as "their proper level" (*WN* 132).

Why is it "natural" for human societies to grow and "improve?" What does it mean to say that equilibrium market outcomes are "natural"? And why, notwithstanding, may "natural" outcomes be delayed or altogether frustrated? Some light is thrown on these matters by Smith's use of the adverb "naturally."

In chapter VII of Book I, "Of the Natural and Market Price of Commodities," the average frequency of "naturally" rises to 0.70 per page as against 0.29 for *WN* as a whole. The purpose of that chapter is to show how "the quantity of every commodity brought to market *naturally* suits itself to the effectual demand" (*WN* 74; my italics). An important part of the explanation postulates a "natural price" as "the central price, to which the prices of all commodities are continually *gravitating*" (*WN* 72, 75, 77; my italics). This powerful Newtonian image is invoked in order to suggest a process whereby a "natural" state of affairs not yet in existence might (or must; Smith was characteristically ambiguous) come about "naturally" – that is, according to the *laws of nature* – after a lapse of time. Similar usage occurs throughout *The Wealth of Nations* in such ideas as "that balance which naturally establishes itself among the various employments of society" (*WN* 499; see also 453, 465, 504, 604, 606, etc.) and that "the mercantile stock of every country ... naturally courts the employment which in ordinary cases is most advantageous" (*WN* 629). Yet deliberate human action may delay such processes "for ages together" or at any rate "for many centuries" (*WN* 79). The laws of nature govern a disequilibrium adjustment process that must – or only might? – bring about "natural" outcomes, but humans can frustrate or impede the operation of those laws.

This equivocation is obviously unsatisfactory. The laws of nature, if they really are that, must operate of necessity. The laws of Newtonian physics are ineluctable, as Smith well knew. At many points in *WN*,

Smith used "naturally" and "necessarily" as synonyms or near synonyms, (e.g. *WN 19*, 98, 104, 277, 417, *446*, 454, 455, 456, 476, 598, *627*, 715, 767, *802)* but where I have italicized page numbers he suggested a distinction between the two that seems to imply that what might happen "necessarily" may not always be what happens "naturally." Moreover, two different kinds of ambiguity are introduced with the adverb. In the normative proposition, "Commerce ought naturally to be ... a bond of union and friendship" (*WN* 493), "naturally" is redundant unless, by adding color and strength to "ought," it is able to suggest that what happens "naturally" is both optional and to be preferred. The same identification of "naturally" with "ought" occurs even more clearly in IV.vii.c.87: "that ... which ought to take place, and which naturally does take place" (*WN* 629). But the near juxtaposition in IV.vii.b of "cattle naturally multiply" with "The prohibition of exporting ... naturally tended to lower the price" (*WN* 577, 579–80) seems to imply, somewhat tendentiously, that market processes happen *in the same kind of* way as biological processes.

In IV.vii.c.86, Smith belatedly acknowledged that one may speak only figuratively of market processes as occurring "naturally." "The mercantile stock of every country, it has been shown in the second book, *naturally* seeks, if *one may say so*, the employment most advantageous to that country." But "one may say so," of course, only because of the motives and actions of "the owner of the stock" who "*necessarily* wishes to dispose of as great a part of [his] goods as he can at home" and who "*naturally* ... endeavours ... to turn his carrying trade into a foreign trade of consumption" (*WN* 629–30; my italics). Nature has endowed "the owner of stock" – like all other human beings – with "the propensity to truck and barter;" whence comes a division of labor "not the effect of any human wisdom" and the associated institution of market exchange (*WN* 25). The adjective "human" in this context seems to imply a contrast with divine wisdom – "the wisdom of nature" (*WN* 674). Moreover it would seem to be something like a law of nature that humans should exhibit this propensity in such a way as to maximize profit: "The consideration of his own private profit, is the sole motive which determines the owner of any capital to employ it" (*WN* 374). It is therefore admissible for Smith to write figuratively of market processes as "natural" because they are the consequences of "natural" human behavior. Yet it would seem that for Smith in *WN*, using "natural" in this positive sense, some "natural" outcomes are good ("natural" in a normative sense) and others bad ("unnatural" in a normative sense). How can we make sense of this?

In III.i.8–9, Smith wrote of "a natural order of things" according to which capital is directed first to agriculture, then to manufactures, and finally to foreign commerce:

> But though this natural order of things must have taken place in some degree in every such society, it has, in all the modern states of Europe, been, in many respects, entirely inverted. The foreign commerce of some of their cities has introduced all their finer manufactures.., and manufactures and commerce together have given birth to the principal improvements of agriculture. The manners and customs which the *nature* of their original government introduced and which remained after that government was greatly altered, *necessarily* forced them into this *unnatural* and retrograde order. (*WN* 380; my italics)

This is complicated and sophisticated. The "nature" of their original government "necessarily" subverted a "natural order of things" and has brought about an "unnatural" order. One is tempted to see in this an example of, or at any rate a secular parallel to, a Scotch Calvinistic, predestinarian account of the inescapable consequences of primordial human lapse. However, if we recall Smith's recognition that aggregative, social phenomena are the unintended consequences of individual human agency, an even more complex – and theologically more interesting – explanation begins to emerge.

The "original principles in human nature" (*WN* 25) – though often obscured by "levity and inconstancy" (*WN* 92–3), "folly and injustice" (*WN* 378, 674), "sloth and intemperance" (*WN* 674), "avidity" (*WN* 563, 567), "gross ignorance and stupidity" and cowardice (*WN* 788) – are evidently taken by Smith to be essentially good. For all who exhibit these defects are said to be "deformed in [some] ... essential part of the character of human nature" (*WN* 788). Such language is clearly teleological. Human beings were meant, or intended, or created to be better than we actually are. That "nature" within which humans exist, and according to whose laws we behave, possesses a "wisdom" that humans can never attain (*WN* 674; see also pp. 25, 564, 626, 687). Though human reason is "feeble" (*WN* 803), there is to be discerned in part an "order of nature and reason" (*WN* 145) by which we ought to govern ourselves. Insofar as we do, we may be said to be acting "naturally" in the normative sense, which I will distinguish as "naturally"(1). And this will produce social outcomes that are "natural" in the normative sense (clearly contrasted by Smith with "unnatural") and that I will designate as "natural"(l) and meaning "good, as intended by Nature." But human beings, Smith evidently assumes here, have a free choice between good and evil. We may ignore or evade the order of nature and reason: But insofar as we do, we still must act "naturally" in a merely positive sense "naturally"(2). And the social outcomes that result will be natural(2) because everything in "nature" must be rule-governed in some way and hence must occur "necessarily." Therefore, when Smith distinguished "naturally" and "necessarily," he would seem to have had in mind that what has to happen within "nature" by human agency, though always "necessary," is not always {"natural" = natural(1)}.

In order to make further progress, we must examine in more detail the textual evidence in W for the inequality *naturally(1) ≠ naturally(2).*

6.3 Interest

"Commerce ... ought naturally to be, among nations, as among individuals, a bond of union and friendship." But because of "the capricious ambition of kings and ministers" and "the impertinent jealousy of merchants and manufacturers," it has become "the most fertile source of discord and animosity." Neither the "violence and injustice of the rulers of mankind" nor "the mean rapacity ... of merchants and manufacturer" can be "corrected" (though the latter "may very easily be prevented from disturbing" the tranquility of others) because they are evils for which "the nature of human affairs can scarce admit of a remedy." (*WN* 493)

In this passage, especially the phrase "the nature of human affairs," Smith came as near as he ever does in *The Wealth of Nations* to the traditional Pauline, Augustinian doctrine of Original Sin. And in describing its putative consequences, Smith's language is almost always more highly colored than usual. He assails the "savage injustice" of Europeans who colonized the New World. (*WN* 448 etc.) In general, "All for ourselves, and nothing for other people, seems, in every age of the world, to have been the vile maxim of the masters of mankind." (*WN* 418) Feudal proprietors were actuated by "the most childish, the meanest and the most sordid of all vanities." (*WN* 419) The "great lords" beheld the growing prosperity of the burgesses – their former serfs – "with a malignant and contemptuous indignation." (*WN* 854) The protective legislation "which the clamour of our merchants and manufacturers has extorted ... may be said to have been all written in blood." (*WN* 648) Even the relatively minor vices of profusion, prodigality and extravagance, draw strong moral condemnation. The "prodigal" perverts capital "from its proper destination." "Like him who perverts the revenues of some pious foundation to profane purposes, he pays the wages of idleness with those funds which the frugality of his forefathers had, as it were, consecrated to the maintenance of industry." (*WN* 338, 339) Such language is intelligible only on the assumption that "the original principles in human nature" are, and can be known to be, good; that traces of a "common humanity" remain (*WN* 587, 648); that humans have genuine free will; and that deviation by individuals from what is natural(1), when voluntary, is culpable.

What impels individuals to act, whether naturally(1) or simply naturally(2)? *The Theory of Moral Sentiments*[16] expounds a detailed psychology according to which human action is motivated by a less-than-perfect balance of the "sentiments:" self-love, justice, and beneficence or benevolence. Because I am concerned in this paper solely with what one may read in *WN*, I shall resist the temptation to reconstruct Smith's complete theological thinking and concentrate rigorously on what is contained in that text alone. "Self-love" occurs only twice in *WN*, "beneficence" not at all. There are 89 occurrences of "justice" – of which more later. But only one of these (*WN* 711) refers to individual virtue, the remainder either to institutional arrangements (80 examples) or abstract principle (eight examples). What drives individuals to act either naturally(1) or naturally(2) in *WN* is what they perceive – correctly or incorrectly – to be their "interest."

"Interest" in *The Wealth of Nations* is a morally neutral term. It is natural(2) for humans to be led by it. The "regard that all men have for their own interest" (*WN* 350) is taken for granted throughout *WN* and illustrated in many ways. "The desire of bettering our condition ... comes with us from the womb." (*WN* 341; see also 540) "Interest" and its cognates occur 557 times in *WN* (Table 6.1), and though perhaps as many as 100 instances may refer to "interest" as the price of loans, the word means "self-advantage" in the great majority of cases. In I.ii, the phrase "their self-love" is twice used as synonymous with "their own interest," (*WN* 26, 27) and this is highly significant theologically. For "self-love" had already been explicated as a Christian duty by Joseph Butler and Josiah Tucker in response to Mandeville's reductionist apologia for "vice."[17] In principle, therefore, an action driven by {"interest" = "self-love"} can also be natural(1). It is evidently for this reason that Smith was so willing, throughout *The Wealth of Nations*, to recommend that "The law ought always to trust people with the care of their own interest." (*WN* 531) It has often been remarked, correctly, that Smith's doubts about the efficacy and beneficence of legislation are consistent with a belief he shared with Hume in the fallibility and myopia of "human wisdom." (*WN* 564, 626) Thus, the "duty of superintending the industry of private people, and of directing it towards the employments most suitable to the interests of the society," is one "for the proper performance of which no human wisdom or knowledge could ever be sufficient." (*WN* 687) It is also the case that Smith saw that "Civil government ... is in reality instituted for the defence of the rich against the poor" (*WN* 715) and therefore that laws and courts must normally be biased in favor of "the rich and the powerful" against the "poor and the indigent." (*WN* 644; see also 157–8 and IV viii *passim)* But even if neither of these objections existed, the ethical question would remain: Why *ought* individuals to be trusted to pursue their own interests? Smith's implicit answer was typical of eighteenth-century Christian moral theory: Private interest may be a reliable guide to right conduct. As Samuel Johnson, oracle of high-church, Tory piety once put it, "There are few ways in which a man can be more innocently employed than in getting money."[18]

Yet *The Wealth of Nations* is filled with examples of the maleficent consequences of interest, sometimes but by no means always linked to the moral corruption of agents. (*WN* 336, 585, 649, 717, 722, 780, 808, 826, 897) It is "private interest" that animates "the spirit of monopoly," (*WN* 474) and it is the latter that creates so very sharp a conflict between the interest of "merchants and manufacturer" and that of "the great body of the people." (*WN* 493–4) For though "to narrow the competition, is always the interest of the dealers," it is "an absurd tax upon the rest of their fellow-citizens." (*WN* 267) The "Conclusion of the Mercantile System" (IV.viii) explains with many examples how, under the mercantile system, "the interest of the consumer is almost constantly sacrificed to that of the producer." (*WN* 660) The interest of that "superior order of people" rich enough to "brew for their own private use" has probably prevented a fairer and more efficient "system of excise duties." (*WN* 893, 891) In universities where "the teacher" receives a salary, his "interest is ... set ... in opposition to his duty". (*WN* 760) The "clergy of every established church ... pursue their interest upon one plan and with one spirit," and their "interest as an incorporated body is never the same with that of the sovereign, and is sometimes directly opposite to it." (*WN* 797) Sometimes indeed, conflict of interest exists within the same person or body. The "interest of the East India company considered as sovereigns" is "the reverse" of "their interest as merchants." (*WN* 638) And quite apart from all this, the *class* interest of "the employers of stock" can never be "connected with the general interest of the society" as is those of landlords and laborers. For the rate of profit is "naturally low in rich, and high in poor countries, and it is always highest in countries which are going fastest to ruin." (*WN* 265, 266)

Therefore, quite independently of any moral corruption that may afflict some *but not all* agents, the world as it actually is evinces many different and sometimes ineradicable conflicts of interest. At the very least, the perfectly natural(2) collusive pursuit of their interest by powerful individuals can – or will or must – harm the interest of weaker individuals. (e.g. *WN* 644) Laws that "trust people with the care of their own interest" invite, or at any rate permit, a wide range of social evil. Yet though the law ought not to foster or encourage collusive interest seeking, it cannot and ought not to prevent it:

> People of the same trade seldom meet together, even for merriment and diversion, but the conversation ends in a conspiracy against the public, or some contrivance to raise prices. It is impossible indeed to prevent such meetings, by any law which either could be executed or would be consistent with liberty and justice. (*WN* 145)

If agents act with regard to their own interest, they will collude whenever that seems likely to improve their relative position. And when they do so, the outcome of their actions will not be natural(1).

Some legislative reform, of course, will do good, in particular, the dismantling of legal monopolies and other trade restrictions. Moral reform may help, too. Conditions of work in a commercial society induce "a torpor of mind" in the laborer, which "renders him ... incapable of conceiving any generous, noble or tender sentiment, and consequently of forming any just judgment concerning many even of the ordinary duties of private life."(*WN* 782) Public education in parish schools is a partial remedy and well worth the expense, for an "instructed and intelligent people ... are always more decent and orderly than an ignorant and stupid one." (*WN* 788) And where – as under a Presbyterian polity – economic incentives conduce to "exemplary morals" in the clergy of an established church, the parish minister may exhibit "that system of morals which the common people respect the most ... . He becomes careful to instruct them, and attentive to assist and relieve them" and so gains an "influence over the minds of the common people," which produces a general moral improvement of the lower orders. The Church of Scotland is exemplary in maintaining not only "the uniformity of faith, the fervour of devotion," but also "the spirit of order, regularity, and austere morals in the great body of the people." (*WN* 813)

Yet it seems clear that in *The Wealth of Nations*, legal and moral reform can never be sufficient to take away or prevent completely that evil which results from the unimpeded operation of private interest. Humans have free will. Though they need not, and perhaps ought not, to intend "the publick good," (*WN* 456) their choices ought to result in social outcomes that are natural(l). Yet even with good laws (or no laws) and good – or at any rate not individually corrupted – morals, their interest may lead them to produce outcomes that are contrary to "the general interest of the country" (*WN* 613) and hence *not* natural(1). This is a consequence of "the nature of human affairs;" perhaps (though Smith nowhere says so in *The Wealth of Nations*) of some initial derangement of "the original principles in human nature." The way in which Smith deals with all this – guarded as he always is and to the modern, secularized ear often seemingly ironical – could easily be taken by his friends and associates in the Moderate party of the Church of Scotland to be consistent with some broadly conceived notion of Original Sin.

6.4 Theodicy

Smith's advocacy in *The Wealth of Nations* of "the natural system of perfect liberty and justice" (WN 606) depends crucially on the possibility of justice. "Every man, *as long as he does not violate the laws of justice,* is left perfectly free." (*WN* 687; my italics) Because private interest will frequently lead to such violation when it appears profitable, there must be a sovereign with "the duty of protecting, as far as possible, every member of the society from the injustice or oppression of every other member of it." (*WN* 687) And in fact, the "equal and impartial administration" of British justice, "by securing to every man the fruits of his own industry, gives the greatest and most effectual encouragement to every sort of industry." (*WN* 610) Yet where institutional arrangements provide incentives to judges and plaintiffs to evade or bend the laws, "the corruption of justice" will occur "naturally." (*WN* 718) And in general throughout *WN*, we see that it is the possibly quite innocent, natural(2) pursuit of private interest *within the wrong kind of institutional framework* that leads to social outcomes that are {"unnatural" = inefficient and unjust}.

But the institutional framework is described, in V.i.a,b and in III, as coming gradually into existence, over very long periods, as the largely unintended and certainly unforeseen outcome of a myriad of private – or at least local – decisions, all of which take place in response to two fundamental, *biological* laws of nature announced in I.xi.b:

(i) "men ... naturally multiply in proportion to the means of their subsistence;"
(ii) "land ... produces a greater quantity of food than what is sufficient to maintain all the labour necessary for bringing it to market." (*WN* 162)

By means of natural law (i), the human race is propelled through the first two "states of society" – those of hunters, "the lowest and rudest state," and shepherds, "a more advanced state" (*WN* 689, 690) – to the third, that of husbandmen. At that point, when all fertile land has been appropriated, the effect of natural law (ii) provides that "the surplus produce of the country" becomes "the subsistence of the town" (*WN* 377). "Conveniency and luxury" become possible, and their development, fostered by the effect of two fundamental, *psychological* laws of nature;

(iii) the "propensity to truck and barter" (*WN* 25) and
(iv) the "natural effort of every man to better his condition" (*WN* 540), a consequence or concomitant of "interest,"

brings about the fourth and highest stage of development, a commercial society. Social and political institutions are associated with, perhaps in some sense determined by, each mode of production. And for reasons explored in III.iv, "commerce and manufactures gradually introduced order and good government." (*WN* 412) "Interest" therefore plays an important part in bringing about an eventual improvement in that institutional framework within which "interest" works for good or ill. The better are social and political institutions, the greater the likelihood of justice. And the more pervasive is justice, the more probable it is that human behavior that is natural(2) may also be natural(1). The broad outline of a theodicy in *WN* is now apparent.

Every religion that acknowledges a God who is all-powerful, all-wise, all-knowing, and perfectly good faces the so-called problem of evil. How is the abundant evidence of unwilled suffering in sentient beings (physical evil), and of human wickedness and its consequences (moral evil), to be reconciled with the divine attributes? (Why does God allow cancer, war, injustice, and so on?) Answers that diminish any of the attributes are rejected as heterodox. To suggest, as Malthus did for example, that the perfectly benevolent God is less than all-powerful to prevent evil is Manichaean; and Malthus was promptly accused of that heresy by Ricardo and James Mill.[19] An inquiry that seeks to demonstrate the possible coexistence of all the divine attributes is known as *theodicy*. The term seems to have been coined by Leibniz,[20] but within the Judeo-Christian tradition, the enterprise goes back at least to the Book of Job.

By far the most influential theodicy in the Christian West is that of St Augustine of Hippo (354–430), whose voluminous and powerful writing set the theological agenda for more than a thousand years. Augustine began with the Pauline doctrine of Original Sin and the Fall of Man and attributed all moral evil, and most if not all physical evil, to that single cause. What then does God do about it? Augustine's answer was complex and not entirely satisfactory.[21] But his account of political society is suggestive. The state and its institutions are a self-inflicted *punishment* of human sin. Augustine had no illusions about the human cost of maintaining internal peace and external security. Moreover, without *justice*, the state is an unmitigated evil: "Remota itaque justitia, quid sunt regna nisi magna latrocinia?"[22] And because of human sin, true justice is never fully obtainable: "vera autem justitia non est nisi in ea republica cuius conditor rectorque Christus est."[23] Yet some degree of justice remains possible; therefore, God allows the self-regarding acts of sinful human beings to bring the state into existence because its institutions – especially those of private property, marriage, and slavery – are also a *remedy* for sin. By means of the state, the evil in human life may be constrained to that minimum that must result from freedom of the will in fallen humanity.[24] I wish to suggest that there are parallels between this aspect of St. Augustine's theodicy and the account we may read in *WN* of the way in which "the wisdom of nature" provides that natural(2) human behavior may bring about natural(1) outcomes.

Though human nature is – or was – natural(l), it is frequently if not always deformed in some "essential part." In particular, "human wisdom" is feeble and unreliable by comparison with "the wisdom of nature." Though "to cultivate the ground was the original destination of man" (*WN* 378; cf. Genesis 2:15), human life is faced with a recurrent possibility of "natural scarcity arising from soil and climate" (*WN* 466; cf. Genesis 3:17–19; in the next paragraph, Smith refers to "barrenness of the earth" as a "curse"). Given its "laws and institutions," the opulence of any country is constrained by what "the nature of its soil, climate, and situation might admit of." (*WN* 112) Thus, human fecundity (natural law [i]) creates competition for resources and so induces the "natural effort of every individual to better his own condition." Human action must therefore be driven by "self-love," or "interest," which may be a reliable guide to right conduct and which may produce natural(1) social outcomes. But where the institutional framework is defective, interest will naturally(2) produce outcomes that are "unnatural." However, the institutional framework itself is the result of an evolutionary process in which human interest, manifesting itself as natural law (iv), operates in face of natural laws (i), (ii), and (iii). And "the natural progress of improvement" (*WN* 708) brings about institutions that are more conducive to justice.

By such a process, for example, the "silent and insensible operation of foreign commerce" (driven by natural laws [iii] and [iv]), acting on "the meanest and most sordid of all vanities" in the nobility, caused the end of feudal dependency. For as objects of luxury become available, an ever larger proportion of the surplus is spent on these, rather than on courtiers, private armies, and domestic retainers. "A revolution of the greatest importance to the publick happiness, was in this manner brought about by two orders of people, who had not the least intention to serve the publick:" the "great proprietors" and "the merchants and artificers" acting "merely from a view to their own interest, and in pursuit of their own pedlar principle of turning a penny wherever a penny was to be got." (*WN* 419, 418–22) In another example, we read of that effect of the "pride of man which makes him love to domineer" and "to prefer the service of slaves to that of freedom" but which is gradually offset by the fact that slavery is inefficient and relatively unprofitable. Hence, to "the slave cultivators of antient times, gradually succeeded a species of farmer." (*WN* 386–9) Similarly, the "invention of firearms, an invention which at first sight appears to be so pernicious, is certainly favourable both to the permanency and to the extension of civilization." (*WN* 708) In cases such as these, and indeed throughout *WN*, we see that what Christian theology identifies as the deadly sins of "pride," "fury" or anger, "avarice," "envy," "sloth," and "intemperance" or gluttony (*WN* 388, 674, 709) are co-opted by the wisdom of nature to become the means of human amelioration. (The benign consequences – if any – of the seventh deadly sin, "lust," are decently veiled in *WN*. But Smith's first great disciple, the Rev. T R. Malthus, wrote in similar fashion of "the passion between the sexes.")

Above all is this so of the benefits conferred on mankind in a commercial society by "the natural system of perfect liberty and justice," which it is the chief purpose of *WN* to recommend. It is indeed the case that

Smith means to include by "justice" not only the courts and the legislature but even more important the built-in "moral sentiments" that induce individuals to play the game according to the rules even when the umpire is not looking. But even where such perfection is yet to be achieved, nature finds ways of effecting outcomes that are natural(1):

> If a nation could not prosper without the enjoyment of perfect liberty and perfect justice, there is not in the world a nation which could ever have prospered. In the political body, however, the wisdom of nature has fortunately made ample provision for remedying many of the bad effects of the folly and injustice of man; in the same manner as it has done in the natural body, for remedying those of his sloth and intemperance. (*WN* 674)

For "the uniform, constant, and uninterrupted effort of every man to better his condition," we read in II.iii.31, "is frequently powerful enough to maintain the natural progress of things toward improvement" notwithstanding either individual corruption or perverse policy and institutions. "Like the unknown principle of animal health, it frequently restores health and vigour to the constitution in spite, not only of the disease, but of the absurd prescriptions of the doctor." (*WN* 343) Institutions in *WN*, above all those of "the political body" and the market, may therefore be seen – by those with theological eyes to see – as an Augustinian *remedium peccatorum*. They supply the means by which the potentially destructive "passions" of fallen human nature are harnessed to good and creative ends.

When we read *WN* theologically as a work of theodicy, we perceive that Smith reserved his masterstroke, with nicely calculated art, to the midpoint of his final book. For when viewed from this standpoint, the literary and rhetorical climax of *WN* is to be seen in the grand sweep of ecclesiastical history outlined in V.i.g. 20–39.

Like all good Protestants of his day, Smith professed to regard the Church of Rome as "the most formidable combination that ever was formed against the authority and security of civil government, as well as against the liberty, reason and happiness of mankind, which can flourish only where civil government is able to protect them." (*WN* 802–3) The temporal power of that church rested on "the great landed estates which the mistaken piety both of princes and of private persons had bestowed" on it. As with the barons, the "immense surplus" from these estates – resulting from natural law (ii) – could only be employed "in the most profuse hospitality, and in the most extensive charity," which also "increased very much the weight of their spiritual weapons." Hence, "the grossest delusions of superstition [eighteenth-century code word for "popery"] were supported ... by the private interests of so great a number of people as put them out of all danger from any assault of human reason." (*WN* 800–3)

However, "that immense and well-built fabric ... was *by the natural course of things* undermined." For the "gradual improvement of arts, manufactures, and commerce, the same causes which destroyed the power of the great barons, destroyed in the same manner, throughout the greater part of Europe, the whole temporal power of the clergy." (*WN* 803–4; my italics) Temporal sovereigns moved in to fill the vacuum, nationalized the church where they could, and tolerated reformation. And the consequence of this process for ecclesiastical polity was wholly benign. Where "episcopal government" survived the emancipation from popery, as in England, this "system of government was from the beginning favourable to peace and good order, and to submission to the civil sovereign."(*WN* 807) Better still was "the equality which the presbyterian form of church government establishes among the clergy." (*WN* 809) For equality among the clergy supplies incentives that reward "learning," "irreproachable regularity" of life, and "diligent discharge" of duty. (*WN* 809) Consequently, "There is scarce perhaps to be found any where in Europe a more learned, decent, independent, and respectable set of men, than the greater part of the presbyterian clergy of Holland, Geneva, Switzerland, and Scotland." (*WN* 810) Smith concludes his remarkable, Providentialist account of "the natural course of things," which by enlisting "interest" can bring good out of evil in a manner never possible by "the feeble efforts of human reason" (*WN* 803), with that ringing tribute to his own national church already cited in part:

> The most opulent church in Christendom does not maintain better the uniformity of faith, the fervour of devotion, the spirit of order, regularity and austere morals in the great body of the people, than this very poorly endowed Church of Scotland. All the good effects, both civil and religious, which an established church can be supposed to produce, are produced by it compleatly as by any other. (*WN* 813)

6.5 Conclusion

I have attempted to show that one may construe the text *WN* as containing, and possibly even as shaped by, a quasi-Augustinian account of the way in which God responds to human sin by using the consequences of sin both as a punishment and as a remedy.

It is evident that the theology of *The Wealth of Nations* is entirely "natural theology," that is, putative knowledge of God arrived at by the study of nature alone, without any reliance on the "revelation" supposed by the faithful to be recorded in sacred scripture. "Nature" is almost always viewed teleologically in W. It exists for and with a purpose, and part of that purpose is human welfare. And to say that must imply either a transcendent, Newtonian God *of* Nature or an immanent, Leibnitzian God *in* Nature.[25] Smith does not say which, and though his text is capacious of either interpretation, it usually easier to read it in the second of these ways. Even on the first interpretation, however, Smith's putative God/Nature does not merely wind up "the great machine" and leave it ticking, as the Deists were held to have believed. She continues to act in various ways, but always wisely and well, so as to make creative use of human folly and wickedness in ways that bring good out of evil. Such redemptive activity, we may assume, is needed at all only because humans have been endowed by God/Nature with freedom to choose and, though intended by God/Nature to choose well, suffer from some universal failing that may be primordial and that often impels them to choose ill.

Natural theology is ecumenical in a way that revealed theology can never be. For its truths are available to all who will read "the Book of Nature," whatsoever their religious tradition. Yet it is a mistake to imagine that natural theology would have been regarded, in eighteenth-century Britain, as in any way opposed to or even inconsistent with Christianity. Smith's great exemplar was Newton, and the Newtonian character of *WN* has been remarked from the first.[26] Newton published his *Principia* in 1687 "with an Eye upon such Principles as might work with considering Men for the belief of a Deity."[27] Throughout the 18th century, Newton was read by Cambridge men preparing for Holy Orders in the Church of England as part of their *theological* training. By Malthus's day, they did so with the help of Colin Maclaurin's popularizing textbook. According to Maclaurin, we learn from Newton that "Our views of Nature ... represent to us ... that mighty *power* which prevails throughout, ... that *wisdom* which we see ... displayed ... the perfect *goodness* by which they are evidently directed."[28]

Much has been made by some recent authors of the fact that there is no mention in *WN* of "Jesus," "Christ," or "the Son."[29] But neither is there mention of that Person in the *Principia*, yet it would never have occurred to any of Newton's readers for that reason to describe his work as an "atheistic science."[30] The whole point of natural theology is to show by means of a scrupulously positive ("objective," "secular," "ecumenical," and so on) inquiry that knowledge of God may be had *without any resort to revelation whatsoever.* To introduce scriptural categories would subvert the enterprise. And it is perfectly clear, from our knowledge of the theological training of the Christian clergy in eighteenth-century Scotland and England, that natural theology formed an important prolegomenon to the more doubtful and more controverted mysteries of revealed religion. Smith himself taught natural theology as Professor of Moral Philosophy at Glasgow.[31] Before doing so, he publicly subscribed his assent to the Calvinist Confession of Faith before the Presbytery of Glasgow.[32] It is entirely proper, therefore, to regard *WN* as a work of "science" in exactly the same sense as that of Newton, and in exactly the same way to identify its theological content. Smith drops a useful hint in V.i.g.14, where he declares that "Science is the great antidote to the poison of superstition and enthusiasm." (*WN*, p. 796) "Superstition" meant popery. "Enthusiasm" meant in general the religion of any sect claiming direct illumination by the Holy Spirit and in 18th-century England was specially associated with Methodism.[33] Anglican divinity of that period is full of warnings against, and refutations of, the errors of superstition and enthusiasm. The position of Church of Scotland Moderates approximated in this matter to that of the Church of England.

It has been no part of my intention in this paper to suggest that a theological reading is the only way, or even the only *right* way, to study economic literature. My claim is merely that if one puts on one's "rhetoric-of-economics" spectacles, one can see that "economic theology" has been with us from the first. Of course, economics is "science." But that does not in the least rule out the possibility that economics is also "theology:" indeed, quite the reverse. *The Economist as Preacher* needs to be very sure of the facts. The more "scientific" economics is, the more valuable it becomes as theology.

7
The Sudden Separation of Political Economy

The history of "economics" strictly speaking is of short duration. The most defensible *terminus a quo* is the decade beginning in 1885, which saw the foundation of the *American Economic Association*, the first numbers of the *Quarterly Journal of Economics* and the *Economic Journal*, and the publication of Marshall's *Principles*.[1] But "economic thought" of some kind is a detectable feature of every civilized society. "Economic thought" – like all other "thought" is and must be conceived and developed within the metaphysical and theological presuppositions of its time and place. In the Christian West, from the thirteenth to the eighteenth century, "economic" thought can be regarded for the most part as a specialized branch of moral theology: in particular, of casuistry. *Mutatis mutandis*, much the same can be said of Talmudic economic doctrine during the same period.[2]

However, "economic thought" in eighteenth-century Europe can also be regarded, and often is regarded, as having become a largely "secular" inquiry, separated from theology by a well-defined epistemological boundary. The subsequent mutation of "political economy' into "economics" at the end of the nineteenth century left that boundary unchanged. In arguing that this view is somewhat anachronistic I shall present a strong thesis for debate:

> The origin of "political economy" as a distinct inquiry, clearly to be demarcated from Christian theology, is the publication of Malthus's first *Essay on Population* of 1798.

First I shall describe the theological matrix of economic thought in eighteenth-century Britain; second, the crucial importance of the first *Essay* in my story; and, finally, the resulting theodicy of scarcity and the appearance of a boundary between "political economy" and Christian theology.[3]

7.1 Economic thought in eighteenth-century Britain

The Enlightenment in Britain was regarded as an opportunity rather than as a threat to established religion. Anglophone "economic thought" in the eighteenth century was congenial to, and to some extent intertwined with, Christian theology. And the canonical text of eighteenth-century economic thought, *The Wealth of Nations*, may be read as congruent with the theological assumptions of Anglican orthodoxy.

Enlightenment and religion

It is still sometimes supposed that a so-called "Enlightenment" in the eighteenth century, by raising new doubts about the Christian religion, began a "secularization" of European thinking which manifested itself in all branches of philosophy, science and politics.[4] But if the term "secular" and its cognates be taken to refer to culture as a whole, it is evident that "secularization" is a consequence of a widespread industrialization and urbanization which did not begin until the next century. "Enlightenment was of the few. Secularization is of the many."[5]

Quite apart from this, it appears that whereas the Enlightenment of the French *philosophes* was indeed associated with infidelity or at any rate with deism, this was decidedly not the case in Britain. For though, in general, the Enlightenment may be understood as an attempt to extend the method of Newtonian science into all branches of inquiry,[6] that enterprise was not perceived as subversive of orthodox religion either in Scotland or in England.[7] Why this should have been we may learn from Colin Maclaurin's *Newton*, first published in 1748 and read by Adam Smith almost immediately after.[8] Maclaurin became required reading for undergraduates at universities in both countries for most of the rest of the century.

For British students at any rate, "natural philosophy is subservient to purposes of a higher kind, and is chiefly to be valued as it lays a sure foundation for natural religion and moral philosophy." But this may not be so in papist countries, for Maclaurin warns against the "superstition" (eighteenth-century code word for Roman Catholicism) which "discourages inquiries into nature, lest, by having our views enlarged, we may escape from her bonds."[9] In papist France Newton may be an enemy of established religion: in Protestant Britain he is an ally. For not only does natural science demonstrate the unity, omnipotence, omniscience and goodness of God: it also "disposes us to receive what may otherwise be revealed concerning him" and thus provide a proper introduction to the study of scripture.[10] Moreover we are led by science to a belief in the afterlife.[11] Finally, we learn from science "to consider our present state ... as a state of preparation or probation for farther advancement"[12] – a central tenet of orthodox, Anglican moral theology of that time.[13]

The Scottish Enlightenment is therefore largely free of that bias against Christianity which characterized the *philosophes*; and David Hume's untypical infidelity may have been acquired in France. As for the English Enlightenment, to the extent there was any such thing at all it took the form of an "ideology of politeness"[14] which united Christian, specifically Anglican orthodoxy, both with modern science and with an unprecedented freedom of opinion, and has been described today as "conservative, clerical and Magisterial."[15] Economic thought that flourished in Scotland in the writing of Hume, Steuart, and Adam Smith, and in England in that of Berkeley, Josiah Tucker, Paley and even Malthus, must be understood in this intellectual context.

Religion and economic thought

The first and (and perhaps the greatest) of these, the philosopher George Berkeley (1685–1753), Bishop of Cloyne in the Church of Ireland, sought not only to relieve but also to cure endemic poverty in his diocese. *The Querist*[16] addresses the problem of economic development upon the assumptions, first, that the attitudes of the population are the most important determinant of wealth-creation; second, that it is the "aim of every wise State ... to encourage industry in its members." Berkeley apprehended "some censure ... for meddling out of my profession; though to feed the hungry and clothe the naked by promoting an honest industry, will perhaps be deemed no improper employment for a clergyman who still thinks himself a member of the commonwealth."[17] We may see in Berkeley's writing what, from the standpoint of this paper, are three of the four most important characteristics of eighteenth-century economic thought: first, that wealth is a good thing; secondly, that wealth-creation is always feasible; and, thirdly, that measures to increase the wealth of nations are consistent with, and may actually belong to, the Christian religion.

What is absent is the theme, first clearly stated by Bernard Mandeville,[18] which came to dominate the economic thought of

Berkeley's successors: that general prosperity occurs only when individuals vigorously pursue their own private economic ends without much, if any, consideration of the common good. Mandeville's provocative slogan, "Private Vices, Publick Benefits," appeared to create a conflict between wealth-creation and Christian morality. Many were deceived and *The Fable of the Bees* was denounced from the pulpit, indicted by the Grand Jury of Middlesex as a public nuisance, burned by the hangman in France, and placed by the Vatican on the *Index Librorum Prohibitorum*.[19]

It was apparent to shrewder minds, however, that Mandeville's shocking paradox depended upon an improper construal of *self-love* as "vice." The first, and also the most powerful, analysis of self-love in relation to "private good" and "public good" was supplied by Joseph Butler (1692–1752), successively Bishop of Bristol, Dean of St Paul's and Bishop of Durham. Butler's sermons at the Rolls Chapel[20] were preached between 1723 and 1725, in the immediate aftermath of the public outcry aroused by the 1723 edition of the *Fable*.[21] Although Mandeville is never mentioned, it seems clear that his seemingly subversive doctrines were vividly in the minds of preacher and congregation, and that they supplied both the motivation and the agenda for the sermons.[22] Butler showed that the ends of private good and public good "do indeed perfectly coincide;" that "self-love is one chief security of our right behavior towards society;" that under Providence much unintended social good is produced by self-regarding actions; and that "there is seldom any inconsistency between what is called our duty and what is called interest."[23] Sermons XI and XII, "On the Love of our Neighbour"[24] recognize that self-love is a duty commanded by Christ himself.

It has been conjectured by F. A. Hayek[25] that Hume studied Mandeville at the time he was planning the *Treatise*;[26] and it seems probable that the essay "Of Luxury"[27] would have been regarded by its readers as one of the many responses to Mandeville that appeared from time to time for several decades after 1723. At any rate Hume acknowledged Mandeville in the Introduction to his first work – along with Locke, Shaftesbury, Hutcheson and Bishop Butler – as one of those "who have begun to put the science of man on a new footing."[28] Though Hayek[29] has argued that Mandeville "made Hume possible," it would seem from Hume's reference to Butler that it was the latter who made it possible for Hume so to generalize Mandeville as to produce the "theory of spontaneous order" now seen as the characteristic contribution of the Scottish Enlightenment to social theory.[30] The multifarious activities of any large human society, most notably its economic activities, arise *and can only arise* in a gradual, unplanned, accidental, piecemeal fashion in response to the incentives to individual, self-regarding actions created by others' needs, wants and desires. A decade or so before Hume began to write, Butler had established that this putatively providential outcome might arise from a wholly virtuous attention by all individuals to their "interest" as determined by the Christian duty of self-love.

Bishop Butler's chaplain, the Reverend Josiah Tucker (1713–99) later Dean of Gloucester, became – through the good offices of Lord Kames – a correspondent and friend of Hume, with whom he conducted a successful dispute over comparative advantage in international trade.[31] Karl Marx later acknowledged that Tucker, though a Tory *(sic)* and a parson, was "an honourable man and a competent political economist,"[32] and it is certainly the case that in Tucker, more than any previous author considered in this essay, we see the concepts of "self-love" and 'interest" employed in specifically economic discourse. His *Essay on Trade*[33] describes the study of commerce as "this noble and *interesting* science; on which the *Riches,* the *Strength,* the *Glory,* and I may add, the *Morals* and *Freedom* of our Country, so essentially depend." His uncompleted *Elements of Commerce and Theory of Taxes* notes that:

> as our present *secular* Happiness appears to arise from the Enjoyment of superior Wealth, Power, Honour, Pleasure, or Preferment, SELF-LOVE, the great Mover of created Beings, determines each Individual to aspire after these *social* Goods, and to use the most probable Means of obtaining them.[34]

And in *Instructions for Travellers* Tucker clearly sets out what was to become, two decades later, the central message of *The Wealth of Nations*:

> let the Legislature but take Care not to make *bad Laws,* and then as to *good ones,* they will make themselves: That is, the Self-Love and Self-Interest of each Individual will prompt him to seek such Ways of Gain, Trades and Occupations of Life, as by serving himself, will promote the public Welfare at the same Time.[35]

Tucker's own bishop (of Gloucester), the polemical William Warburton, once derided him for making "trade his religion,"[36] and the slur was tendentiously – and inaccurately – revived by R. H. Tawney.[37] But it is clear from the whole of Tucker's life and writings that this is to misunderstand the theological climate of the day. It is correct, as Tawney[38] noted, that by Tucker's day economic thought had ceased to be concerned primarily with casuistry, at least in Protestant countries.[39] But contrary to Tawney's astonishingly successful propaganda, *it had not ceased to be Christian.* Tucker was a faithful disciple of his first bishop (of Bristol), patron and friend. He viewed the unintended but beneficent economic outcomes of "interested" action in a purely Butlerian way, as examples of divine Providence and as congruent with that Newtonian, natural theology which characterized the Scottish and English Enlightenments.

7.2 *The Wealth of Nations* and Christian theology

Though Adam Smith seems never to have met Josiah Tucker, he acquired Tucker's economic writings for his own library.[40] Moreover, he would have known of Tucker and his ideas from his friends David Hume and Lord Kames, and also from the "oeconomists" he met on his visits to France. Tucker's Butlerian view of the human condition can be discovered both in *The Theory of Moral Sentiments*[41] (hereafter *TMS*) and in *The Wealth of Nations*[42] (hereafter *WN*), though Smith may well have learned of Butler's doctrines in the first instance from his old teacher, Francis Hutcheson.[43] In *TMS* (I.iii.1), Butler is referred to as "a late ingenious and subtile philosopher."

From the standpoint of this paper, the two most important philosophic ideas in early eighteenth-century debate that Smith assimilated to his system were those of self-love and human ignorance.

Mandeville is dealt with, in *TMS* (VII.iv), in a way that resembles Butler's, and self-love explicated, following Aristotle and the Stoics, as both integral to human nature (*TMS* II.ii.1) and as at least possibly virtuous (*TMS* VII.ii.3.16). Most recent commentators[44] agree that Smith – who in 1751 had subscribed to a Calvinistic Confession of Faith upon accepting the Chair in Logic at Glasgow[45] – moved away from "orthodox Christianity" in later life towards Stoicism and natural religion. This may well have been the case.[46] But the relevant passages in the final (1790) recension of *TMS* remain consistent with Butler's treatment of self-love, and differ chiefly in omitting any reference to Dominical command.

The last of Butler's Rolls Sermons was "Upon the Ignorance of Man,"[47] and in this at any rate he agreed with Mandeville[48] who had written of "the narrow Bounds of human Knowledge." Hume followed both Mandeville and Butler in propounding a modest, not to say pessimistic, view of the possibility of human knowledge, and in particular of any prior knowledge about the social consequences of individual human acts. The polemical message of *WN* is founded upon this conviction. For the duty of "superintending the industry of private people" assumed by "the sovereign" is one "for the proper performance of which no human wisdom or knowledge could ever be sufficient"(*WN* IV.ix.51). The implicit contrast with *divine* wisdom is left open.

It has lately been asserted by Peter Minowitz[49] that "*The Wealth of Nations* is an atheistic and anti-Christian work" and that Smith's "campaign against religion ... emerges – indirectly – from his wholly secular examination of 'nature and causes'." Much is made of the fact that there is no mention in *WN* of "Jesus," or "Christ," or "the Son," and no direct reference to "God" or "Providence." But in truth there is little difference between Smith's language of social explanation and that of Josiah Tucker.[50] Moreover, it would seem from the literature reviewed so far that Smith's conceptual apparatus in *WN* was far from novel, and that at least two of his most powerful ideas were developed fifty years before, in an explicitly Christian context, in the profoundly influential writings of Bishop Butler. Neither Butler nor Tucker appear in the index to Minowitz.[51]

In my opinion it is more plausible to regard *WN* as a work of Newtonian natural theology, based on a complex and sophisticated exploitation of the ambiguous term "nature."[52] Ian Ross has correctly remarked that Smith's "philosophy of explanation involves final explanations, couched in terms of a purposeful nature or God, and this variety of theism is an integral party *[sic]* of his approach to social phenomena."[53] As a social-scientific equivalent of Newton's *Principia* it could be regarded by the Anglican Establishment of that period – but not perhaps by the rigorously Calvinist party within the Presbyterian Establishment[54] – as wholly compatible with orthodox Christianity. According to my reading, moreover, *WN* may also be thought to contain, and possibly to have been shaped by, an Augustinian account of the way God responds to human sin by using the consequences of sin both as a punishment and as a remedy.

Whether or not I am right in this interpretation, it is certainly the case as we shall see, that *WN* was never regarded at the time as "hostile to religion:" whereas the "political economy" of Malthus and Ricardo most certainly was.

7.3 The *Essay on Population*

An important feature of the economic thought considered so far is Berkeley's tacit assumption that continual wealth-creation is always, or at any rate normally, feasible: an assumption subsequently protected by Smith's account of *increasing returns*. But Malthus's *Essay on Population* created the presumption that *diminishing returns* would dominate increasing returns, and so retard or even extinguish economic growth. And the consequent metamorphosis of "political oeconomy" – from an "Inquiry into the Nature and Causes of the Wealth of Nations" into the "Dismal Science" of rational response to scarcity – raised for the first time the possibility of strong dissonance between political economy and Christian theology.

Unbounded wealth creation

The emphasis in *WN* on increasing returns, both from division of labor and from endogenous technical progress is too well-known to require comment. It has been suggested by some[55] that we may also discern in *WN* some recognition of diminishing returns, and hence that Smith's system may be subsumed under the "canonical classical model of political economy."[56] The best recent attempt to formalize Smith's growth theory[57] accepts the possibility of diminishing returns in *WN* but associates it solely with the agricultural sector, locates increasing returns in the manufacturing sector, shows that Smith assumes that an endogenous taste for "luxury" increases the relative share of manufacturing; and hence that increasing returns must come to dominate diminishing returns. Other recent Smith growth models[58] abstract altogether from diminishing returns.

Therefore, though Smith – like every serious economic thinker, at least since Plato[59] and the author of *Ecclesiastes* (5:11), understood that "Every species of animals naturally multiplies in proportion to the means of their subsistence, and no species can ever multiply beyond it" (*WN* I.viii.39), his optimistic vision of increasing returns caused him to regard capital, rather than land, as the chief constraint upon growth. Since the capital constraint can be removed by "parsimony," the way is open for more or less continuous population growth *and this is a good thing*. The "demand for men ... necessarily regulates the production of men;" demand for labor increases with capital, the "continual increase [of] which occasions a rise in the wages of labour," and "the liberal reward of labour ... is the cause of increasing population." Hence "it is in the progressive state, while the society is advancing to further acquisition ... that the condition of the ... great body of the people, seems to be the happiest and the most comfortable." Indeed "The progressive state is in reality the chearful and hearty state to all the different orders of society." (*WN* I.viii, 40, 18–21, 22, 42, 43) This state of affairs will come about "naturally" – if capitalists are parsimonious – under "the obvious and simple system of natural liberty." (*WN* IV.ix.31)

There were, of course, authors whose works suggest some modification of this rosy picture. Malynes (f1. 1586–1641) had recognized that continual population increase might be stopped by the "positive check;" "divested of non-essentials the 'Malthusian' Principle of Population sprang fully developed from the brain of Botero in 1589;" Quesnay believed that pressure of population in France "was actually present around 1750;"[60] and Sir James Steuart[61] had observed that "the generative faculty resembles a spring loaded with a weight, which always extends itself in proportion to the diminution of resistance; when food has remained some time without augmentation or diminution, generation will carry numbers as high as possible." Moreover it was Steuart,[62] so pointedly ignored by Adam Smith, who was among the first to formulate an account of diminishing returns which are necessary for this to happen. According to Schumpeter[63] the *locus classicus* of the law of diminishing returns is Turgot's *Observations sur le Mémoire de M. de Saint-Péravy en Faveur de l'Impôt Indirect,*[64] which "suffices in itself to place Turgot as a theorist high above A. Smith."

But despite these outliers the great majority of eighteenth-century economic thinkers – including those like Cantillon[65] who noted that "*Les Hommes se multiplient comme des Souris dans une grange, s'ils ont le moïen de subsister sans limitation*" – believed that for all practical purposes there were no physical limits to capital accumulation and population growth, and that these should be objects of national policy. Berkeley and Hume were populationists;[66] Wallace,[67] who first formulated what was to become Malthus's antiperfectibilist argument, believed that in practice no limit would appear "till the whole earth had been cultivated like a garden;"[68] Tucker[69] proposed to tax bachelors and childless widowers to encourage them to add their "*proper* Increase to the *publick* Stock of Inhabitants, in which the *Riches* and *Strength* of a Nation do consist;" even Steuart,[70] viewing the matter from the standpoint of the nation state, believed that the possibility of importation removes any physical barrier to growth.

Perhaps the clearest and most complete account of the *un*-canonical, *pre*-classical model of political "oeconomy" common to all the authors so far mentioned is to be found in the long chapter "Of Population and Provision" in Paley's celebrated textbook on *Moral and Political Philosophy*.[71] The Reverend William Paley (1743–1805), Archdeacon – later Chancellor – of Carlisle, whom Keynes[72] thought was perhaps "the first of the Cambridge economists," cited Berkeley and seems from context to have been familiar with the work of Mandeville, Hume, Steuart, Tucker, *TMS* and possibly *WN*.[73] Paley agreed with his predecessors in regarding population as a good, but rejected their quasi-mercantilist reasons. Though willing to assent to the standard of Anglican orthodoxy,[74] he was a utilitarian and a methodological individualist, and for him population was an index of social welfare to be maximized. He used his sophisticated two-sector model ("provisions" and "luxuries") to analyse the determination of output and population in a world without scarcity. As with virtually all before him, aggregate effectual demand determines aggregate supply in the long period because – implicitly – population, fully employed work-force and output are all produced under constant costs.[75]

Diminishing returns, misery and vice

The Reverend T. Robert Malthus (1766–1834) was required to read Paley's textbook as a Cambridge undergraduate, and, together with *WN* and Hume's political essays it is the prime source of his economic and political ideas.[76] In 1796 Malthus agreed with Smith, but disagreed with Paley, that it is the rate of increase in population, rather than its level, which determines the "happiness and prosperity of a state."[77] And in 1798 he disagreed further with Paley, who – like Wallace and (by implication) most other eighteenth-century economic thinkers – had assumed that "the number of people have seldom, in any country" arrived at the limit set by "all the provisions which the soil can be made to produce."[78] For:

> At every period during the progress of cultivation, from the present moment, to the time when the whole earth was become like a garden, the distress for want of food would be constantly pressing on all mankind, if they were equal. Though the produce of the earth might be increasing every year, population would be increasing much faster, and the redundancy must necessarily be repressed by the periodical or constant action of misery or vice.[79]

It is made clear at the outset of the first *Essay* that the dismal consequences of resource scarcity are not merely an hypothetical result of the egalitarian communism that Godwin had advocated, but are actually present in all times and places. "Necessity, that imperious all-pervading law of nature" restrains "the germs of existence ... within the prescribed bounds."

> The race of plants, and the race of animals shrink under this great restrictive law. And the race of man cannot, by any efforts of reason, escape from it. Among plants and animals its effects are waste of seed, sickness and premature death. Among mankind misery and vice. The former, misery, is an absolutely necessary consequence of it. Vice is a highly probable consequence, and we therefore see it abundantly prevail.[80]

Although the institutions of private property, marriage and wage-labor – by raising *average* per capita income above subsistence – can shelter the propertied classes from the worst effects of these in "the civilized state," the great mass of the lower orders must normally live at or near the margin of subsistence. And even their betters must often endure the misery of compulsory celibacy (or give way to the temptation to vice) as Malthus himself well knew.

The seeming necessity of "misery" or "vice" in all human existence is an entirely new element in economic thought. Almost immediately it created a conflict – and therefore a distinction, not previously apparent – between Christian theology and economic thought.

Malthus was, and remained for the whole of his life, a faithful clergyman of the Church of England. It was therefore obvious to him that the economic reasoning of his *Essay*, though brilliantly successful in disposing of Godwin's Jacobin attack on private property, had only succeeded at the cost of creating a serious theological problem. Why should a God who is believed to be perfectly good and wise, all-knowing and all-powerful, have created a world in which men and women must live in misery or vice? He therefore attempted to "vindicate the ways of God to man" in the last two chapters of the *Essay*. Unfortunately Malthus's talents as a theologian were not equal to those as an economist and his theodicy was seriously defective.[81] Paley, who was converted to Malthus's population theory by the *Essay*, sketched a more satisfactory theodicy of scarcity in his last work, *Natural Theology* (1802). And in 1816, J. B. Sumner's definitive *Treatise on the Records of the Creation* abstracted the "principle of population" from the Problem of Evil by showing that it might instead be regarded, following Paley, as an example of divine wisdom and "contrivance."[82]

In the 1817 recension of the *Essay* Malthus[83] paid tribute to Mr Sumner's "masterly developement and completion" of his views.

Economists versus human beings

Meanwhile, two other developments were occurring which were to identify "political economy" in the public mind as an intellectual enterprise altogether distinct from Christian theology. The "principle of population" gave rise to explicit formulations by Anglophone economists (unaware of Turgot's pioneering work) of diminishing returns, so producing the "canonical classical model" and converting what was hitherto a study of wealth into the new science of scarcity. And almost from its birth, "political economy" in the new, nineteenth-century sense was perceived and denounced by many as "hostile to religion."

Though doubting whether Malthus ever intended that his "ratios" should be integrated to afford a diminishing-returns production function,[84] Samuel Hollander[85] has lately lent his authority to the view that many other passages even in the first *Essay* do indeed imply diminishing returns. An explicit formulation is to be found at the very latest in Malthus's *Inquiry into the Nature and Progress of Rent* (1815), appearing in the same year as essays by Ricardo, West and Torrens which expounded essentially the same doctrine. The classic formulation of what became known as the "Ricardian" theory of rent – and the most complete explication of the "canonical classical model" – was supplied two years later by David Ricardo's *Principles of Political Economy, and Taxation*. The friendship between Malthus and Ricardo dated from 1811. In 1821 they combined with James Mill, Sumner, and most other leading practitioners of the new science, to found the *Political Economy Club*.

Well before this, Malthusian theory and the "political economy" which grew out of it had become objects of suspicion and hostility to many theological critics. The first *Essay* passed largely unnoticed and its eccentric theodicy incurred only gentle mockery.[86] But the 1803 recension brought down a torrent of rage and execration upon its author for his unwelcome speculations, especially his brutal rejection of a traditional doctrine Paley[87] had maintained: that "the poor have a claim founded in the law of nature" upon the resources of the rich. In a notorious passage – expunged in the next (1806) edition – Malthus had written, with obvious allusion to Paley's metaphor of a "banquet:"

> A man who is born into the world already possessed ... has no claim of *right* to the smallest portion of food, and in fact has no business

> to be where he is. At nature's mighty feast there is no vacant cover for him. She tells him to be gone, and will quickly execute her own orders if he does not work on the compassion of some of her guests.[88]

Southey's vitriolic review (*Annual Review* 1803) inaugurated what Donald Winch[89] has lately called "one of the enduring fault-lines in British cultural debate ... separating economists from the self-appointed spokesmen for human beings." A series of pamphlets and articles appeared denouncing Malthus for heresy and hardness of heart;[90] specifically for his "impious and blasphemous assertion, *that the Almighty brings more beings into the world than he prepares nourishment for.*"[91] The wound went deep. Many of the greatest luminaries of nineteenth-century Britain – Coleridge, Hazlitt, Southey, Wordsworth; Carlyle, Ruskin, Morris, Toynbee – took the "human" side in "the bitter argument between economists and human beings."[92] A full century after Southey's review, a leading Christian Socialist of the day, W. E. Moll (1857–1932), decried "the false political economy which 'teaches men to say that there are those for whom God has placed no plate at the banquet of life.'"[93]

It cannot be emphasized too strongly that the rift, or "fault-line," between economic thought and Christian theology opened up very suddenly in the decade or so after 1798 and is quite without precedent. Eighteenth-century high-church men of impeccable orthodoxy such as Samuel Johnson could believe that "there are few ways in which a man can be more innocently employed than in getting money."[94] As late as the 1790s Edmund Burke, who once said that *WN* was "probably the most important book ever written,"[95] could boldly declare – with possible allusion to Warburton's sneer at Josiah Tucker – that "the laws of commerce, which are the laws of nature, [are] consequently the laws of God."[96] But the *Essay on Population*, by deflecting both popular and scientific attention from the benign effects of wealth and wealth-creation to the seemingly malign consequences of resource scarcity and diminishing returns, created a wholly new climate of opinion. In 1832 a reputable journal could remind its readers that the writings of Malthus and Ricardo had "tended to lead the public far away from the true path of inquiry," and to make of political economy "a hideous chain of paradoxes at apparent war with religion and humanity."[97] More than two decades later an influential American economist and protestant cleric contrasted the "harmony" apparent in *WN* with "the great law of discord, promulgated by Malthus and Ricardo."[98]

7.4 The boundary between political economy and Christian theology

Although by 1820 Paley, Sumner and Coplestone had worked out an accommodation of political economy that was acceptable to Christian orthodoxy, the good effect of their work was undone by an attempt of the Philosophic Radicals to hijack the new science to their avowedly atheistic program of reform. In responding to this new challenge Richard Whately was led to propound an epistemological *boundary* between "scientific" and "religious" knowledge, later popularized by Nassau Senior, which laid the foundation of methodological orthodoxy in political economy – and in what is now called "economics." And precisely because of Whately's demarcation between what came to be seen as distinct and non-competing inquiries, the possibility was created of fruitful *exchange at the boundaries*.

Ideological crisis in the 1820s

As a result of Sumner's *Treatise* (1816), Malthusian political economy was able to retain the anti-utopian aspect of Malthus's original polemic against Godwin whilst replacing its amateurish theology with a more acceptable theodicy. What had appeared to Malthus as a nasty case of the Problem of Evil was shown by Sumner to be an example of the Argument from Design, so successfully developed by Paley in his *Natural Theology* (1802). This, in turn, enabled Sumner to reintroduce the orthodox, Butlerian doctrine of human life on earth as a "state of discipline and trial" that Malthus had denied in 1798. Malthus himself had strengthened the argument for private property in 1803 by appropriating Paley's concept of "moral restraint" – which also afforded a legitimate escape from "misery or vice." And "moral restraint," which must be taught and learned, required that attention to institutional reform which all Whigs took for granted and which "liberal Tories" were beginning to tolerate. All of these improvements were incorporated in the 1817 recension of the *Essay*.[99]

Though Edward Copleston's two *Letters to Peel* (1819) were much briefer and narrower in scope than Sumner's *Treatise*, they added intellectual content to the Malthus–Sumner argument and also helped to make it more widely respectable. For Copleston, then Provost of Oriel College, was one of the first Oxford men, certainly the most influential, to take political economy seriously; and Oxford was the spiritual home of the English Establishment.[100] Through the combined efforts, therefore, of the Simeonite evangelical, Cambridge Whig

Sumner (later Archbishop of Canterbury) and the pre-Tractarian high-church, Oxford Tory Copleston (later Bishop of Llandaff), political economy had been made, if not palatable, at least digestible to orthodox members of the Church of England by 1820. Only the most intransigent Romantics – the Lake Poets and their circle – persisted in their hostility.

In the early 1820s however the good work of Sumner and Copleston in reconciling English Christians to political economy was seriously endangered. Bentham, James Mill, and their allies – including Ricardo until his death in 1823 – having established the *Westminster Review* (1824) and the University of London (1826) to propagate their views, were alarmingly successful in harnessing the new science of political economy to the cause of radical "reform." Bentham had lately published his widely noticed pamphlets attacking the Church of England, natural theology and St Paul,[101] and Mill's hostility to Christianity was well known. The Philosophic Radicals made no secret of the fact that reform was to be guided only by the criterion of the greatest good of the greatest number, without reference to the traditions and structure of a Christian society; and that political economy was to be the instrument of designing reform. All the worst fears of Oxford Tories, ever ready to suppose "the Church in danger," were reawakened. As Whately observed in 1828, many once again regarded political economy "with a mixture of dread and contempt – as a set of arbitrary and fanciful theories, subversive of religion and morality." It had become essential to reassure those who feared "the pursuit of knowledge of any kind, as likely to be injurious to the cause of religion" that "truth ... can never be at variance with truth;" and that "the Bible ... was not intended to teach men Astronomy or Geology, *or*, it may be added, Political Economy, but Religion."[102]

Richard Whately (1786–1863), former Fellow of Oriel, Principal of St Alban's Hall, was ideally situated for performing this task. A pupil and friend of Copleston, logician, Christian apologist and ecclesiastical statesman, he occupied a position of great power and influence in Oxford of the 1820s.[103] His first move was to engineer the election of his former pupil, Nassau Senior, as Drummond Professor of Political Economy in 1826. When Senior's term expired Whately accepted the chair himself, for it seemed to him, he wrote to a friend in 1829:

> that there is a sort of crisis for the science in this place, such, that the occupying of it by one of my profession and station may rescue it permanently from disrepute. Religious truth ... appears to me to

> be intimately connected, at this time especially, with the subject in question. For it seems to me that before too long, political economists, of some sort or other, must govern the world; ... now the anti-Christians are striving hard to have this science to themselves, and to interweave it with their own notions.[104]

Whately was duly elected to the chair and delivered his first and only lectures[105] during the Easter Term of 1831. His academic career was abruptly terminated three months later by his appointment as Archbishop of Dublin.

Whately's demarcation

Though Whately had intended to treat political economy in the manner of Paley's *Natural Theology*, his first object was "to combat the prevailing prejudices against the study, and especially those which represent it as unfavourable to religion."[106] I have elsewhere described Whately's argument in detail[107] therefore a brief summary will suffice.

Whately was much influenced in his approach by another former pupil, Samuel Hinds, whose important book on *Inspiration and the Authority of Scripture* (1831) had just appeared. Hinds argued that "sacred knowledge" (of God) and "secular knowledge" (of nature) are distinct. Though the former "should have a due share of our intellect', its truths must – and can only – be 'spiritually discerned." But the latter is autonomous, requiring only reason to comprehend. Hence the Bible can not be the source of all truth, "but only of such truth as tends to religious edification" and in this alone is the Bible infallible.[108] Whately used these ideas to argue that it is erroneous to appeal to scripture "on questions of physical science," for "Scripture is not the test by which the conclusions of Science are to be tried."[109] Since political economy resembles physical science in that it consists of "theory" in relation to "observable phenomena," we may expect:

> That Political-Economy should have been complained of as hostile to Religion will probably be regarded ... with the same wonder, almost approaching to incredulity, with which we of the present day hear of men's having sincerely opposed, on religious grounds, the Copernican system.[110]

Moreover, not only does the Bible contain no authoritative *scientific* knowledge; even its account of "*moral* truths ... must be received with considerable modification." For biblical doctrine of virtue and vice is such "as seems to presuppose a natural power, or capacity for acquiring that power, to distinguish them."[111] A moral sense of some kind is therefore necessary for right conduct. Thus Whately opposed head-on the consequentialism which Bentham (the enemy) shared with Paley (the ally).

By arguing that political economy and theology are distinct, incommensurable and non-competing fields of inquiry, Whately was able to assuage the doubts and fears of his Oxford colleagues about the new science. And by arguing that a consequentialist ethic is defective, he countered the intellectual imperialism of the Philosophic Radicals who sought to impose "scientific" reform in the name of "utility." The latter were correct in regarding political economy as a valuable instrument for implementing the social values which guide public policy. But they were wrong to suppose that the hedonistic calculus can be a reliable source of (or substitute for) those values. Only a moral sense, preferably illuminated by holy scripture, can determine those ends to which political economy is only the means.

Whately's insistence that political economy is merely a value-neutral, "positive" study of means was embedded in a philosophy of science, derived from Dugald Stewart, that emphasized the priority of theory in all scientific inquiry, and the essentially analytical nature of such theory.[112] As developed and popularized by Senior[113] it became the foundation of methodological orthodoxy later expounded by J. N. Keynes[114] and Lionel Robbins,[115] recognized in present-day textbooks as "positive economics." The boundary between political economy and Christian theology, now so obvious and impermeable, was thus erected for the first time in 1831 by Whately – in response to an ideological crisis that was ultimately traceable to the publication thirty years before of Malthus's first *Essay on Population*. It is important to realize that this boundary, which is a corollary of Whately's rigorous specification of the scope and nature of political economy, is the *fons et origo* of all subsequent boundaries between economics and other inquiries.

Exchange at the boundary

It is only when the existence of a boundary has been generally recognized that there can be any "exchange at the boundaries." Whately himself was the first to undertake any such "exchange" consciously and knowingly.

Before accepting his appointment as Drummond Professor, Whately had thought of "making a sort of continuation of Paley's 'Natural

Theology', extending to the body-politic some such views as his respecting the natural."[116] Though most of the *Introductory Lectures* were taken up with methodology, Whately did include one celebrated passage, eventually incorporated into Samuelson's[117] famous textbook, which describes the market solution of "the problem of supplying with daily provisions of all kinds such a city as our metropolis" by individuals "who think each of nothing beyond his own immediate interest."[118] The radicals' quarterly described it sarcastically as "one of the most beautiful pieces of Sunday reading it ever fell to the lot of the Westminster Review to recommend."[119]

Yet it would appear that Whately was doing nothing different in this case from what Tucker, Sumner, and possibly even Smith had done; and in a sense this is so. The first two at any rate had intended to show that the unintended result of self-love in economic life may be socially beneficent, and that this can be taken as evidence of "contrivance" in a wise and benevolent Author of Nature – whom they identified with the Judaeo-Christian God. Were they not, then, "exchanging" at a boundary already in existence before 1831?

It is certainly the case that the purpose of natural theology is to show from the results of a strictly positive, scientific inquiry that knowledge of God may be had without any resort to putative "revelation" (sacred scripture, tradition and the beatific vision). It is therefore of the essence that the data brought in evidence should be, and should be recognized to be, those which may occur in the ordinary course of nature. If "the laws of commerce" are indeed "the laws of God" they are so in exactly the same way as are the laws of celestial mechanics discovered by Newton. But to Paley, Malthus and Sumner, Newtonian mechanics had been taught (as it had also been taught to Hume, and presumably by Smith in his Glasgow lectures on natural theology) almost as a branch of theology, and certainly as an adjunct to theology. Hence the distinction between "science" and "theology" was far less clear in their minds – and in the minds of their contemporaries – than it could ever be again after the intellectual turmoil created by the *Essay on Population.* Malthus and Sumner lived through that turmoil. But Malthus, though indeed a "Christian moral scientist"[120] was first and foremost an "economist" – so-called as early as 1804[121] – with little taste or talent for theological niceties. Sumner was primarily a theologian, with no methodological interest in the new science. Whately was neither, but a logician with a keen appetite for conceptual refinement. To him, therefore, belongs the credit for first recognizing and defining a boundary which he then deliberately crossed.

At least two kinds of "exchange at the boundaries" were now possible between political economy – or "economics" as it eventually became – and Christian theology. In the first place, the latter could be used, as it already had been, to supply evidence of Divine contrivance in the "spontaneous" order – Hayek's terminology is carefully untheological[122] – which arises unintentionally out of the self-love of individuals. Second, in societies which continued to acknowledge Christianity as the public religion, economics could be combined with theology in the construction of normative social theory.

Natural theology continued to flourish in nineteenth-century Britain and America, undisturbed by Hume's posthumous *Dialogues Concerning Natural Religion*, until the appearance of *Origin of Species*. Darwin's great work[123] compelled Victorians – who assumed hitherto that Hume had been adequately answered by Paley, Whately, and others – to acknowledge that nature may afford no evidence of "design." It was thus in the 1860s, Keynes believed, that "Christian dogma fell away from the serious philosophical world of England, or at any rate of Cambridge."[124] Until that time political economy was frequently pressed into theological service, as, for example, in Frédéric Bastiat's *Harmonies* (1850), or textbooks by the American, Henry Carey (1837) – who like Bastiat preferred to contemplate the "harmony" apparent in *WN* rather than the "discord" introduced by Malthus and Ricardo.

The other respect in which there has been border-crossing between economics and theology is of more recent date, and is the consequence of a belated recognition by ecclesiastical authorities of Whately's claim that economics may be regarded as an ethically neutral science of means, autonomous with respect to theology. At least since the papal Encyclical *Quadragesimo Anno* (1931), but especially since a revival of interest in economic policy by many churches in the 1970s, it has come to be thought that economic science may be employed as a means to social ends proposed by Christian theology. Many conferences between economists and theologians have taken place in the last twenty years;[125] a flourishing *Association of Christian Economists* was founded in the USA during the 1980s; and numerous books by economists and others have appeared on various aspects of "Christianity and Economics."[126]

Though Whately's demarcation has become so much a part of methodological orthodoxy in economics that few now realize how or when it came into being, it must not be supposed that it is, or has been, universally accepted. For although French economists such as J.-B. Say appear to have taken the autonomy of economics for granted, those whose sympathies lay with the *ancien régime* were still inclined to regard economics as subservient to, or as part of, Christian theology. The most considerable of these, Alban de Villeneuve-Bargemont, published his *Economie Politique Chrétienne* in 1834. At the outset of his study, Villeneuve records:

> *un horizon vague et immense s'était offert à mes regards; peu à peu, à l'aide surtout du phare lumineux du christianisme, il me sembla que l'on pouvait distinguer nettement les causes des désordres moraux et matériels des sociétés: les fairs se classèrent naturellement.*[127]

An Irish reviewer described Villeneuve's work as "catholic in its faith, and catholic in its manner of conceiving science."[128] It influenced subsequent French and German catholic social theorists, and through them the famous Encyclical of Leo XIII, *Rerum Novarum*.[129] Among protestant economists, a similar unwillingness to recognize the autonomy of economic science has been evinced by those of the Dooyeweerdian, Neo-Calvinist tradition associated with the Free University of Amsterdam. According to this school of thought, there can and ought to be a "Christian economics" which might in principle differ both in method arid results from "secular economics."[130]

It is evident that "Christian economics," whether of the catholic or protestant variety, obliterates the boundary between economics and theology. In such work, therefore, there can be no "exchanges at the boundary."

8
Methodology of Classical Political Economy

> The English writers ... or followers of Dr Smith ... define their science as that of the laws which regulate the production and distribution of wealth ... . The foreign school ... holds that it is the office of the political economist to point out in what way social happiness may best be attained through the medium of national wealth ... . We contend that the study is purely a science: our opponents, that it includes the practical adaptations of the science to existing circumstances. In other words, that it is at once a science and an art. According to them it is a *deontology:* according to us, an *ontology* only.[1]

The *Edinburgh's* notice of Nassau Senior's *Outline of the Science of Political Economy* (1836) pinpoints a distinction between "ought" (deontology) and "is" (ontology) which characterizes the method employed by the "followers of Dr Smith" down to the present. Long before Adam Smith indeed, Petty had sought to separate science, viewed as a means to an end, from moral problems which arise in the selection of ends.[2] And Richard Cantillon – "foreign" though he was – had made "a conscious separation of pure theory from normative policy declarations and value judgments."[3] By the end of the nineteenth century, J. N. Keynes was in a position to isolate "a *positive science* ... as a body of systematized knowledge concerning what is" from either "a *normative ... science* as a body of systematized knowledge relating to criteria of what out to be," or "an *art* as a system of rules for the attainment of a given end."[4] The most complete account of Petty's distinction between "means" and "ends" in social science was supplied by Lionel Robbins in the 1930s. Since "Economics ... is concerned with that aspect of behaviour which arises from the scarcity of means to achieve given ends" it follows that "Economics is entirely neutral between ends."[5] The argument of his book converges on a corollary of that proposition: "Economics cannot pronounce on the validity of ultimate judgments of value."[6] The entire tradition is summed up for the first-year student in a well known textbook:

> The distinction between positive and normative allows us to keep our views on how we would like the world to work separate from our views on how the world actually does work. We may be interested in both. It can only obscure the truth however, if we let our views on what we would like to be bias our investigation of what actually is. It is for this reason that the separation of the positive from the normative is one of the foundation stones of science and that scientific inquiry, as it is normally understood, is usually confined to positive questions.[7]

Subject only to such other criteria as may be necessary in defining "positive," economics is regarded by all such writers as a positive science.

Possibly because of verbal confusion, this venerable tradition has lately been smeared by association with the modern doctrine of "positivism" and made to share in the blame so freely bestowed by "postmodernist" writers on the latter.[8] Without wishing to deny a "rhetorical" significance to the separation of "fact" from "value," I shall focus my attention in this chapter on that aspect of it identified by Robbins: the putative inability of economics to throw any light on "ultimate judgments of value." I do so without wishing to recommend any such a view to the skeptical, or to even to associate myself with what is undoubtedly a prevailing orthodoxy among modern economists. My object is intellectual history. Why is the modern orthodoxy what it is? How did it emerge from theological and philosophical debate in the late eighteenth and early nineteenth centuries? And what light does this throw on the relation between economics and theology which is the subject of this book?

The modern view that economics can only tell us about facts (if even that) and not at all about values has its origin, as I have suggested, in the dichotomy between "science" and "art" drawn by classical and pre-classical political economists. But the distinction was first made sharp and explicit by Richard Whately and subsequently canonized by his friend and former pupil, Nassau Senior. Whately was responding to a particular ideological emergency in the third decade of the nineteenth century: an attempt by the "Philosophic Radicals" to hijack political economy for their (avowedly) godless program of "reform." It was essential, as Whately saw it, to rescue political economy from the hands of Bentham and his disciples and restore it to the political and ecclesiastical establishment. But the establishment needed reassurance that political economy was not "hostile to religion." Whately had, therefore, to create a strategy for immunizing religious "knowledge" against encroachment by the scientific "knowledge" afforded by economics, whilst at the same time preserving the authenticity of the latter. He did so by extracting from the vague distinctions of his predecessors a clear-cut, epistemological demarcation of "religious" from "scientific" knowledge. It is therefore a somewhat paradoxical feature of my exposition in this chapter that a methodological principle designed to protect religion from economics and vice versa should have its origin in a particular theological doctrine invoked to meet a particular religious need.

In what follows I shall first outline the circumstances which caused Whately to develop his methodology; secondly, describe that methodology; and thirdly, trace its incorporation into the mainstream of modern economics by way of Senior's influential work.

8.1 The ideological crisis of the 1820s

An ideological alliance of political economy and Christian theology[9] was first constructed by Malthus[10] to provide a defense of the British *ancien régime* against the revolutionist polemic of the English Jacobins, in particular, of William Godwin.[11] Certain deficiencies in Malthus's original argument were gradually corrected over the next twenty-one years by Paley,[12] John Bird Sumner,[13] Malthus himself,[14] and Edward Copleston.[15] By 1820 a coherent body of political doctrine existed. Poverty and social inequality are the inevitable outcomes of scarcity. Human life is a state of "discipline and trial" for eternity. The institutions of private property, marriage, and the market economy are necessary and inevitable, suited to the human condition, and consistent with scripture. It is impossible to achieve social progress by legislation, and this is evidence both of design – in God's creation of the self-regulating economy – and of the moral and religious need of Christians to practice charity and compassion.[16] By 1820 political economy had in this way been made wholly acceptable to Lord Liverpool's implacably counter-revolutionary cabinet.

When the Political Economy Club was founded in 1821, the discipline still united all who studied it. But almost from the first a schism began to appear between those later called "radicals" and the rest.

During the previous decade Bentham and James Mill had gradually moved away from the whiggism of Holland House and the *Edinburgh*, and under the influence of Francis Place had allied themselves with "radical" causes whilst clearly distinguishing their position from the plebeian radicalism of Cobbett and Hunt. Though Halévy[17] exaggerates in saying that "all the actions in Ricardo's life, after 1811, were willed by James Mill," it is certainly the case that Ricardo tended to side with the Philosophic Radicals until his death in 1823.

Philosophic Radicalism is the term coined in 1837 by John Stuart Mill to identify the political doctrine developed by his father and Jeremy Bentham from the second decade of the nineteenth century. At its core is the peculiar version of utilitarianism invented by Bentham: Paley's "theological utilitarianism" with God left out. The doctrine is defined by the union of Benthamite utilitarianism with *political economy:* in particular with the extreme version of the deductive, *a priori* approach to political economy typified by Ricardo's *Principles* (1817).[18] Its popular character derived less from these purely intellectual components, however, than from a puritanical hatred of the arts and of literary culture with which James Mill managed to invest it. The attack by Philosophic Radicalism on British political culture in the 1820s was launched in the *Westminster Review* (founded in 1824) and from University College, London (founded in 1825).

It was this third, essentially accidental, feature of Philosophic Radicalism which first attracted the hostility of the literary world. The *Westminster's* attacks on Sir Walter Scott and the Lake Poets provoked a fierce response from the leading Romantic authors, all of whom were already predisposed against political economy from their horror of Malthusian population theory. The "New School of Reform" was attacked for breaking up the old reformist alliance:

> The obvious ... effect of the Westminster tactics is to put every volunteer on the same side *hors de combat*, who is not a zealot of the strictest sect of those they call Political Economists ... to leave nothing intermediate between the Ultra-Toryism of the courtly scribes and their own Ultra-Radicalism.[19]

It was also attacked for godlessness, and in a way that associated godlessness with political economy. In the works of Malthus and Ricardo, the *Eclectic Review* supposed, political economy had become "a hideous chain of paradoxes at apparent war with religion and humanity."[20] Wordsworth[21] perfectly expressed the detestation felt by Christian

Romantics for the New School in his famous sonnet on King's College Chapel:

> Tax not the royal Saint with vain expense,
> With ill-matched aims the Architect ...
> ... high Heaven rejects the lore
> Of nicely-calculated less or more;
> So deemed the man who fashioned for the sense,
> These lofty pillars ...

By the second half of the 1820s, this is to say, the aggressive propaganda of the New School had undone much of the good work performed by Paley, Sumner, and Copleston in winning acceptance of political economy by the ecclesiastical and political establishment.

Oxford was the spiritual home and intellectual center of that establishment; and by the mid-1820s Richard Whately, Warden of St. Alban's Hall, was its most powerful and articulate spokesman. Whately was fully alive to the threat presented by the Philosophic Radicals but, unlike the Tory romantics, was wholly persuaded both of the scientific value and the ideological efficacy of political economy. He therefore saw his task as that of rehabilitating political economy in the eyes of the establishment by wresting it from the grasp of the radicals. For:

> ... if the cultivation of this branch of knowledge be left by the advocates of religion, and of social order, in the hands of those who are hostile to both, the result may easily be foreseen.[22]

There can be no doubt that he was behind the election of Nassau Senior to the newly founded Drummond Chair in 1825, nor that Senior's *Introductory Lecture* (1827) was composed, in part, with Whately's objects in view. A letter of Whately to Senior written in 1826 shows that the latter submitted a draft of his lectures for the approval of his old tutor before he began the series.[23]

Senior's first lecture was delivered in December 1826. It defined political economy as:

> ... the science which teaches in what wealth consists, ... by what agents it is produced, – and according to what laws it is distributed, – and what are the institutions and customs by which production may be facilitated and distribution regulated, so as to give the largest amount of wealth to each individual.[24]

The study is thus both practical and theoretical. The former is to discover what institutions are most favorable to wealth and requires "induction from phenomena" and must "take into account the influence of every human passion and appetite."[25] The latter rests "upon a very few general propositions ... which almost every man, as soon as he hears them, admits" and is "capable of all the certainty that can belong to any science, not founded exclusively on definitions."[26] Though the pursuit of wealth may be regarded as inferior to the pursuit of virtue or knowledge, and though it is not *sufficient* for happiness, yet it is "a great source of moral improvement" and national wealth (suitably distributed) is *necessary* for morals and happiness and even for "all real religion."[27]

Whately (1828) reviewed Senior's first four lectures for the *Edinburgh* and warmly welcomed both the establishment of the chair at Oxford and the productions of its first occupant:

> A study which has so far received the sanction of that learned and orthodox body, stands some chance of being rescued from uninquiring contempt; and no set of men, we presume, could more safely be entrusted to appoint such professors as shall at least be untainted with extravagant anarchical principles.[28]

Nevertheless it is clear that he felt that Senior's justification of political economy, though "perfectly satisfactory" so far as it went, was "far from exhausting the subject."[29] He therefore devoted a few hundred words to each of two issues, the first of which Senior had ignored, and the second only briefly dealt with.

Many at Oxford (Whately knew) regarded political economy "with a mixture of contempt and dread – as a set of arbitrary and fanciful theories, subversive of religion and morality ... ." It was essential to reassure those who feared "the pursuit of knowledge of any kind, as likely to be injurious to the cause of religion" that "truth ... can never be at variance with truth;" that holy scripture "should be appealed to only in respect of matters beyond the reach of unassisted reason;" and that "the Bible ... was not designed to teach men Astronomy or Geology, or, it may be added, Political Economy, but Religion ... ."[30]

In the second place, there was more to be said about the vexed relation between wealth and virtue than in Senior's easy dismissal of the problem. Mandeville's argument that a society may have "either national wealth or national virtue" *but not both*[31] is too deeply rooted in the very nature of economic theorizing to be brushed aside. Though

Whately made no attempt to supplement Senior's treatment of this matter in his review it is clear from what followed that he had decided to do something about it himself.

In 1829 Senior's first term as Professor expired, and Whately allowed himself to be nominated to the chair. For:

> ... some of my old friends have persuaded me that this is a sort of crisis for the science in this place, such, that the occupying of the office by one of my profession and station may rescue it permanently from disrepute. Religious truth ... appears to me to be intimately connected, at this time especially, with the subject in question. *For it seems to me that before too long, political economists, of some sort or other, must govern the world;* ... now the anti-Christians are striving hard to have this science to themselves, and to interweave with it their own notions ... .[32]

Whately was duly elected (unopposed) and delivered his first and only set of lectures in the Easter Term of 1831. For his academic career was abruptly terminated on 14 September 1831, by his nomination to the See of Dublin, making Whately the only economist in history to move directly from a professorial chair to an Archbishopric without intervening stages.

8.2 Demarcation of "scientific from religious"

Whately's first object, in his *Introductory Lectures* (1832), was "to combat the prevailing prejudices against the study [of political economy], and especially those which represent it as unfavourable to Religion."[33] His method was to contrast the epistemological character of religious and scientific "knowledge," to argue that political economy consisted only of the latter, and to demonstrate that it was a merely instrumental study of the *means* to ends which must be sought elsewhere.

For English protestants of Whately's time religious knowledge came both by "reason" and by "revelation." "Reason," meaning in particular, Newtonian science, was supposed to demonstrate, or at any rate corroborate, the existence and attributes of the deity: including His power, wisdom, and perfect goodness.[34] "Revelation" meant the putative evidence of God's self-disclosure contained in holy scripture: yet as Whately was careful to note in his review of Senior, "The Scriptures should be appealed to only in respect of matters beyond the reach of unassisted reason."[35] In his development of this point in the *Introductory Lectures* Whately was much influenced by another former pupil, Samuel Hinds, who became his first domestic chaplain (1831–33) and, like Senior, remained a lifelong friend and correspondent.

Hinds published an important book in 1831 on *Inspiration and the Authority of Scripture* which identified more clearly than ever before in Anglican discourse the distinction between "secular" and "sacred" knowledge. The two are not mutually exclusive, as we have seen, for the secular knowledge which comes by science can enlarge or confirm our knowledge of God. The domain of "natural theology" is precisely their intersection. But though revelation "should have a due share of our intellect," its truths must be "spiritually discerned:"[36] that is, by "faith." Faith and reason are not in conflict, therefore, nor do they operate over wholly different fields: but the former is needed to give us knowledge not accessible to reason alone. The question is: what kind of knowledge does faith, operating on revelation, produce?

According to Hinds, the Bible contains its own hermeneutic rule in many passages declaratory of the criterion by which its inspiration must be judged:

> It is not ... truth of all kinds that the Bible was inspired to teach, but only such truth as tends to religious edification; and the Bible is consequently infallible as far as regards this, and this alone[37]

Thus when scripture refers to other matters of an historical, scientific, or social-theoretic nature, its authority is not divine and therefore absolute, but only "that which attaches to the work of an honest and sincere author, and varies according to his individual circumstances, and the circumstances of the country and age in which he wrote."[38] The Mosaic account of the Creation, for example, has divine authority for "the religious truth involved in it ... that God created and appointed the sun its sphere;" but its "authority for the astronomical truth is only human,"[39] and therefore provisional and corrigible. It is clear from this that there can in principle be no conflict between the secular knowledge afforded by astronomy, geology, biology and the like, and the sacred knowledge to be "spiritually discerned" in scripture.

Whately's second lecture treated at length these matters. "Some there are who sincerely believe that the Scriptures contain revelations of truths the most distinct from religion." But it is an "erroneous principle" to appeal "to Revelation on questions to physical science" for "Scripture is not the test by which the conclusions of Science are to be tried." Now political economy resembles "physical science" in that it consists of "theory" in relation to "observable phenomena:"[40]

> That Political-Economy should have been complained of as hostile to Religion will probably regarded a century hence (should the fact be then on record) with the same wonder, almost approaching to incredulity, with which we of the present day hear of men's having sincerely opposed, on religious grounds, the Copernican system.[41]

Not only does the Bible contain no authoritative information about *scientific*, matters: even its account of "*moral* truths ... must be received with considerable modification."[42] For:

> God has *not* revealed to us a system of morality such as would have been needed for Beings who had no other means of distinguishing right from wrong. On the contrary, the inculcation of virtue and reprobation of vice in Scripture are in such alone as seems to presuppose a natural power, or a capacity for acquiring the power, to distinguish them.[43]

A moral sense of some kind, an ability to perceive a "natural law," is necessary for right conduct: thus Whately opposed head-on the consequentialist – utilitarian – ethic which Paley (an ally) had maintained in common with Jeremy Bentham (the enemy).

So much for sacred knowledge, which comes by *faith* – the "spiritual discernment" of "revelation" – and which, though capable of being confirmed by *reason*, cannot be disconfirmed by it. What, then, is *scientific* or secular knowledge, and how is it to be sought?

It has recently been persuasively argued by Pietro Corsi (1987) that the chief source of Whately's approach to this question – an approach that he made in common not only with Copleston, Baden Powell, Hinds, Senior, Hampden, and the entire "Noetic" circle but also with John Stuart Mill – is the second volume[44] of Dugald Stewart's *Elements of the Philosophy of the Human Mind*.[45] The powerful influence that Stewart exercised not only over what became the *Edinburgh Review* circle but also over the first generation of Cambridge economists has long been attested.[46] Corsi has shown that from the time that Stewart's pupil and friend, John William Ward (1781–1833, first Earl of Dudley), moved to Oxford and became Edward Copleston's tutorial pupil at Oriel (1799) there was continual communication, mutual respect, and mutual criticism between the two men who dominated the intellectual lives of Edinburgh and Oxford, respectively. Copleston was the chief formative influence upon Whately; thus it is no surprise that "the philosophical strategy adopted by Copleston and Whately in their polemical works was greatly indebted to *Stewart's ... Elements of the Philosophy of the Human Mind*."[47]

Like most philosophers of his time and place Stewart worked to uphold "the Inductive Philosophy of the Newtonians" from "the hypothetical systems of their predecessors" (meaning Descartes). Volume II of his *Elements* was written, in part, to meet the objections raised by Francis Jeffrey[48] against naive or crude inductivism. A "chaos of insulated particulars" must be distinguished from "a knowledge of connected and well ascertained facts." But this can only be had by *theory*, for "without theory ... experience is a blind and useless guide." Indeed there can be no "facts" without theory, and "so deeply rooted in the constitution of the mind is that disposition on which philosophy is grafted, that the simplest narrative of the most illiterate observer involves more or less of hypothesis." Hence "when a scientific practitioner quits the empirical routine ... in quest of a higher and more commanding ground," he looks for "general laws" which enable him "to separate accidental conjunctions from established connexions."[49]

Stewart, a pupil, friend, admirer, and authoritative expositor of Adam Smith, illustrated his argument from *Political Economy*, a scientific (or philosophic) discipline contrasted with the crude observationism of the "*political arithmeticians, or statistical collectors*." It was "acknowledged by Mr. Smith, with respect to himself, that he had 'no great faith in political arithmetic.'" Smith's aim was to identify "the *general laws* which regulate the course of human affairs." These are few, are "practical maxims of good sense," and may be known by introspection: "we have only to retire within our own bosoms, or to open our eyes to what is passing around us" to be assured of their truth. Chief among them are: (1) that "Political Order" is the spontaneous outcome of individual decisions and is "less the effect of human contrivance than is commonly imagined;" (2) that "every man is a better judge of his own interest than any legislator can be for him;" (3) that "private interest may be safely trusted to as a principle of action universal among men;" and (4), as a consequence of these, that "there is a reasonable presumption in favour" of maximizing economic freedom subject to "the security of property."[50] Though Stewart presented these as "conclusions" deduced from premises it is evident that (1), (2) and (3) might be, and in fact were by Adam Smith and his successors, used as axioms in a *deductive* system of economic theory.

Thus Stewart contrived to defend science in general and political economy in particular as a theoretical structure deduced from axioms, essential for the organization and even the correct identification of "facts," whilst continuing to pay lip-service to the Baconian inductivism mythopœically attributed to the immortal Newton. The third axiom, first stated as that "*the desire of bettering our condition appears ... to be the masterspring of human action*"[51] was crucial. Together with the resemblance which Stewart noted[52] between the "general laws" of political economy and those of physics, it formed the centerpiece of that Anglican, Oxonian reconstruction of the methodology of political economy which ever since has defined orthodoxy in our discipline. As Copleston put it in his first *Letter to Peel,*[53] "Profit is in mercantile dealings, what *gravitation* is in the system of the universe: and no problem is worth listening to, which supposes the absence of that universal principle."

Whately applied Dugald Stewart's account of the inescapably theoretical way in which "the human mind" organizes "experience" into scientific knowledge to Senior's definition of political economy as the science of *wealth.*

A narrower definition would have suited Whately better. For since the "propensity to truck and barter" noted by Adam Smith distinguishes human from other animals, and since "it is in this view alone that Man is contemplated by Political-Economy," it would be better to speak of "catallactics, or the science of exchanges." But if "the term Wealth is limited to *exchangeable* commodities" (as Whately supposes it is in Smith), then the more usual definition will do.[54] Whately noted that some have criticized political economists for "confining their attention to the subject of Wealth." But for his polemical purposes (as against the Philosophic Radicals who wished to base policy recommendations on political economy alone) this is entirely proper. For the seeming recommendations of economists must be understood "*in reference to national wealth* alone; and not as giving any decision as to its *absolute* expediency." An economist who presumes to recommend upon the assumption "that wealth constitutes the sole ground of preference of one thing to another" will be "deserving of censure:" for his "proper inquiry" is "as to the means by which wealth may be preserved or increased. To inquire how far wealth is *desirable,* is to go out of his proper province."[55]

It is, of course, the abstract nature of political economy which isolates it from the welter of particular circumstances as interpreted by human values – each of which is historically and culturally specific –

which must largely govern public policy formation. Whately's exposition of the abstract, theoretical, ahistorical nature of economic science closely follows Dugald Stewart.

All observation is theory-laden; hence there can be no escape from theory, only a choice between explicit and implicit theorizing. The latter is almost always inferior:

> When you find anyone contrasting ... . Experience with Theory, you will usually perceive ... that he is in reality comparing the results of a *confined*, with that of a *wider*, experience; – a more imperfect and crude theory, with one more cautiously framed ... [56]

Theory may be true or false. This is where observation plays its part in scientific inquiry: for "if we really are convinced ... of the falsity of any theory ... we must needs believe that that theory is also at variance with observable phenomena."[57] We must therefore begin with theory and search for the "facts" which may confirm or disconfirm it:

> ... it is remarkable ... that the revolution brought about in philosophy by Bacon was not the *effect* but the *cause*, of increased knowledge of physical facts:..men ... discovered new phenomena in *consequence* of a new system of philosophizing.[58]

In modern parlance, the "logic of scientific discovery" proceeds deductively, not inductively. Each individual:

> ... has in his mind certain *major premises* – i.e., principles – relative to the subject in question; that observation of what actually presents itself to the senses, supplies *minor-premises;* – and that ... (... which is reported as a thing experienced) consists ... of the *conclusions* drawn from the combination of these premises.[59]

Economic theory consists of "collecting, arranging, and combining whatever general propositions on the subject can be well established,"[60] and these are few. Its "fundamental principles" are based on "very little information beyond what is almost unconsciously, and indeed unavoidably, acquired by every one."[61] Man is "by nature a social Being" and human society is not preconceived but is the unintended outcome of "instinctive propensity."[62] "Private interest," especially the "desire of wealth" and "emulation" (neither of which in themselves are "either virtuous or vicious"), are a constant and universal feature of human behavior.[63]

Humans are to be regarded as "rational free agents."[64] From such simple ingredients we can construct a model of spontaneous market order which solves, for example, "the problem of supplying with daily provisions of all kinds such a city as our metropolis" by individuals "who think each of nothing beyond his own immediate interest."[65] The illustration was, as Whately meant it to be, of service to natural theology. He had intended, if appointed to the chair, to make "a sort of continuation of Paley's 'Natural Theology,' extending to the body-politic some such views as his respecting the natural."[66] His example of the beneficent operation of the Invisible Hand, eventually incorporated in Samuelson's textbook,[67] afforded "one of the most beautiful pieces of Sunday reading it ever fell to the lot of the Westminster Review to recommend."[68]

The *method* of economic theory is analytical: "we are more likely to advance in knowledge, by treating one subject at a time." "Human propensities," and the "general laws" which can be deduced from them, enable us to model social processes as they would be if no "disturbing cause" were present.[69] Thus Malthus's theoretical principle – that population has a *tendency* to increase beyond subsistence – means that "there are in Man propensities which, if unrestrained, lead to that result." "Propensities" are the behavioral assumptions used to formulate *ex ante* relations between economic variables (e.g. food; population) which are the starting point, but only the starting point, of a scientific account of economic phenomena. Because it is essential, in constructing any such account, to be absolutely clear about the concepts employed, good *definitions* are of the essence:

> The terms which may be considered as forming the technical language of Political-Economy, being all taken from common discourse ... stand more in need than those of almost any science, of accurate definition, and rigid confinement to their defined sense ... [70]

Whately had requested Senior, upon the latter's first becoming Drummond Professor, to supply the Appendix on "Definitions in Political Economy" to his famous textbook on *Logic*.[71]

8.3 The methodology of political economy

Whately's second object was to weaken and if possible to refute the "fundamental doctrine," first advanced by Mandeville and widely supposed to discredit political economy, of "the incompatibility or discordancy of national Wealth and Virtue."[72] He addressed this first by insisting on the ethical neutrality of political economy, and secondly by enlarging Senior's defence of Wealth with a theological interpretation of the market economy.

Because the "strict object" of political economy "is to inquire only into the nature, production and distribution of wealth; not, its connexion with virtue or with happiness," its usefulness is unaffected by the moral nature of wealth. For "whether wealth be good, or an evil, or partly both, the knowledge of all that relates to it is not the less important."[73] But in fact, as Whately goes on to show (Lecture II), both private and public wealth are "in themselves goods:"[74] it is the human use of wealth which may be good or evil. His Lecture IV, as I have noted above, applied to human social phenomena the teleology of Paley's *Natural Theology*:[75] by means of market exchange "Man is, in the same act, doing one thing, by choice, for his own benefit, and another, undesignedly, under the guidance of Providence, for the service of the community."[76] The argument was strengthened in Lecture VI by a consideration of "emulation:" what we should now describe as "positional goods" are valuable as incentives to economic growth:

> The race never comes to an end, while the competitors are striving, not to reach a certain fixed goal, but, each, either permanently to keep ahead of the next, or at least, not to be among the hindmost.[77]

But apart from the claim that political economy is merely a study of wealth whether wealth be good or bad, Whately's treatment of wealth and virtue has no direct bearing upon methodology.

We may therefore summarize the methodology expounded in the *Introductory Lectures* as follows:

1. Scientific or secular knowledge is sharply distinct from theological or sacred knowledge.
2. The former comes by "experience:" that is to say, through the interpreting of observational data by "theory," where "theory" consists of "general laws" or "principles" which may begin as *inductive* generalizations but which function as the axioms of a *deductive* system of putative, falsifiable inferences.
3. The latter comes by "faith:" that is to say, by the "spiritual discernment" of the strictly *religious* truths contained in "revelation" (meaning the canonical texts of the Christian Bible) and which are and must be "beyond the reach of unassisted reason."

4. "Reason" may make use of scientific knowledge in corroborating religious knowledge (natural theology).
5. But reason cannot be at variance with faith for it is precisely the function of the latter to generate knowledge where the former cannot operate.
6. Thus science cannot falsify the *religious* truths contained in scripture, and scripture cannot falsify the *scientific* truths obtained by "theory" and "observation."
7. *Political economy* is a scientific study of the "nature, production and distribution of wealth."
8. Like all scientific inquiry its method is analytical: the whole of human "experience" is deliberately ignored except only those elements abstracted for the purpose of formal study.
9. Because it is a science of the social phenomena produced by the actions of "rational free agents," its "principles" are either generalizations about the nonhuman constraints on human action (e.g. diminishing returns) which may be known by common observation, or generalizations about goal-directed human behavior which may be known by introspection.
10. Because scientific method in general abstracts from religious considerations which are the prime source of values, and because political economy in particular abstracts from all ethical aspects of its subject matter, political economy can provide guidance only with respect to the *means* of obtaining certain social ends, and none at all about whether those *ends* ought to be pursued.
11. Which is to say, political economy attempts to answer questions of what *is*, leaving strictly alone questions of what *ought* to be: *per se*, an economist can give no advice about public policy.

It is evident from this summary first, that Whately's methodology was highly serviceable for his ideological counter-attack on the Philosophic Radicals; secondly, that it supplied the first extended account of the "orthodox" or "mainstream" economic methodology developed and restated by Senior, J. N. Keynes, Robbins, and their successors; and thirdly, that *a sufficient insulation of "economics" from "religion" is an integral part of that methodology.*

The Philosophic Radicals had sought to capture the "reform" movement and make it subject to political economy. Whately showed that political economy is indifferent as to the ends of policy, that the latter require value-premises, and that the utilitarian basis of Benthamite values is philosophically inadequate.[78]

When Senior accepted a second term as Drummond Professor (1847–52) he greatly amplified his treatment of methodology in *Four Introductory Lectures*,[79] closely following the Stewart-Whately line. Political economy is a "mental study,"[80] a "science," not an "art,"[81] and as against J. S. Mill,[82] "a positive, not an hypothetical science."[83] McCulloch,[84] as might be expected, had strongly objected to Senior's insistence, in the Introduction to his *Outline*,[85] that "the economist 'is not to give a single syllable of advice.'" Senior responded with a clear distinction between "science" which permits us merely to instruct, and "art" which permits us to "advise."[86] Though Marian Bowley[87] maintained that Senior changed his mind more than once, between 1827 and 1847, about whether there could be an "art" of economics, she rightly insisted that "it was Senior, rather than Cairnes, who was the most important writer on scope and method among the classical economists;" and the one whose work was most influential for the twentieth-century development of economic methodology.

Finally it is clear that the orthodox view of economics as an instrumental study of *means*, based upon deductions from quasi-axioms supposed to be inductive generalizations about human motivation and physical constraints, can have nothing to say about "ultimate judgments of value." If there can be any such thing as "sacred" knowledge, then all "ultimate" judgments, all consideration of human *ends*, must lie wholly within its field.

This does not mean that economics has nothing to say about God, for it is possible, as Whately, Sumner, Bastiat, and many others believed, that the scientific knowledge of human societies afforded by political economy may be co-opted by natural theology. But the apologetic value of such an enterprise depends precisely upon the fact that such knowledge should indeed be "scientific" in the sense of being publicly accessible to all who will submit to its discipline *regardless of religious belief or theological preconceptions*. It is for that reason I believe, that present-day orthodox methodology, descending as I have shown from the work of Whately and Senior, takes it for granted that "economic analysis" is indeed "as far removed" from theological considerations "as are the techniques of any other science."

9
Peasants, Population and Progress

The word "peasant" is now an anthropological concept designating a particular mode of production and its concomitant culture in terms so general that we may use it with equal propriety of societies in eighteenth-century Poland, medieval China and contemporary India. Alan Macfarlane[1] has summarized the evolution of this concept since the Second World War and concluded, on the basis of his own historical investigations, that:

> England has been inhabited since at least the thirteenth century by a people whose social, economic and legal system was in essence different not only from that of peoples in Asia and Eastern Europe, but also in all probability from the Celtic and Continental countries of the same period.[2]

A market economy, capitalist production and individualistic attitudes and institutions are to be found as far back as detailed records go,[3] and in this respect "England stood alone."[4] The implication for traditional old-fashioned, quasi-Marxian historiography[5] is drastic, for "one of the "most thoroughly investigated of all peasantries in history," turns out to be not a peasantry at all."[6] Putting it even more strongly:

> one of the major theories of economic anthropology is incorrect, namely the idea that we witness in England between the sixteenth and the nineteenth centuries the "Great transformation" from a non-market, peasant society where economics is embedded in social relations, to a modern market, capitalistic, system where economy and society have been split apart.[7]

Yet the word "peasant" was current in English at least since the fifteenth century, and though Macfarlane may be correct in suggesting that it was at first used only of foreigners[8] this was no longer the case by the eighteenth century. Samuel Johnson (1755)[9] defined *peasant* as "A hind; one whose business is rural labour" and *peasantry* as "Peasants; rusticks; country people." It seems to have been in this sense that Malthus wrote in the first *Essay* of "The sons and daughters of peasants" (who are not "such rosy cherubs in real life, as they are described to be in romances"); or of "the Scotch peasants" whose customary standard of living was then so far below that of "labourers of the South of England."[10] In his Scandinavian journal of 1799[11] Malthus used the words "peasant" and "boor" synonymously, and in one passage[12] "boor" and "farmer" are equated. "Boor" was cited in Johnson"s *Dictionary* as a synonym for "hind," itself a synonym for "peasant" as we have seen. It is evident from these passages and others in the journal[13] that Malthus meant to include under the general term "peasant" both "slaves" (i.e. serfs) and freemen, "farmers," "farmers' sons," "housemen" and "labourers."

Though Thomas Chalmers, being a Scotchman, would have known a rural culture less obviously exceptional than that of England, he too seems to have used the word "peasant" in much the same way as Malthus. In Chapter I of *Political Economy*[14] "agricultural labourers" suddenly became "peasantry" in the context of a Malthusian analysis of population growth and diminishing returns. But the word "farmer" is seldom used in this work and when it is,[15] it is synonymous with "capitalist." To an even greater extent than Malthus, it would seem, Chalmers understood "peasant" to mean simply a landless, agricultural laborer employed for wages by a capitalist farmer.

Though, as I shall show (see Appendix), their analysis subsumed the case of the self-employed peasant proprietor, it was with the peculiarly English, or at any rate British conception of "peasant" and "peasantry" that Malthus and Chalmers were chiefly concerned. In what follows I shall consider first, why they should have been concerned at all; second, how they constructed within the general framework of classical political economy an analytical schema for dealing with their concerns; and finally, the inferences for public policy which they drew from that schema.

9.1 Why did "peasants" matter to Malthus and Chalmers?

At some risk of oversimplification it may be said that most classical economists were interested primarily in trade and markets, money and capital. From Adam Smith to Marx the chief focus of attention was upon merchants, capitalists, entrepreneurs and bankers. Malthus and Chalmers are somewhat unusual in the degree of attention they devoted to agriculture and the "peasantry." This has nothing to do with the claim, sometimes advanced by the ignorant or disingenuous,[16] that they were apologists for the landed interest as against merchants and manufacturers. It is simply because each was a clergyman of an established church.

On Trinity Sunday (7 June) 1789 Robert Malthus was ordained to the Diaconate by the Bishop of Winchester, and the following day licensed to serve the curacy of Okewood, a chapel-of-ease in the Surrey parish of Wotton, a few miles from his parents' house in Albury. At his ordination the bishop would have instructed him, according to the Anglican ordinal, in what "appertaineth to the Office of a Deacon in the church," which includes:

> where provision is so made, to search for the sick, poor and impotent people of the Parish, to intimate their estates, names, and places where they dwell, unto the Curate, that by his exhortation they may be relieved with the alms of the Parishioners, or others.[17]

The Perpetual Curate of Okewood, John Hallam, was non-resident and Malthus therefore performed all the duties of the cure from the first, at an annual stipend of £40. He was ordained Priest in February 1791 and remained as residentiary assistant curate until preferment to the Rectory of Walesby in 1803. Malthus became Professor at the East India College in 1805 and resided at Haileybury thereafter, but retained Walesby as non-resident Rector, appointing an assistant curate and visiting several times a year. In 1824 John Hallam died and Malthus also became Perpetual Curate of Okewood, here too employing a stipendiary assistant curate.[18]

The English Ordinal was compiled in 1549, half a century before the Elizabethan Poor Law of 1601, and it was the latter which made statutory "provision" for parochial relief of the "sick, poor and impotent." A great majority of the population at that time lived *par-oikia* (literally "round the *house*," meaning house of God, or *church*) in communities of seldom more than five or six hundred souls.[19] Under the Act of 1601 churchwardens and vestrymen were empowered to levy a tax upon their fellow-parishioners, using the proceeds to provide useful work for the able-bodied poor, and maintenance for the sick and impotent. Accounts were to be rendered annually to a Justice of the Peace, and these officials came eventually to dominate the administration of the old Poor Law.[20]

The Poor Law overseers and guardians were thus almost entirely lay. But it would be a mistake to infer from this that the English parochial clergy was to any significant extent detached from the perennial problems of rural poverty. As one of three or four of the richest and most powerful members in a village society of some two hundred families, it is inevitable that the clergyman should have played a large part, formal or informal, in all that affected the local economy. As pastors whose normal ecclesiastical duties – public worship, catechizing the young; baptizing, marrying, burying; churching of women and visitation of the sick – brought them into daily contact with their parishioners, it is impossible that clergymen should not have been intimately acquainted with the circumstances of every family. As registrars of births, marriages and deaths they were uniquely placed to take a strategic, or at any rate a statistical, view of society. And as the local and contemporary representatives of a church that for more than a thousand years had united the village, enjoining charity upon all and caring no less for the temporal than for the spiritual welfare of the faithful, they were expected to be – and in many cases actually were – the Christian *persona* or "parson" of their flock: the "ensample of Godly life" as prescribed in the *Book of Common Prayer*. The diaries of Parson Woodforde from 1759 to 1802 are filled with references to the needs of his humble neighbors and the measures he or others had taken for their relief; and they form an eloquent, but artless, witness to a unique blend of worldly, even coarse realism with authentic piety, which characterized the culture of that eighteenth-century Anglicanism in which Robert Malthus was formed and nurtured.[21]

In Scotland, as in England, statutory provision "For Punishment of Masterful Beggars and Relief of the Poor" (1579, 1592, 1649, 1661) was parochially administered. Each Kirk Session was required to implement not only the Poor Law but also the Vagrancy provisions: ministers, elders and deacons were authorized to nominate commissions, and in 1661 Justices of the Peace were instructed to appoint two or more persons of standing in every parish to be overseers of the poor. There was little change in the Scottish Poor Law between the Union of Parliaments (1707) and the Poor Law (Scotland) Act of 1845.[22]

Thomas Chalmers was educated at St Andrew's for the Ministry of the (Presbyterian) Church of Scotland and was ordained as Minister of the Fifeshire parish of Kilmany in May 1803. For the first eight years of his incumbency he ignored his parish in furtherance of academic ambition, military duties in the St Andrew's Volunteers, a disastrous love affair, and the composition of his first – totally unsuccessful – work of political economy. After serious illness and a nervous breakdown, Chalmers experienced an evangelical conversion in the winter of 1810–11 from which he emerged a new man. Quickly becoming one of the most famous preachers in Scotland, he now entered wholeheartedly into the work of his long-neglected parish: preaching and teaching; visiting each of the families in his cure of about 800 souls, and taking responsibility with his Kirk Session for the collection and distribution of the poor-relief fund.[23] Though he quickly exchanged his obscure, country living for the challenge of a large, industrial parish in Glasgow, and though he went from Glasgow in 1823 to a series of university appointments in St Andrew's and Edinburgh, his latest and best biographer has argued convincingly that the view Chalmers formed, in his last years in Kilmany, of the proper interdependence of a Christian community remained definitive for his understanding of social policy for the rest of his life.[24]

As clergymen, both Malthus and Chalmers were confronted at an early stage of their ministry with the pervasive reality of rural poverty, Malthus at Okewood in the last decade of the eighteenth century, Chalmers at Kilmany during the second decade of the nineteenth. Though each devoted much of his later life to academic work, early experience was decisive. Moreover its effects would undoubtedly have been reinforced by the general culture – the attitudes and expectations, assumptions and common experience – of the clerical order to which each belonged to the end of his life. The question which must now be asked is, what did they believe ought to be done about the condition of those they sometimes referred to as the "peasants"?

According to Marx, and others who have set out to grind his polemical axe, the answer is very simple. Malthus and Chalmers were "parsons:" *ergo* they must have been "servitors" of "the conservative interest." Malthus "idolized" the "interests of the ruling class with all the fervor which was fitting in a parson." Malthus, "like his pupil Parson Chalmers, glorifies, as an economist, the class consisting of pure buyers or pure consumers." For that reason he defends the importance of "the landed aristocracy, the placemen of the state, the beneficed clergy" against that of "the capitalist actually engaged in production." Malthus, moreover, is "a master of plagiarism;" and in a long footnote, so deliciously vitriolic as to arouse the suspicion that Marx may actually have had a sense of humor, "Parson Wallace, Parson Townsend, Parson Malthus, and his disciple, the arch-parson Thomas Chalmers" are condemned for their "bungling interference." Chalmers is reported to have had "his suspicions as to Adam Smith's having invented the category of 'unproductive laborers' solely for the Protestant parsons, in spite of their blessed work in the Lord's vineyard."[25]

It is of course problematic just how much of all of this Marx himself really believed, and how much is merely a rhetorical scarecrow to deflect his readers' attention from those features of the Malthus-Chalmers analysis which seriously threaten his own enterprise. But true believers down to the present day have taken his lightest word for gospel truth. Thus the latest textbook in what is described as the "critical (i.e. Marxist) perspective" faithfully repeats the absurd idea that Malthus was an apologist for "the landlord class;" assures innocent undergraduates that "in Malthus's theory, the ultimate difference between the rich and the poor was the high moral character of the former and the moral turpitude of the latter;" describes Malthus as "an outspoken champion of the wealthy;" and interprets the first *Essay*, with flagrant anachronism, as an attack on schemes to "promote the welfare and happiness of workers" victimized by "the industrial revolution."[26] It may not be altogether redundant, therefore, to put the record straight.

It is certainly true, as I have shown in detail, that the first *Essay* was written to defend "the established administration of property" against the attacks of Condorcet and Godwin; that Malthus argued for the inevitability of a class division in society between a minority who owned land or capital and a majority who did not; and that his work inaugurated an influential tradition of social theory (to which Chalmers amongst others contributed) which sought to demonstrate the optimality of such arrangements.[27] But having described what he believed was the unavoidable reality of economic life in late eighteenth-century Britain, Malthus went on to distinguish between Godwin's unfeasible plan "to promote the welfare and happiness of workers" (by the abolition of government, marriage and property rights), and his own more modest, but more practical proposals:[28]

> Mr Godwin seems to have but little respect for practical principles; but I own it appears to me, that he is a much greater benefactor to mankind, who points out how an inferior good may be attained, than he who merely expatiates on the deformity of the present state, and the beauty of a different state, without pointing out a practical method, that might be immediately applied, of accelerating our advances from the one, to the other.[29]

From the beginning Malthus's own recommendations turned upon economically efficient ways of achieving a permanent increase in the relative share of "the peasantry of England"[30] by raising the real agricultural wage – strange indeed in an apologist for the landlord class and "outspoken champion of the wealthy." After his clear specification of "moral restraint" in 1803, the *Essay on the Principle of Population* became an "inquiry into our Prospects respecting the future Removal or Mitigation of the Evils which occasions." There is no reason to doubt Malthus's sincerity in wishing for "the future improvement of society," for though:

> We have every reason to believe that it will always consist of a class of proprietors and a class of labourers; ... the condition of each, and the proportion which they bear to each other, may be so altered as greatly to improve the harmony and beauty of the whole.[31]

To the manner in which Malthus and Chalmers proposed to alter the condition of "the class of labourers" we now turn.

9.2 How to ameliorate the condition of the "peasants"

In a nutshell, Malthus and Chalmers argued that the working class had some degree of command over the surplus.

The "subsistence wage," which determines that family income at which population and work-force is stationary, is conventional and socially conditioned. If workers choose the age of marriage and (therefore) the rate of family formation in order to maintain a conventional living standard, they determine the aggregate supply of labor. In an agricultural economy, with a given state of technique and diminishing returns to labor-and-capital applied to land, aggregate production of "the means of subsistence" is therefore determined. If the return to the (variable) labor-and-capital unit is governed by competition it will equal the marginal product of the variable factor at equilibrium. If surplus be conceived as the *agricultural* surplus – that is, the excess of the currently produced "means of subsistence" over the irreducible minimum of those "means" necessary to (re-) produce them – then it is clear first, that workers actually do appropriate part of the surplus; second, that they can vary this amount at will. For the excess of the *conventional* subsistence wage over the *biological* subsistence wage is the amount per capita that each employed worker appropriates in stationary equilibrium. And by increasing the former through a lengthening of the pre-marital period, workers as a class can increase their relative share of the surplus at *the expense of landlords* but not of capitalists. There is no clear statement of this crucial theorem in any of Malthus's writing that I have so far examined, but it is made explicit by Chalmers[32] in a sophisticated analysis of the power of technical progress to offset the distributional effects upon rents of an increase in the subsistence wage.

The analysis so far outlined is an early example of comparative statics. What is now known as the "stationary state" of the "canonical classical model"[33] is a stable equilibrium determined by available land, the state of technique, the subsistence wage, and a corresponding, quasi-"subsistence" rate of return to capital at which net saving-and investment is zero. A once-for-all change in one of these parameters will lead (conceptually) to a new stationary equilibrium of population, production and income distribution – with the same rates of return to labor and capital as before unless the change occurred in one of these.

It must be pointed out first, that even the comparative statics analysis was more complex than I have described it above; and second, that Malthus and Chalmers, like all other classical economists, were fully conscious of the dynamic properties of the model and their practical importance. For in the first place each author, but especially Chalmers in his first work, included in the static model a sophisticated analysis of the "unproductive" (or "secondary" and "disposable") population supported by expenditure out of the agricultural surplus.[34] And in the second, each author, but especially Malthus in the *Principles*,[35] explored the behavior of the model when either or both wages and profits were in excess of subsistence levels for appreciable periods, considered both a "ratchet" effect upon the subsistence wage and the possibility of endogenous technical progress, and adumbrated a quasi-Keynesian – perhaps more properly, "quasi-Harrodian" – theory of deficient aggregate demand.[36]

One thing neither did, however, was to include a manufacturing sector in the model. So far as Malthus and Chalmers were concerned, "the industrial revolution" had not yet happened. The economy of Great Britain that they analysed was rural and agricultural. Capitalist "cultivators" or "farmers" employed (uniquely British, landless) "peasants" for market-determined wages on land owned by the "class of proprietors," paying the latter a rent, the expenditure of which – together with any profits not saved – supported an "unproductive" population which afforded "everything ... that distinguishes the civilized, from the savage state."

It is important to keep in mind that the "model" I have so far referred to is a rational reconstruction by modern authors based on a series of works beginning in 1798 with Malthus's first *Essay* and continuing at least until the third volume of Chalmers's *Christian and Civic Economy*.[37] Though it can be argued[38] that all are latent or implicit in the first *Essay*, the full range of its analytical possibilities only very gradually became clear to Malthus, Chalmers, Ricardo and West over the next two decades. Not until the simultaneous discovery by Malthus, West, Ricardo and Torrens of the "Ricardian" theory of rent could it truly be said that anyone really "understood" the Malthusian model.[39] Nevertheless, both Malthus and Chalmers often wrote as if they understood it, and from the first their policy recommendations were fully consistent with its logical structure.

Having previously established the textual evidence for, and explained the formal properties of, the Malthus-Chalmers version of the "canonical classical model,"[40] I shall only summarize below those features which are directly relevant to their concern for the welfare and progress of the "peasantry."

Let F units of (homogeneous) "means of subsistence" be produced annually with given land and technique by the application of N units of what Samuelson[41] calls a "labor-*cum*-capital" composite factor of production. Under suitable assumptions N is also a measure of the current population of "peasants:" the class of landlords and capitalists is negligibly small, and the non-agricultural population ("unproductive," "secondary," "disposable" or even "manufacturing" if any) does not enter directly into the production of the "means of subsistence." It is evident that Malthus's famous "ratios" of hypothetical food and population growth imply that the function relating F to N is logarithmic.[42] Let L be an index that summarizes the availability of land of various qualities, the state of agricultural technique, and the current set of capital–labor ratios. Then

$$F = L.\ln N \tag{9.1}$$

Let the real wage rate be W, the socially determined "subsistence wage" be S, the rate of return to capital be r, and the stationary rate of return be p. Let the capital stock be one year's advance wages, and let the real return to labor-and-capital be brought by competition to equal the marginal physical product, $\partial F/\partial N = L/N$. Capital, wages, profits and output are all measured in the same ("corn") units. Then

$$W(1+r) = L/N \tag{9.2}$$

Let agricultural population and the capital stock grow, respectively, at the rates

$$n = a(W - S);\ a > 0, \tag{9.3}$$
$$k = b(r - p);\ b > 0. \tag{9.4}$$

Then in stationary equilibrium when $n = k = 0$, and $N = N^*$

$$L/N^* = S(1+p), \tag{9.5}$$

and equilibrium population of the "peasants:"

$$N^* = L/S(1+p). \tag{9.6}$$

Given N^*, the equilibrium production of the "means of subsistence," equilibrium income of the "peasants," and relative shares are all determined:

$$F^* = L.\ln\,[L/S(1+p)], \tag{9.7}$$
$$N^*S = L/(1+p), \tag{9.8}$$

the relative share of "peasants,"

$$N^*S/F = (1+p)^{-1}\ \ln\,[L/S(1+p)]^{-1}, \tag{9.9}$$

of "cultivators":

$$N^*Sp/F = p(1+p)^{-1}\ \ln\,[L/S(1+p)]^{-1}, \tag{9.10}$$

and of "proprietors":

$$[F - N^*S(1+p)]\ /F = 1 - \ln\,[L/S(1+p)]^{-1} \tag{9.11}$$

Equations (9.6) to (9.11) show that all aspects of the stationary state of the Malthus-Chalmers model, including those which affect the absolute and the relative welfare of the "peasants" are determined by three parameters: S, which is under the control of the "peasants" themselves, p, which is controlled by the class of "cultivators," and L. Though "cultivators" may play some part in improving techniques, L is largely under the control of the class of "proprietors," who may bring about an increase in its value either by reclamations and improvements which augment the amount of land available for cultivation (for example, the draining of the Fens) or by sponsoring improvement in agricultural techniques (for example, horse-hoe husbandry).

We may draw two general sets of inferences from these results: the first, concerning the invariance of equilibrium to any changes but those of the three parameters; the second, concerning the distributional effects of changes in those parameters.

It is clear, to begin with, that if the population grows according to (9.3), any expedient for increasing the income of "peasants" by transfers from the other classes can have no permanent effect. N would temporarily grow as W exceeded S (and the "unproductive" population would decline as a smaller amount of the agricultural surplus was spent on its employment), but as the marginal product fell the excess of W over S would diminish until N was again stationary with per capita income as before. A larger agricultural population would exist but its condition would be no better. The Poor Laws "may be said therefore in some measure to create the poor which they maintain."[43] For the same reason emigration, or a sudden decline in population caused by plague or famine, can have only a transitory effect. A temporary excess of W over S will be reversed as population returns to the level determined by S, p and L.

In the second place we can specify exactly those changes which will have a permanent effect on the welfare of the "peasants," and on the distribution of income between classes in general. It is clear from inspection of equations (9.9), (9.10) and (9.11) that an increase in S will raise the share of "peasants" *and of "cultivators,"* and lower that of "proprietors;" and that an increase in L will lower the share of "peasants" and "cultivators" and raise that of "proprietors." It is also obvious that an increase in p will lower the relative share of "proprietors:" but the effect on those of the other two classes needs to be elucidated. By partial differentiation of (9.9) and (9.10) with respect to p we obtain:

$$\partial/\partial p\ [N^*S/F] = (1+p)^{-2}\ [(\ln N^*)^{-2} - (\ln N^*)^{-1}] < 0 \qquad (9.12)$$

and

$$\partial/\partial p\ [N^*Sp/F] = p(1+p)^{-2}\{(\ln N^*)^{-2} - [p^{-1}(1+p) - 1](\ln N^*)^{-1}\} > 0 \qquad (9.13)$$

The distributional effects of once-for-all changes in each of the three parameters, the other two remaining constant, are summarized in Table 9.1.

Table 9.1 **Effect of parametric change upon stationary equilibrium values of relative shares in the Malthus–Chalmers model**

Relative share of:	*Peasants*	*Cultivators*	*Proprietors*
Increase In:			
S	increases	increases	decreases
p	decreases	increases	decreases
L	decreases	decreases	increases

If "cultivators" increase *p*, the minimum return on capital at which they simply maintain the stock and accumulate no further, they increase their relative share at the expense of the other two classes. If "proprietors" increase *L* by land improvement or the introduction of superior farming methods, they too increase their relative share at the expense of the other two classes. But if the "peasants" increase *S* by deferring marriage and reducing procreation, they increase both their own relative share *and that of their employers* at the expense of the "proprietors" alone. The reason for this asymmetry is that the total return to capital and labor at all times is determined by (9.2) at *L*. The equilibrium value of the "wages fund," what Malthus[44] called "the fund appropriated to the maintenance of labour" is shown by (9.8) to be $L/(1+p)$. The distribution of *L* between wages and profits therefore depends upon the rate of profit alone. It follows that any relative gain the "peasants" can make by raising *S* will be shared equiproportionately with the "cultivators" provided the latter maintain the value of *p*.

The implications of equations (9.1) to (9.11) are illustrated in Figure 9.1, which represents stationary equilibrium of food and population when the marginal product of the variable factor – the slope *L/N* of the tangent CG-extended – is equal to the "subsistence" factor price, the slope $S(1+p)$ of the ray 0E. Production is $0F^*$, divided between wages N^*D, profits DE and rent EG. Agricultural population is $0N^*$. If landlords spend all of their rents on "government, the administration of justice, national defence," and on personal services, home-produced "ornamental luxuries" and the like; and if "cultivators" and "peasants" similarly expend any excess of their per-capita incomes over a *biological* food requirement, $f < S$; and if any excess of these wage payments over *f* is in turn spent and re-spent until it disappears, the total population at equilibrium, $P = F^*/f$, is represented by the amount 0P. and "unproductive" population, $P - N^*$, by the amount N^*P.

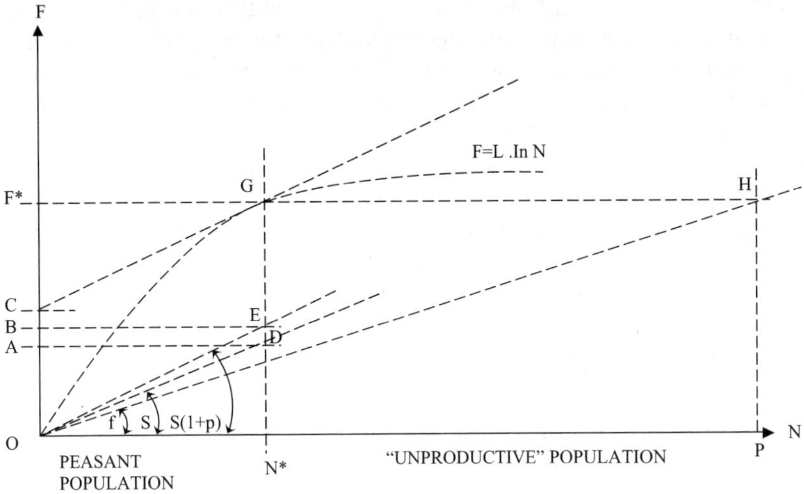

Figure 9.1 Stationary equilibrium of agricultural production and total population

Now if *S* is increased the rays 0D and 0E will rotate anti-clockwise; the point of tangency, G, moves to the left along the *F*-curve; *N**, *F** and *P* will fall; and the absolute share of landlords, EG, declines. Given the logarithmic production function, the absolute amounts *N**D and DE will remain constant, hence relative shares of capital and labor will rise at the expense of that of land. In the more general case of a non-logarithmic, diminishing-returns production function – for example, $F = F(N)$; $F' > 0$; $F'' < 0$ – this result is ambiguous, for the effect of rising factor prices need not outweigh that of falling factor inputs. Nevertheless Chalmers, who made no analytical use of Malthus's ratios, asserted baldly as early as 1808 that "by exalting the taste of the people [i.e., by increasing *S*], you add to the extent of the population who work for their secondary enjoyment. You therefore trench on the disposable population" that is, increase the relative share of labor at the expense of land.[45] I have analysed these cases in the context of Chalmers's three-sector version of the model.[46]

The effect of a once-for-all increase in available land or improvement in technique is captured in the figure by an anti-clockwise rotation of the $F = L.\ln N$ curve. It is obvious that the effects upon equilibrium population and production, and the distribution between classes, will be opposite to those produced by a similar rotation of the rays 0D and 0E. It has been shown in this section that a coherent version of the "canonical classical model" in stationary state may be reconstructed from the writings of Malthus and Chalmers with important implications for social policy respecting the welfare of the "peasants." In striking contrast to the analysis of Marx, the equivalent in this model of a "working class" has the power to appropriate part of the surplus, *and to do so without encroaching on the relative share of capital.* Policy to improve permanently the absolute and relative condition of the "peasants" must therefore be at the expense of "proprietors" alone. In that Malthus and Chalmers strongly advocated such a policy, we see the implausibility of Marxian propaganda which represents them as "servitors" of the landlord class. Armed with this conceptual apparatus, we are now in a position to examine more closely the opinions of Malthus and Chalmers on peasants, population and progress.

9.3 Inferences for public policy

> Political economy aims at the diffusion of sufficiency and comfort throughout the mass of the population ... Now, we hold it to be demonstrable, *on its own principles*, that, vary its devices and expedients as it may, this is an object which it can never secure, apart from a virtuous and educated peasantry.[47]

On Political Economy in Connexion with the Moral State and Moral Prospects of Society (1832) was written to "demonstrate the futility of every expedient, which a mere political economy can suggest for the permanent well-being of a community" and to recommend schemes for the flourishing of a "moral and intelligent peasantry, imbued with a taste for the respectabilities of life, mixing prudence and foresight with every great practical step ... holding it discreditable to enter upon marriage without the likelihood of provision for a family."[48] The first of these amounts to a demonstration, *by means of economic analysis*, that W (or F/N for which W is a proxy) cannot permanently exceed S; the second, to a consideration of the *exogenous* (religious, cultural, sociological) determinants of S. This two-pronged strategy, consisting of an economic-theoretic attack on futile policies to raise W, and a socio-religious advocacy of efficacious policies to raise S, was adumbrated in Malthus's first *Essay on Population* and gradually developed by Malthus himself, J. B. Sumner and Chalmers, whom Malthus once described as his "ablest and best ally."[49] Each of these, and in particular Malthus, was aware that S might be to some extent *endogenous*, for tastes are affected by a long-lived change in the actual living standard for better or for worse. But each also believed that parametric shifts could be effected by "moral and religious teaching." Though for completeness therefore we might write $S = S(M, W)$, where M stands for the state of "moral and religious teaching," we may abstract from the influence of the second argument of the S function for a consideration of this particular policy debate. In what follows I shall briefly consider each line of argument in turn.

The first *Essay* was written to show the futility of Godwin's plan to achieve human perfection by pure anarchy. Abolition of all the institutions of society, Malthus argued, would be self-reversing. Before too long – as the result of rational, optimizing actions of individuals coping with scarcity – property rights, government, marriage, inheritance laws and wage labor would reappear. So successful was Malthus in routing Godwin that this particular argument needed no further amplification. But in developing his case Malthus was led to consider more practical expedients for improving the welfare of the poor, such as the Poor Laws and emigration.

Chapter V of the first *Essay* begins Malthus's lifelong involvement in Poor Law reform,[50] and contains each of his four principle objections: first, the Poor Laws tend to increase the population of paupers; second, they restrict the mobility of labor; third, their implementation by means of monetary transfers will increase food production proportionately less than population; and finally, they undermine the attitudes and expectations essential to a productive and self-reliant working class.

The "first obvious tendency" of the Poor Laws "is to increase population without increasing the food for its support."[51] It is this axiom, which Ricardo, J. S. Mill and Senior accepted as fully as Malthus and Chalmers, that united all who finally succeeded in obtaining the Poor Law Amendment Act of 1834. Moreover, because the operation of the old Poor Law made each parish responsible for its own poor only, the mobility of labor was greatly impeded. The "want of freedom in the market for labour" resulting from this together with "the more general cause of the facility of combination among the rich, and its difficulty among the poor," acts as a brake on real wages in times of expansion when k exceeds n and labor is relatively scarce: "perhaps, until a year of scarcity, when the clamour is too loud, and the necessity too apparent to be resisted."[52] Furthermore, in that transfers to the indigent are made as cash payments, no immediate increase in total food will occur, prices rise, and "the lowest members of society" must still "live upon the hardest fare, and in the smallest quantity." If inflation should "give a spur to productive industry," "the spur that these fancied riches would give to population, would more than counter-balance it, and the increased produce would be divided among a more than proportionably increased number of people."[53]

The last and most important objection to the Poor Laws is their moral effect. "Fortunately for England, a spirit of independence still remains among the peasantry. The poor-laws are strongly calculated to eradicate this spirit."[54] Not only do they degrade the poor by creating attitudes of welfare-dependency and subjecting them to "a set of grating, inconvenient and tyrannical laws, totally inconsistent with the genuine spirit of the constitution." They also degrade those who administer the laws.

> The tyranny of Justices, Churchwardens, and Overseers, is a common complaint among the poor: but the fault does not lie so much in these persons, who probably before they were in power, were not worse than other people; but in the nature of all such institutions.[55]

In the second *Essay* (1803) and in all subsequent recensions, Malthus amplified and refined this attack on the Poor Laws.[56]

Although, as was declared at the time, "the poor-laws of Scotland are not materially different from those of England," they were "very differently understood and executed" and in effect, the poor had "no claim of right to relief."[57] Chalmers devoted less attention than Malthus to attacking the (Scottish) Poor Laws therefore, and contented himself in *Political Economy* [58] with theoretical demonstration of the futility of "a compulsory provision for the indigent." Both he and Malthus also considered emigration as a remedy for poverty, recognizing it, as Chalmers put it, to be "a safety-valve in the boiler." Malthus, moreover, acknowledged that in long swings labor could be in excess supply for "ten or twelve years together." "It is precisely in these circumstances that emigration is most useful as a temporary relief. Nevertheless, "every resource ... from emigration must be of short duration."[59]

Chalmers went further than Malthus in ramming home the merely negative point that nothing but an increase in *S* can permanently improve the condition of the "peasants," and in at least one respect parted company with him. Chapters VI, VII and XII of *Political Economy* argue that foreign trade and land redistribution can do nothing permanently to raise living standards. Chapters VIII, IX and X, building upon the analysis first worked out in *National Resources*,[60] argues that the burden of all taxation falls upon landlords, and therefore that changes in the tax laws can neither harm nor benefit the "peasants." In 1827 Malthus wrote to Chalmers: "I agree with you in much of what you say about the wealth derived [by international trade] from manufacturers, but I think ... you have pushed your principles too far." And again in 1832 after reading *Political Economy:* "Have you not pushed too far the non-importance of foreign commerce?" The same letter contains a long paragraph carefully analyzing the incidence of taxation and distancing Malthus from Chalmers's reductionist claims that the whole burden of all forms of taxation falls eventually upon landlords.[61]

Where Malthus and Chalmers agreed more completely was in the measures they advocated for increasing *S*, though here too Chalmers went much further than the more cautious and realistic Malthus. These measures were the removal of pernicious institutions such as the Poor Laws, the fostering of beneficent institutions such as village savings banks, and above all, the moral and religious education of the "peasantry."

I have already alluded to Malthus's lifelong campaign against the English Poor Laws and his proposal for their gradual abolition.[62] He and Chalmers argued strongly for the importance of voluntary charity in its place, and the latter devised – and to some extend implemented – ambitious schemes for the parochial organization of such voluntary charity.[63]

For theological and ethical reasons, and possibly because of the rudimentary state of contraceptive techniques, Malthus and Chalmers ruled out birth control within marriage. The "preventive check" must therefore be implemented by *moral restraint:* meaning abstinence from marriage "which is not followed by irregular gratifications." And only if large numbers of the "peasantry," *acting by individual choice,* practice this virtue can *S* be raised with its attendant benefits. In order to meet the "free-rider" objection first raised by Lloyd and reiterated since by many including Samuel Hollander[64] it is essential for the Malthus-Chalmers position that it should be rational for peasants to act in this way, or at least that institutions could be reformed or created to make it rational.

Malthus believed that the significance of the preventive check was very widely recognized. It has lately been realized that:

> North-western Europe possessed a distinctive marriage pattern whose two major features were a late age at marriage and a relatively

> high proportion of individuals who never married at all. Analysis of parish records suggests that this pattern was already well established for the majority of the population in Elizabethan and Stuart England.[65]

There is abundant evidence that the age of marriage was strongly correlated with economic circumstances and expectation.[66] It is clear from his writing that Malthus was well aware of all this not only in England, where it was most marked, but also on the Continent. His travel diaries make frequent reference to the marital and child-bearing propensities of the northwest European "peasantry," and he gave prominence in the second *Essay* to the example of a Swiss peasant he met in 1802 who "appeared to understand the principle of population almost as well as any man I ever met with."[67]

He also believed that although "The happiness of the whole is to be the result of the happiness of individuals and to begin first with them" yet "No co-operation is required... . He who performs his duty faithfully will reap the full fruits of it, whatever may be the number of those who fail."[68] Because it is "clearly his interest" for an individual "to defer marrying til ... he is in a capacity to support the children,"[69] social institutions can undermine or reinforce this rational self-interest. The Poor Laws do the former; savings banks the latter:

> Of all the plans which have yet been proposed for the assistance of the laboring classes, the savings banks, as far as they go, appear to me much the best ... . By giving to each individual the full and entire benefit of his own industry and prudence, they are calculated greatly to strengthen the lessons of Nature and Providence.[70]

Malthus devoted some attention in the second *Essay* to savings banks, in which he was followed by John Bird Sumner (later Archbishop of Canterbury) who wrote on the subject in the *Quarterly Review*, and who formulated an arithmetical illustration of Malthus's argument in his most famous book.[71]

Even more important than savings banks were parochial schools, and the principal positive element of the Malthus–Chalmers social program was "moral and religious instruction" in general. Malthus cautiously approved the new Sunday Schools in England, advocated a national system of education, rebutted the arguments of those who feared that this would encourage the lower orders to read Tom Paine, and held up the example of universal literacy in Scotland: "The quiet and peaceable habits of the instructed Scotch peasant, compared with the turbulent disposition of the ignorant Irishman, ought not to be without effect in every impartial reasoner." A system of parochial schools in England, Malthus believed, would "have the fairest chance of training up the rising generation in habits of sobriety, industry, independence, and in a proper discharge of their religious duties, and approximate them in some degree to the middle classes of society."[72]

Chalmers's arguments started where Malthus's left off. "The maintenance of parish schools," he pointed out in an important tract written whilst still in Kilmany, "is a burden upon the landed property of Scotland, but it is a cheap defence against the poor-rates."[73] The religious element in such education is crucial. "The exemption of Scotland from the miseries of pauperism is due to the education which their people receive at schools, and to the Bible which their scholarship gives them access to."[74] Rapidly increasing pauperism in the Southern kingdom must lead to some "mighty convulsion." "If anything can avert this calamity from England, it will be education of their peasantry, and this is a cause to which the Bible Society is contributing its full share of influence."

This is more than just a matter of St Paul's teaching respecting marriage, or his dictum, which both Chalmers and Malthus reiterated, that "if a man will not work, neither shall he eat."[75] The general diffusion of Christian culture – provided it be that of "a Protestant education, and Protestant habits" rather than "the dark and degrading popery which prevails in Ireland" and on the Continent – is the best guarantee against "that impetuous appetency, which leads first to early marriages, and afterwards lands in squalid destruction, the teaming families that spring from them."[76] For "A disciple of the New Testament ... has gotten a superiority over the passions ... a reach of perspective to distant consequences ... and, withal, a refinement and elevation of taste." Wherefore, "An individual Christian is generally in better comfort and condition than other men. A whole parish of Christians would be a parish of well-conditioned families."[77]

Malthus and Chalmers were as concerned for the welfare of the "peasantry" – almost certainly more so, given their clerical profession – as any other economists of their generation. Like all economists of the day they held the view, which Malthus himself had done most to establish and develop, that population is the fundamental determinant of that welfare. Their criticism of existing policy was directed to the fact, as Chalmers correctly perceived it, that "we are still looking objectively to the enlargement of resources in the outer world of matter, instead of looking subjectively to the establishment of habit and principle in the inner world of mind."[78] Classical political economy, to which each contributed so largely, demonstrated both the futility of the former and the efficacy of the latter. But only the Christian religion, each believed, had the power to bring about the necessary changes in "the inner world of the mind."

Appendix: The case of land-owning peasants

Suppose all land is divided into family-sized holdings owned by peasants, who also own the capital necessary to cultivate them. Assume each family is identical and redefine units so that family income is F/N. If this exceeds the minimum acceptable return to the family unit, $S(1+p)$, then $n > 0$, $k > 0$ until $F^*/N^* = S(1+p)$, whence

$$(\ln N^*)/N^* = (1+p)\, S/L. \qquad (9A.1)$$

It is immediately obvious that population and production are determined as before by the three parameters S, p and L, all of which are now under the control of the peasants. Family real income, F^*/N^*, is determined by S and p alone. As before, therefore, the living standard of the peasantry depends upon the (demographic) "target" family income, which in this case is $S(1+p)$.

If $S(1+p)$ exceeds the biological food requirement, peasant families will spend the excess on "unproductive" labor (which supplies what Chalmers preferred to call "the second necessaries of life"). Hence, when all such income is spent and respent, the total population will be F^*/f and "unproductive" population

$$F/f - N^* = N^* \,[S(1+p)/f - 1]. \qquad (9A.2)$$

In Figure 9.1 the ray 0E would be extended until it cut the $f = L.\ln N$ locus: the equilibrium population of peasants would lie vertically below that intersection.

All the negative implications of the model remain in force so long as the parameters a and b in equations (9.3) and (9.4) remain sufficiently large to be relevant. This Chalmers invariably assumed. Hence in Chapter XII of *Political Economy* he considered and rejected the abolition of primogeniture and any other measure for "the more equal diffusion of property."

10

Property Rights in Christian Social Teaching

The "Philosophical Foundations of Economic Science" have afforded occasion of sporadic inquiry for most of this century.[1] Since the appearance in 1953 of Friedman's famous essay on "The Methodology of Positive Economics"[2] scholarly interest has increased greatly[3] and has advanced considerably in sophistication by the attempt to apply Lakatos's generalization of the Kuhn–Popper–Feyerabend debate[4] to the methodological problems of our discipline.[5] It is now become commonplace, within the "mainstream" tradition at any rate, to appraise putative contributions to economic knowledge in terms of "progressive" or "degenerating" "Scientific Research Programmes"[6] and, presumably, to accept the epistemological consequences of that view. Knowledge of the economy, like knowledge of other parts of the physical universe, is socially created within a community of experts bound together by common allegiance to a set of criteria for distinguishing truth from falsehood. Though these criteria are historically conditioned, and may well be ideologically tainted, commitment to the – generalized – falsification principle keeps the fresh air of critical scrutiny circulating in the community, so preserving it from permanent lapse into dogmatism and error.

It is a clear implication of the "mainstream" view that scientific knowledge can and ought to be detached from any knowledge which may be had of ethical, aesthetic or theological matters. Economics can only be a "Moral Science" in the sense of providing us with instructions about what we *ought* to do if its practitioners are provided with a set of exogenously determined value axioms. Division of labor is pushed to the furthest possible extent. Save only that they set the agenda and supply the axioms, Ethics and a *fortiori* Theology have nothing to contribute to "Economic Science." Whether Economics has anything to contribute to Ethics or Theology is a matter of indifference to the economist.

It is impossible not to admire the elegance of this view. Yet a conviction persists that social reality may be too messy to be fully explained, still less prescribed for, by such means. Respectable economists persist in meddling with questions of "justice."[7] The objections of Marxists and other radicals to the "mainstream" enterprise are well known.[8] Adherents of the neo-Calvinist, Dooyeweerdian philosophical school have not yet despaired of constructing a "Christian Economics" which might in principle differ in analytical technique from any existing variety of "secular" Economics.[9]

It is not my intention to intervene directly in this many-sided dispute. I propose instead a more oblique approach. Whatever may be the orthodox position now, it is well known that the concepts of economic analysis were intertwined with those of Ethics and Theology in the past. In this essay I shall consider one such concept, that of "property rights," still very much in use by "mainstream" analysts[10] though its original, normative signification has almost entirely evaporated. The first part outlines the theory of a somewhat equivocal "natural right" to private property classically formulated by John Locke. The second part contains an account of the temporary importation of something like that theory into Catholic social teaching by Pope Leo XIII in his famous encyclical *Rerum Novarum* of 1891. In a concluding section I raise certain questions about the intellectual connection between Economics and Theology suggested by these data.

10.1 Locke's doctrine of property rights

Locke's theory of property was intended to explain "how men might come to have a property in several parts of that which God gave to Mankind in common."[11] It stands at the end of a long tradition of natural law discussion of "rights" and "property;" and is unintelligible in abstraction from its theological presuppositions. The explanatory significance of "labor" in the assignment of property rights and the creation of value has provided support for two rival political views: a "liberal" theory of private property and a "socialist" theory of bourgeois appropriation. And Locke's achievement has been widely misunderstood in our own time and earlier as an ideological justification of capitalism. I address each of these in turn.

The *Two Treatises of Government*[12] first published in 1689, were composed in the early 1680s, probably in response to the Exclusion crisis[13] to refute the Tory propaganda of Sir Robert Filmer's *Patriarcha*[14] so "justifying resistance to arbitrary government and legitimizing its dissolution."[15] Filmer had argued from Adam's "dominion" over the earth[16] that all sovereignty, authority and proprietorship come from patriarchy. Private property rights are thus natural, unlimited and arbitrary As against this doctrine Locke sought to show that Adam's dominion represents "the Dominion of the whole Species of Mankind"[17] and that the rights individuals may acquire to land and other goods *in political society* are "conventional and based on consent."[18]

Chapter V of the *Second Treatise*, "Of Property" begins with two propositions already established in the *First Treatise:* Scripture reveals that the earth is a gift of God to mankind in common;[19] natural reason that men have a right to the means of subsistence.[20] The latter follows from a natural right to self-preservation[21] which in a formal sense can be treated as Locke's first axiom.[22] Before the common stock of goods can be of any use to any particular man, however, there must be a means for him to appropriate what he needs. Locke now introduces his second axiom:[23] "every Man has a Property in his own Person… . The *Labour* of his Body, and the *Work* of his Hands, we may say, are properly his."[24] It follows that, "Whatever he then removes out of the State that Nature hath provided, and left it in, he hath mixed his Labour with, and joined to it something that is his own, and thereby makes it his Property."[25] The exclusive right so acquired is subject to three conditions: first, that after the act of appropriation, there should be "enough, and as good left in common for others;"[26] secondly, no-one may engross more than "anyone can make use of to advantage before it spoils;"[27] and thirdly, by implication, "rightful appropriation appears to be limited to the amount a man can produce by his own labour."[28] Within these limitations individuals may have a "property in" goods, including land, in a state of nature.

Now when the first of these conditions can no longer be met and goods – particularly land – become scarce, claim rights conflict and the enjoyment of property becomes "very uncertain and constantly exposed to the invasion of others."[29] Men seek political community as a remedy, forego their natural power of self-preservation[30] and submit their possessions to the community.[31] All goods revert to common ownership under the state, which now determines the assignment of property rights which are merely conventional.[32] The mechanism of this assignment is "civil law," or human positive law to use the terminology of Richard Hooker, whom Locke cites extensively. Yet there is nothing arbitrary or unjust about this, for the legislature is bound in all its enactments by natural law.[33] Hence the function of the analysis of natural appropriation set out in Chapter V of the *Second Treatise* is to stand "as a normative model to guide society in the prudential application of this law of reason."[34]

Locke's *Two Treatises* mark the culmination of a tradition of natural rights theory which began at Bologna in the Twelfth century with the rediscovery of the *Digest* of Roman law.[35] The distinction between *jura in re* and *jura ad rem* originated by Bassianus opened the way for the recognition by Accursius (fl. 1210–30) of a *dominium utile* to describe the "right in" some good possessed by a usufructuary. "The process had begun whereby all of a man's rights, of whatever kind, were to come to be seen as his property."[36] Though Gratian and other Twelfth century natural law theorists had argued that "common possession of everything, and freedom for everyone" were "received everywhere by natural instinct"[37] it was therefore possible for Thomas Aquinas in the next century to conclude, from the usufruct humans have in the Creation, that "Man does have a natural *dominium* over material things."[38] After another century and more of development and debate, not uninfluenced by the ideological needs of the rival Dominicans and Franciscans,[39] a fully-fledged natural rights theory of property was propounded in 1402 by Jean Gerson.[40]

Gerson asserted that every creature has a *jus* "to do what is naturally good for it," a view of "right" which almost amounts to a propensity to survive. "There is a natural *dominium* as a gift from God, by which every creature has a *jus* directly from God to take inferior things into its own use for its own preservation."[41] Thus if the most effective way of using something is to appropriate it privately, a "natural right" to private property exists. The position was developed by Gerson's successors at Paris in the Fifteenth century, but largely repudiated in the next century under the impact of Reformation theology. It was the medieval tradition, "laboriously rebuilt at the end of the Sixteenth century" by Grotius (1583–1645) which flowered for the last time in the work of Selden, Hobbes, Pufendorf and Locke.[42]

It is not surprising, given the historical context of the natural rights tradition, that the slippery terms *jus* (approximately translatable as "right") and *dominium* (approximately translatable as "property," though often used virtually interchangeably with *jus*) should be if not "value-laden," at least "theology-laden." So far as Locke at least is concerned, the concept of "property" – meaning "any sort of right the nature of which is that it cannot be taken without a man's consent"[43] – is unintelligible apart from its theological underpinnings. This is evident not only of his acceptance of scriptural revelation as authority for the view that the earth is given to mankind in common, but more seriously of the two axioms with which he begins the famous chapter "Of Property."

Why should "natural *Reason*" tell us "that Men, being once born, have a right to their Preservation"?[44] James Tully has traced the process by which Locke could arrive at this belief. In summary: we can know that God exists from evidence of his "design" in the universe; God must have created Man, and must have a purpose for him; Man's reason thus discovers that God intends Man to be preserved; Man may freely choose to act in accordance with God's purpose (by seeking the preservation of God's workmanship); because the act of making confers a right in the product, the fact that Man is God's workmanship gives God a right in Man; God's *right* correlates with Man's *duty* to act in accordance with God's purpose; hence Man has a duty to preserve God's workmanship, which duty coincides with Man's "rational desire;" whence finally, Man's duty to preserve God's workmanship implies Man's right to preservation.[45]

Why is it that "every Man has a *Property* in his own *Person*,[46] even though God is conceded to have property rights? Again I summarize Tully. A person, for Locke, is one who can act freely and can therefore have (non-observational) "intentional" knowledge of his acts. A person is therefore author of his acts, and his actions a species of making: he may thus be said to own his actions including "the *Labour* of his Body and the *Work* of his hands."[47] Hence though Man's life is God's property it is also Man's property in a different sense: Man has a right of use, analogous to usufruct, similar to a tenant's.[48]

The explanatory importance of "workmanship" and "Labour" in the elucidation of Locke's two axioms has given rise to, or at least provided support for, "a quasi-historical model of the development of property relations which could be utilized either to justify the owners of property over and against the propertyless, or to provide a vehicle for the criticism of property owners."[49] On the one hand the conservative, not to say reactionary ecclesiastical establishment of late-Eighteenth century England "assumed as a matter of course that property was an integral part of the natural condition of things, just as they did the Lockean concept that all men possessed it in their labour."[50] Some decades later the "arch-parson" Thomas Chalmers[51] defended property rights because they secure "to each man ..., the fruits of all the labour he may choose to expend upon it," and concluded that the sense of property "implanted in Man from his birth ... bespeaks the immediate hand of God."[52] But on the other hand Locke's insistence that "labour makes the far greatest part of the value of things"[53] seemed to imply, at least to the "Ricardian socialists" of the early Nineteenth century, that "Men who owned property by virtue of their control of the labour of others could be seen ... as usurpers."[54] And though Schumpeter is undoubtedly correct in saying that Locke's doctrine of property "has nothing whatever to do with a labour theory of value"[55] – when the latter is regarded "analytically" as an explanation of the phenomenon of value – Adam Smith's confusion in the matter "inspired by Locke"[56] was the source of Ricardo's theory: and Ricardo's the source of Marx's.[57]

The ideological possibilities latent in Locke's second axiom (that each man "owns" his labor), combined with his frequent assertion throughout the *Second Treatise* that "Government has no other end but the preservation of Property,"[58] have led some to regard Locke as an apologist for nascent capitalism. According to Viner this is indeed "now standard doctrine, originated mostly by socialists and social Christians."[59] The most carefully worked-out interpretation along these lines was that of C. B. MacPherson who attempted to show that "the insistence that man's labour was his own – which was the essential novelty in Locke's doctrine of property ... provides a moral foundation for bourgeois appropriation."[60] "Locke's astonishing achievement was to base the property right on natural right and natural law, and then to remove all the natural law limits from the property right."[61] This he accomplished, according to MacPherson, by arguing that the social contrivance of money obviates the three limitations which Locke conceded must apply to individual appropriation.[62]

MacPherson's interpretation has been shown to be defective on several grounds. The labor theory of appropriation in Chapter V of the *Second Treatise* affords only weak and ambiguous support for a right to private property.[63] In any case, "Labour justified neither the accumulation of nor rights over one's goods; ... it provides a means of identifying something as naturally one's own."[64] Construal of the famous "Turfs" passage[65] as evidence of "the wage relationship" is to misunderstand the master-servant relation in preindustrial society.[66] Locke's theory of property rights does not rest solely upon the admixture of labor, but also upon charity and inheritance.[67] Locke actually denies a natural right to private property in a political society.[68] Finally, as Locke's friend and collaborator James Tyrrell acknowledged, the logic of their case pointed in a radical direction. "There was really no stopping place between the ground he and Locke occupied and logical individualism, final democracy, and the sharing of political power with women, children and servants,"[69] a point which has not escaped the attention of contemporary feminists.[70]

10.2 Property rights in *Rerum Novarum*

"In the *Rerum Novarum*, Leo XIII presented the doctrine that the right of private property was prior to government and rested on natural law, in terms which, in their stress on labor as constituting the original title to property, resembled closely, both in substance and in wording, those used by John Locke."[71] "It seemed to some, therefore, that Pope Leo XIII broke sharply with traditional doctrine when he went beyond St. Thomas in his campaign against nineteenth century socialist doctrine."[72] In this section I shall examine the theory of property rights contained in *Rerum Novarum*;[73] consider the extent to which it may be regarded as Lockean; appraise its putative novelty; and trace its subsequent treatment by Leo's successors.

The celebrated encyclical *Rerum Novarum* is generally acknowledged to mark the beginning of Catholic social teaching in modern times.[74] It was issued on 15th May, 1891, in response to the rise of Marxist "Social Democratic" parties in Western Europe and also to a wave of anarchist violence which culminated in the assassination of Czar Alexander II of Russia on 13th March of that year.[75] The theme of the encyclical was "The Condition of the Working Classes," said to have been reduced to misery by the economic revolution of the past century and also by the "prevailing moral degeneracy."[76] Acknowledging the poverty and wretchedness of the masses, Leo sought to show that a socialist remedy would be unjust and inexpedient. His argument was based on a theory of an inviolable "natural right" to private property which may be summarized in the following propositions:[77]

1. "God has given the earth for the use and enjoyment of the whole human race" (115:8,25).
2. But to acknowledge this "is not to deny that private property is lawful" (115:8).
3. Because "every man has by nature the right to possess private property as his own" (115:6,22,47).
4. This is not merely a right of usufruct but a right "to possess outright" (115:10).
5. Which arises from two considerations:

   (a) Man precedes the State, and has "the right of providing for the maintenance of his body" (115:7,22).
   (b) Man is rational, "links the future with the past," exercises "enlightened forethought," and therefore has need of "stable and permanent possession" of the means of production (115:6,7,22).
6. (a) Right of "outright possession" is acquired "when man thus turns the activity of his mind and the strength of his body towards procuring the fruits of nature" (115:9).
   (b) For "by such act he makes his own that portion of nature's field which he cultivates – that portion on which he leaves, as it were, the impress of his individuality" (115:9).
   (c) It is not just, therefore, that "the fruit of a man's own sweat and labour should be possessed and enjoyed by anyone else" (115:10).
7. Hence, the "first and most fundamental principle" is "the inviolability of private property" (115:15,46).
8. Abolition of private property would therefore be unjust because it would "rob the lawful possessor" (115:4).
9. It is lawful to transmit private property by inheritance (115:13).
10. Because "nature dictates that a man's children ... carry on ... and continue his own personality," and ought to be provided for (115:13).
11. The chief function of the state is "the safeguarding of private property by legal enactment and public policy" (115:38).
12. The "main tenet of Socialism, community of goods, must be utterly rejected" (115:15).
13. Because:
    (a) "it would injure those it would seem meant to benefit" (115: 15).
    (b) it is "directly contrary to the natural rights of mankind" (115:15), and
    (c) it would introduce "confusion and disorder" in society (115: 15).
14. Hence it is beyond the "competence" of the state to abolish private property rights (115:4).
15. Yet the state may "control its use in the interests of the public good" (15:471).
16. Moreover, to reduce class divisions and increase the stability of society private property should be distributed as widely as possible (115:46,47).

In light of the analysis contained in part I of this chapter, it would seem that Leo's doctrine exhibits several distinctly "Lockean" features, whilst departing decisively from Locke at a number of other points.

Propositions 1 and 2 form the starting point of Locke's own enterprise; proposition 5(a) is one of his two axioms. If "outright possession" be construed as equivalent to Locke's "property in," then proposition 4 (and also 3) are consistent with his doctrine, and propositions 6(a), (b) and (c) correspond quite closely to the labor theory of appropriation of the *Second Treatise*.[78] Propositions 9 and 10 are consistent with Locke,[79] and 11 would be if the adjective "private" was omitted. Proposition 15 accords with Locke's view of the purpose of civil society.[80]

Yet if the property theory of *Rerum Novarum* reveals the possible influence of Locke, it also displays considerable divergence both from Locke's intentions and from his conclusions. Locke did not assert the inviolability of private property (proposition 7). Upon Tully's interpretation of the *Two Treatises* moreover,[81] there is no reason to suppose that abolition of private property must always and necessarily be unjust (proposition 8); nor, therefore, that "community of goods must be utterly rejected" (propositions 12, 13); nor that it is beyond the competence of the state to bring this about (proposition 14). The "Locke" who stands behind *Rerum Novarum* would seem to resemble MacPherson's straw man, the apologist for capitalist appropriation.

Whether authentically Lockean or no, the property theory of *Rerum Novarum* has in recent times been regarded as a departure from traditional Catholic doctrine.[82] Though apologists for Leo maintain that he followed Scholastic theory in recognizing private property as a "secondary" right,[83] there is no trace of any such distinction in *Rerum Novarum*. Somewhere between the Reformation and the Nineteenth century, it is alleged, a decisive break in Christian social thought occurs. On this view of things the teaching of Thomas Aquinas, especially that of Question 66 in Part II–ii of the *Summa Theologiae*[84] is taken to be definitive.

Thomas held that "Man has a natural *dominium* over external things, for he has a mind and a will with which to turn them to his own account."[85] It is not natural for man to possess external goods as his own, however: community of goods is part of the natural law, though the distribution of goods by human positive law is not contrary to natural law but an addition to it devised by human reason.[86] Thomas follows Aristotle[87] in arguing that such distribution is expedient, but only because of the self-seeking propensities consequent upon human sin.

The Thomistic theory is evidently at variance with propositions 3, 4, 6, 7, 12, 13(b), 14 and possibly 8. These elements in Leo's doctrine are consistent with the hypothesis that they were derived, as indicated above, from some author possessing only an imperfect understanding of the Lockean argument. The source of the latter may have been the *Encyclopédistes*, who popularized Locke's political writings in the Eighteenth century and whose influence may be detected in the *Déclaration des droits de l'homme* of August, 1789: "*Le but de toute association politique est la conservation des droits naturel et imprescriptible de l'homme. Ces droit sont la liberté, la propriété ... .*"[88] The French Revolution destroyed the culture and seriously disrupted the intellectual life of European Catholicism.[89] The first to attempt a reconstruction of the Scholastic tradition was the Italian Jesuit, Taparelli d'Azeglio, whose "Lockean" treatment of property rights may have been influenced by Lacordaire,[90] and whose Essay on Natural Law marks the "solemn and definitive entry into Neo-Scholastic literature and Catholic teaching" of the thesis of a "natural right to property."[91] It was Taparelli's colleague and fellow-Jesuit, Liberatore, who prepared the first draft of *Rerum Novarum* and "apparently introduced the argument in question into that document."[92]

This account ignores the continuous development in Scholastic theory after Thomas in the work of Ockham, Jean Gerson, John Major, Lessius, Molina and Suarez.[93] Some idea of a "natural right" to property is always present after Gerson, and Molina's "revival of Gersonian rights theory" in the Sixteenth century "looks very much like an attempt to produce an ideology of mercantile capitalism."[94] The de Sousberghe-Viner hypothesis also ignores the famous bull *Quia Vir Reprobus* (1329) of John XXII, which attacked the Franciscan doctrine of holy poverty and clearly anticipated *Rerum Novarum* in asserting a natural rights defense of private property.[95] But it explains what is an unquestioned novelty in Catholic doctrine: the labor theory of appropriation. It explains, moreover, how the "Lockean" elements of *Rerum Novarum* came to be distorted into an unequivocal theory of a "sacred and inviolable" natural right to "private ownership." And finally the hypothesis explains why there are two alternative accounts of property rights in the encyclical: a traditional, Thomistic version derived from human rationality (proposition 5[b]) and a "Lockean" version derived from human labor (proposition 6). For both were contained in Taparelli's attempted reconstruction and synthesis of natural law theory.[96]

In *Rerum Novarum* Leo was thought at the time to have "fixed an impassible gulf between Socialism and the great religious organization over which he presides,"[97] a judgment which still stands.[98] The encyclical *Quadragesimo Anno* of Pope Pius XI,[99] written to commemorate the fortieth anniversary of *Rerum Novarum*, was no less intransigent: "Socialism, if it remains truly socialism, ... cannot be reconciled with the teachings of the Catholic Church."[100] A natural and inviolable "right of private ownership" was said to have been maintained by all "who have taught under the guidance and authority of the Church."[101] and a somewhat equivocal version of the labor theory of appropriation upheld.[102] "Certain kinds of property ... ought to be reserved to the state," however, "since they carry with them a dominating power so great that it cannot without danger to the general welfare be entrusted to private individuals."[103]

This slight concession to "community of goods" opened the way for a fairly sympathetic treatment of "socialization" in *Mater et Magistra* of Pope John XXIII, thirty years later.[104] Pius was reported to have asserted the incompatibility of Christianity and socialism, but nothing was added by way of confirmation or emphasis.[105] The doctrine of Leo[106] and Pius[107] on a natural right to property was similarly reported, and a barely recognizable concession made to the labor theory of appropriation.[108] Two new features were added by John to papal property theory however: first, that private property is an important safeguard of political freedom;[109] and secondly, reflecting a revival of interest in biblical theology in the Church of Rome, that "the right of private ownership is clearly sanctioned by the Gospel."[110] The first of these was reasserted in *Pacem in Terris*, where the right to private property, "including that of productive goods," was said to derive "from Man's nature."[111] A more eirenical attitude towards socialism was evident for the first time in that encyclical.[112]

The retreat from the Leonine position on "socialism" and private property rights perceptible in the reign of John XXIII became much more marked in that of his successor, not least because of the temper of the Second Vatican Council. Its pastoral constitution *Gaudiam et Spes*[113] reaffirmed the Thomistic doctrine of property and quietly buried the Protestant innovations of *Rerum Novarum*.[114] The view that private property is necessary for political freedom was reasserted, but the lawfulness of a "transfer of goods from private to public ownership" – subject only to "adequate compensation – was clearly recognized."[115] The encyclical *Populorum Progressio*[116] went even further. All rights whatsoever, "including the rights of property and free trade" are to be subordinated to the principle affirmed in *Gaudiam et Spes.* that "God intended the earth and everything in it for the use of all human beings and peoples."[117] Ambrose, the great Fourth century contemporary of Augustine, was quoted in support of the anti-Leonine (but not anti-Lockean) doctrine that "Private property does not constitute for anyone an absolute and unconditioned right. No-one is justified in keeping for his exclusive use what he does not need, when others lack necessities."[118] Finally, in the encyclical *Laborem Exercens* of John Paul II (1981), issued to mark the ninetieth anniversary of *Rerum Novarum,* the doctrine of the latter was explicitly repudiated. The customary lip service is paid to Leo's encyclical. But his "first and most fundamental principle, ... the inviolability of private property"[119] is denied: "Christian tradition has never upheld this right as absolute and untouchable."[120] Less than a century after *Rerum Novarum,* Catholic social teaching had reverted to the *status quo ante.*

10.3 Some questions raised by the foregoing

What is the relevance of all this to the ordinary professional concerns of the reader of this *Review;* or even to their holiday concern with the "Philosophical Foundations of Economic Science?" I shall suggest two groups of related issues where some intellectual connection, or at any rate some "interface," may exist between Economics and Theology: first, with respect to the scientific status of the concept of "property rights;" secondly, with respect to the epistemic claims of Christianity in general and the papacy in particular in relation to a "scientific" knowledge of the economy. I am acutely aware of the tentative and conjectural nature of this section, and offer it as a provocation to discussion rather than as a substantive contribution to knowledge.

The concept of "property rights" is used in "mainstream" economic theory in what is intended to be a strictly "positive" way. Property rights are "*sanctioned behavioral relations among men that arise from the existence of things and pertain to their use.*"[121] This is consistent with the standard definition of "right" taught to beginning Political Science students: "a privilege or prerogative conferred by usage or law upon a person or group."[122] Right of ownership is an exclusive right to use an asset, change its form and substance, and to transfer all or some of the rights in it, "limited *only* by those restrictions that are explicitly stated in the law ... ".[123] Assignment of rights is determined in the last resort by the state, hence "a theory of property rights cannot be truly complete without a theory of the state."[124] It is a clear implication of all this that the concept is intended to capture a particular social phenomenon, the political allocation of scarce resources, and so to facilitate its explanation. The latest contribution on the matter to this *Review* affords a clear example. The American economist A. T. Hadley is said to have explained the evolution of property rights in terms of the technical requirements of economic development. "Society required property structures that fostered efficient resource use if it was to advance via capital formation."[125]

It is not altogether certain, however, that the merely "positive" use of the concept can be sustained.

In the first place, to restrict the term "right" to mean only that which *is* allowed by the state, and to ignore or exclude (as Furubotn and Pejovich seem to do) that which *ought* to be allowed is already to commit oneself to one of two possible ethical positions. On the one hand one may take the view that what the state permits is thereby good: either because one believes the state to implement the general will through democracy or some other approved process, or for the more Hobbesian reason that the state is necessary to survival. Or on the other, one may take the view that nothing certain can be known about how the state *ought* to behave. The latter would seem to imply, further, the theological judgment that God does not exist; or that if He exists, He has no particular purpose in mind for human beings and their institutions; or that if He does He is unable or unwilling to communicate it.

In the second place, the positive usage flies in the face of an ancient and still current normative connotation and is, to say the least, potentially misleading. As I have attempted to show in part I, the terms "right" and "property" are inextricably bound up in our culture with theological beliefs and ethical principles, and for many people – even perhaps for economists when they wear some other hat – still resonate to these themes. Stipulative definitions intended to escape from a normative flavor are not easy to construct. That of Furubotn and Pejovich rests upon two terms, "sanctioned" (by whom? on what authority?) and "law" (whose law, God's or Man's? just or unjust?) which are inescapably value-laden.

Thirdly, it is in fact the case that economists do make use of the concept in a normative context. Furubotn and Pejovich cite Alchian: "The allocation of scarce resources in a society is the assignment of rights to uses of resources ... the question of economics, or of how prices *should* be determined, is the question of how property rights *should* be defined and exchanged ... ".[126] Now this may only be an example of what Marshall called the "drudgery" which Mistress Ethics now delegates to her humble servant Economics.[127] Given value axioms, "Economic Science" grinds out answers to the question. "What must we do to be saved?" It would seem, however, that economists have taken for granted a particular set of value axioms by no means invulnerable to ethical or theological criticism. " ... the focus of discussion is on economic efficiency and the conditions under which markets should be, or should not be, extended into new areas." Moreover, "Strong concern is shown for the individualist basis of choice."[128]

Finally, the use of the concept by Hadley and others in historical explanation seems to imply a determinism which prejudges the transcendental aspect of "rights" and links ethics (and by implication their theological rationale) to a particular stage of economic development. Though this need not be inconsistent with a Lockean account of "conventional" property rights in political society, it ignores the Lockean connection between conventional rights and natural rights, and hence deprives the former of their metaphysical and theological basis. To do so is therefore to occupy a theological position.

The second group of issues raised by the first two parts of this article concerns the relation between the epistemic claims of Christianity and those of "scientific" economics. The problem is illustrated – to the point of caricature – by *Rerum Novarum*, in which as I have shown, the pope did not hesitate to assert empirical judgments (e.g., propositions 13[a], 13[b], 16) on all fours with theological pronouncements. "We approach the subject with confidence," declared Leo, "and in the exercise of the rights which manifestly appertain to Us, for no practical solution of this question will be found apart from the intervention of Religion and of the Church."[129] In the slightly mythological account supplied by Pius XI, "all eyes as often before turned to the Chair of Peter, to that sacred depository of all truth," whereupon "that long awaited voice ... taught the whole human family to strike out in the social question upon new paths."[130] A contemporary authorized translation of paragraph 9 was less equivocal: "... the venerable Pontiff taught mankind new methods of dealing with social problems."[131] The "new methods" consisted in recognizing a characteristic of human nature, known by reason and certified by the authority of the Church, but unknown or rejected by many outside the fold: "We have seen that this great labour question cannot be solved save by assuming as a principle that private ownership must be held sacred and inviolable."[132] Property rights, in a transcendent, metaphysical sense quite unlike their analytical use in modern Economics, are asserted to be part of a "solution" to a social problem, the poverty of the working classes.

In what may well have been the last article by a Catholic bishop to appear in a leading journal of our profession (1891), John Keane pointed out that the pope did not mean by this "that the Catholic Church has any system of economics which she would wish to impose upon the nations of the earth;"[133] but "she offers the world ethical and religious principles which ... ought to regulate all economic systems" and therefore "repudiates the school of economics ... of *laissez-faire*" because it was "avowedly without heart and without ethics."[134] "The laws of economics have their very roots in morals" and this intimate connection between economics and morals is "just as true of economics and religion." [135] Hence the "inalienable rights" referred to in the Declaration of Independence, which are "the basis of sound economics," can only be known theologically: "Without the Creator man has not inalienable rights ... . But with the Creator man has inalienable rights which every fellow-creature must respect."[136]

Now it is evident that by "economics" in these passages Keane does not mean "Economic Science" in the modern sense, but what we should now call "economic systems." No respectable practitioner of the former today would presume to prescribe for the latter save in the light of his clients' value axioms. So far there is no difficulty for contemporary "mainstream" theory – subject only to a proper explication of "the laws of economics." But behind his argument (and explicit in the encyclicals) is a claim to epistemic superiority which can only be ignored at the cost of commitment to a particular theological position: that the Christian church does not, in fact, have peculiar access to knowledge of any kind; or that if it does, it is of a kind that is irrelevant to scientific inquiry. If the Chair of Peter is really the "sacred depository of all truth;" or if, for that matter, a similar authority is located elsewhere in the church or else enshrined in sacred scripture: what is the status of "scientific" knowledge of the economy?

Quadragesimo Anno concedes that the church may use her authority "not ... in matters of technique for which she is neither suitably equipped nor endowed by office"[137] but only in "things that are connected with the moral law;" for "economics and moral science employ each its own principles in its own sphere."[138] A similar disclaimer appears in the most recent encyclical.[139] Though seems sufficient acknowledgement of the autonomy of scientific knowledge, it still leaves some loose ends. First, some Christian economists, especially those of the Neo-Calvinist tradition, continue to hold that "the attempted assertion of epistemological autonomy involves ... the denial that the very possibility of knowledge rests on the being of God."[140] Secondly, the concession throws doubt on the authoritative character of that very substantial part of Catholic (or at least Papal) social teaching which consists not of theological and ethical pronouncements, but of empirical judgments about the economy. Finally, it fails to explain what may be the most interesting claim of Christian social thought: that some particular economic problem (e.g., poverty) may require for its solution the general recognition of some truth that could only be known theologically (e.g., that there exists a natural right to private property).

I have not attempted to settle these questions in this paper. My purpose has been the more modest one of persuading the reader that the boundary between "Economic Science" and Theology is neither so well-defined, nor so unimportant, as is commonly supposed. If "Social Economy" is ever to be more than a soft option in the "mainstream" tradition, its practitioners must get to work on the relation between the two.

11 The Intellectual Context of *Rerum Novarum*

The subject of this article is the intellectual isolation of Roman Catholic – specifically papal – social doctrine, an isolation not only symbolized, but in some ways ratified and made permanent, by *Rerum Novarum*. I shall deal with it by posing and attempting to answer three related questions. Why did the Roman Church wait until 1891 before uttering any official teaching about the industrial economy? Why, when it did, was it so hostile to a market order? Why, in view of its lack of confidence in the market, was it so opposed to socialism? The short answer to all three questions is "The French Revolution."

11.1 Why not before 1891?

According to Pius XI, *Rerum Novarum* was "... so unexpectedly in advance of its time that the slow of heart scorned the study of this new social philosophy, and the timid feared to scale its lofty heights."[1] Pius was only voicing the official ideology: *Rerum Novarum* marks the beginning of any serious attempt to apply Christian theological principles to the social problems created by the market economy.[2] Before I ask *why* it was "belated," therefore, I must first show that it was.

The French Revolution destroyed – on the continental mainland *but not in Britain* – an *ancien régime* which in one way or another had determined European culture since the coronation of Charlemagne. Europe was a Christian society: church and state were one, and political theory a branch of ecclesiology. To be sure, there were continual struggles for supremacy over this society. At times the papacy had the upper hand, at others the secular power: first the Emperor, later the princes of the great sovereign states, Spain, France, England and

Austria. The Reformation made little difference. France and Austria remained in communion with the Holy See. England did not. In all three countries the crown dominated the church, determining the appointment of bishops, the existence if any of religious orders, the powers of ecclesiastical courts, and the system of education.[3] But in all three – as everywhere else in Christendom – its divine authority was ratified by the national church in the coronation ritual and in the whole body of Christian teaching enforcing religious obedience to the powers that be.

It was Burke who first saw clearly that the French Revolution, quite unlike the Glorious Revolution of 1689 or the American Revolution of 1776, was a full-scale assault on the Christian foundations of European civilization. By 1793 events had proved Burke right, and the process had begun of constructing an intellectual defence of the *ancien régime* against the ideology of "Jacobinism." But from the first there was a drastic bifurcation between the method – if any – of the continental reaction and that of the British.

In France, Belgium, Italy, Austria and the Iberian Peninsula blow followed blow with such rapidity that there was little opportunity between 1791 and 1814 for more than anguished cries of protest. Pius VI responded to the first assault on the French church with his last encyclical, *Charitas* (1791), which promulgated what was to be the normal Roman position for a century:

> Carefully beware of lending your ears to the treacherous speech of the philosophy of this age which leads to death.[4]

Seven years later French troops entered Rome and arrested Pius, deporting him to Valence there to die obscurely in a government hostel.

During the next decade-and-a-half the new pope and his church suffered many vicissitudes and humiliations. Like his predecessor, Pius VII was arrested and imprisoned by the French. He was compelled to crown Napoleon as Emperor and to ratify a drastic reduction in his temporal power. His one encyclical, *Diu Satris* (1800), issued shortly after his election, could only respond to new ideas by a call to rally round the church and suppress dangerous literature:

> Books which openly oppose the teaching of Christ are to be burned. Even more importantly, the eyes and minds of all must be kept from books which do so more stealthily and more deceitfully.[5]

Though the Papal States were returned in 1815 and papal religion once again tolerated throughout Europe, the Roman communion was never restored to its commanding position as the established church of the West: "After the French Revolution, after the Italian revolutions. neither society nor the Church-in-society could be the same again."[6] Not until the pontificate of Leo XIII (1878–1903), however, did there reign a pope who had not lived through the catastrophic events of 1800–1814. Like the Bourbons they generally supported, Leo's predecessors "forgot nothing and learned nothing." Though Pius VII himself, when Bishop of Imola, had published a pastoral in 1797 interpreting *Liberté, Egalité, Fraternité* in a Christian sense;[7] though militant ultramontanes such as Lacordaire, Montalembert and Lamennais came to see that the church should ally itself with liberal-democratic forces against Erastian absolutism, the Curia preferred to believe with *émigré* intellectuals that all revolution was a "satanic" conspiracy against Christ[8] engineered by the Freemasons.[9] *L'affaire Lamennais* ended in the suppression of *L'Avenir*[10] and the first encyclical of the new Pope Gregory XVI: *Mirari Vos* (1832) which condemned "liberalism." The "absurd and erroneous" doctrine of "*liberty of conscience,*" the "never sufficiently denounced" freedom of the press, and the "unbridled lust for freedom" were explicitly linked to "writings which attack the trust and submission due to princes."[11] Subsequent events in 1848 throughout Europe, and in Italy during the later course of the *Risorgimento*[12] only served to confirm the worst fears of the popes and those who advised them. The *Quanta Cura* (1864) of Pius IX[13] and the annexed *Syllabus Seu Collectio Errorum Modernorum*[14] epitomize papal social thought in the nineteenth century. The eightieth anathema of the *Syllabus*, nuanced though it is by reference to the allocution *Iamdudum Cernimus*, says it all: it is forbidden for the faithful to believe that "The Roman Pontiff can and should reconcile and align himself with progress, liberalism and modern civilization."[15]

Meanwhile in Britain, securely insulated by the Channel and the Royal Navy from any military threat to the *ancien régime*, a totally different strain of counter-revolutionary ideology quickly evolved. In the luxury of a domestic tranquility unique in Europe at that time and facilitated by a long-established freedom of the press, a full-scale literary battle was fought between the English Jacobins and the champions of Church and King. The first shots exchanged were Richard Price's revolutionary *Discourse on the Love of our Country* (1789)[16] and Burke's *Reflections* (1790)[17] written in reply. Tom Paine, (1791)[18] Mary Wollstonecraft, (1790)[19] James MacIntosh (1791)[20] and more than seventy others joined in the fray. The high point of the Jacobin campaign was reached in William Godwin's massive rebuttal of Burke, the deservedly famous *Political Justice* (1793, 1796, 1798).[21] The battle was brought to an end, and the Jacobins finally routed by T. R. Malthus's first *Essay on Population* (1798). Despite propagandist historiography of the past, it is now generally agreed that the English Jacobins were vanquished in fair fight.[22]

It is impossible to overestimate the importance of Malthus in all this, not so much because he administered the *coup de grâce* to revolutionary ideas in Britain as because of the method he chose in doing so. The intellectual problems created by the first *Essay* and the gradual development of their solution over the next three decades are the seedplot of "christian social thought." The papal response to revolution, and that of all continental Christians obedient to the papacy, had been to repudiate the Enlightenment philosophy on which the Jacobins claimed that their revolution was founded. But the Anglican response was to challenge the Jacobins' attempt to hijack the Enlightenment, and to show that Enlightenment ideas, rightly understood, were incompatible with revolution but wholly consistent with orthodox Christianity and the established order in church and state. I have told this story at considerable length elsewhere.[23] What follows is supported by that work and the references therein.

Malthus's weapon against Godwin, Condorcet and the Jacobins in general was *political economy*, meaning by that what we should now call the "canonical classical model"[24] constructed by modifying *Wealth of Nations* in light of the "principle of population." According to Malthus, political economy proves that private property, wage labor, marriage, market institutions and the legal sanctions for these are inevitable. If once removed by revolutionary action, they would reappear within a generation or two. Moreover they are beneficial, for by maximizing the social surplus at equilibrium they afford "all that distinguishes the civilized, from the savage state." Now private property and wage labor create inequality, and the dominance of scarcity in human affairs means poverty for most. But inequality and poverty are beneficent, for they supply the necessary incentives to industry and thrift.

Powerful and sophisticated as this argument was, it could only be of ideological service to the *ancien régime* if it could be shown to be consistent with Christian orthodoxy. How can a God who is good, wise and all-powerful create a world ruled by scarcity in which the many must live in subordination and poverty? Malthus, though a priest of the Church of England, was an amateurish theologian and his own answer to this question was seriously defective. William Paley sketched a solution in his last work, *Natural Theology* (1802). But the definitive treatment was provided by J. B. Sumner (Archbishop of Canterbury, 1848–63) in his immensely influential *Records of the Creation* (1816). "Inequality of Ranks and Fortunes is the condition best suited to the exercise and development of virtue." It is proof of the divine wisdom. Sumner argued, that this is "universally established by a single principle" – the principle of population.

In the post-war period the foundations laid by Malthus, Paley and Sumner were built on by two of the most eminent Oxford men of their generation: Edward Copleston (Bishop of Llandaff, 1827–49) and Richard Whately, the only person in history to move directly from a professorial chair in Economics to an archbishopric (Dublin, 1831–63) without intervening stages. Copleston distinguished "indigence" (income below the socially conditioned "subsistence" level) from poverty ("subsistence" income) and showed that the former is not the equilibrium outcome of Malthusian dynamics but a disequilibrium phenomena – frequently associated with unanticipated inflation – which the state may quite properly correct. He also refined Sumner's theology by showing that the futility of attempting to legislate benevolence (taken to have been proved by economics) can be fitted into Paley's teleology. Whately corrected Paley's theology (of which Sumner's was an enlargement) by showing that "theological utilitarianism" and natural theology must fail without some Butlerian assumption of a "moral sense," but can be rescued by means of it. Even more important perhaps, Whately defended Christian Political Economy from confusion with the "Philosophic Radicalism" of Bentham and James Mill who were attempting, from the mid-1820s, to harness the prestige of political economy to their own avowedly, godless program of "reform." Whately's famous demarcation of "scientific" from "theological" knowledge rationalized the ideological alliance of the two different kinds of knowledge in Christian Political Economy, and guaranteed the integrity both of science and of theology in the decades to follow.

The *ancien régime* in Britain expired peacefully between 1829 and 1832[25] and there was gradual transition to a new age of pluralism, parliamentary democracy and social mobility founded on continual economic growth. By the coronation of Victoria at the latest the first – and only complete – attempt at a Christian social theory had been incorporated into the mainstream of English-speaking culture where, divested of those features peculiar to the 1790s, it became part of the air we all breathe.[26]

For reasons I have already hinted at and will explore more fully in the next two sections of this article, it was impossible for the papacy and those bound by its discipline to recognize and avail themselves of the English ideas. This had nothing to do with their provenance in Anglican divinity. There was precedent for recognizing intellectual contributions of the English church. As long ago as 1686, at Bossuet's motion, the Synod of St. Germain had formally communicated its thanks and appreciation to the Anglican Bishop George Bull (St. David's, 1705–1710) for his *Defensio Fidei Nicaenae*.[27] The causes of papal hostility to a free-market order, lay not so much in theology as in economics and philosophy.

11.2 Hostility to a market social order

As late as 1959 the French, Roman-Catholic economist Daniel Villey could write that:

> ... almost all Roman Catholic theologians and economists repudiate economic liberalism, and do so very frequently in the name of their faith. Very few ... understand the working of the market economy ... Catholic mentality often remains much more pre-Columban, pre-Copernican, pre-Baconian, pre-Cartesian – and in France pre-revolutionary, and in economic matters pre-capitalist – than most Catholics realize.[28]

Villey's view was explicitly confirmed in a recent article in this *Review* by his German co-religionist, the late Goetz A. Briefs who remarked "the continuing hostility with which Catholic social thought confronts even the modern versions of liberalism and capitalism."[29] Villey and Briefs ranged far and wide in their attempts to explain this hostility. Confining my attention to the period between *Mirari Vos* and *Rerum Novarum*, I shall focus upon what seem to me to be the three most important elements of the explanation: a misunderstanding of classical political economy, a misperception of the European economy in the nineteenth century, and philosophical commitment to a purely Scholastic understanding of "liberty."

Classical political economy was supposed to be about the "wealth of nations." Because of the ideological need to distinguish themselves clearly from the Philosophic Radicals, Malthus and his allies were careful to minimize both the heuristic scope of political economy and the range of meaning attaching to "wealth" in the later 1820s: "I cannot agree with you in the meaning which you seem inclined to give to wealth," wrote Malthus to Chalmers in 1827:

> I cannot help thinking that it is more correct ... and in accordance with all our common feelings to say that security, independence, moral and religious instruction, and moral and religious habits, are very superior in importance to what we usually mean by wealth, than to say that they ought to be included in that term.[30]

Whately's *Introductory Lectures* (1831) took the point for granted and pressed home the second stage of the argument: the merely "scientific" (that is "positive" and value-neutral) character of political economy. Though Whately never quite succeeded in converting the profession to his minimalist, "catallactic" account of economic theory, his main point was immediately accepted, and classically enounced by his former pupil and life-long friend N. W. Senior.

"The foreign school," said the *Edinburgh Review* in its notice of Senior's *Outline*:[31]

> ... ask whether all questions concerning the moral and physical well-being of man in society, as far as wealth affects it, are not closely connected with the abstract science of wealth? ... .We contend that the study is purely a science ... a science which neither recommends to do, or abstain from doing ...[32]

For as Malthus insisted to Chalmers:

> We have always been told, and most properly, to prefer virtue to wealth; but if morals be wealth, what a confusion is at once introduced into all the language of moral and religious instruction.[33]

The most distinguished member of the heterodox "foreign school" (and chief target of the *Edinburgh*) was Sismondi. But the most influential for subsequent papal teaching was his disciple the legitimist, ultramontane Vicompte Alban de Villeneuve-Bargemont (1784–1850), "Precursor of Modern Social Catholicism"[34] No economic writing illustrates more clearly than his the "confusion" introduced by "the foreign school" into "all the language of moral and religious instruction." For as the (pro-papal) *Dublin Review* observed in its notice of Villeneuve's *Economie Politique Chrétienne*,[35] "The Catholic school … affirms, that the moral virtues, probity, chastity, and temperance, do truly … come into the catalogue of national riches."[36]

Villeneuve's *Economie Politique Chrétienne* was utterly unlike the "Christian Political Economy" of Malthus, Whately and their colleagues. Both were part of the Christian reaction to the French Revolution: but whereas "Christian Political Economy" – in the spirit of Enlightenment thinking – was merely a contingent alliance of the epistemologically distinct enterprises of Christian theology and classical political economy, *Economie Politique Chrétienne* was a unified enterprise, "catholic in its, manner of conceiving science,"[37] which reasserted the ancient sovereignty of theology over all other inquiry. Villeneuve vividly described his vision in the preface to his great work:

> … je m'attachai … à interroger tour à tour l'économie politique [meaning Smith, Say and Malthus], les théories philosophiques de la civilisation, la statistique, la legislation et les sciences morales … peu a peu, à *l'aide surtout du phare lumineux du christianisme* il me sembla que l'on pouvait distinguer nettement les causes des désordres moraux et matériels des sociétés.[38]

There is no "smoking gun" which undeniably links *Rerum Novarum* to Villeneuve. Sr. Mary-Ignatius Ring's parallel quotations[39] are suggestive but rather strained in some places. There is considerable testimony, however, to Villeneuve's influence upon ultramontane social thinking both before and after the suppression of *l'Avenir:* he was a friend of de Coux and linked through him with Lacordaire and Montalembert; he collaborated with Gerbet in *l'Université Catholique*. It is improbable that Ketteler, whom Leo XIII explicitly acknowledged,[40] should not have known of Villeneuve and his work. Villeneuve's characteristic misunderstanding of classical political economy[41] became the conventional wisdom, part of the intellectual furniture of Leo's drafting committee in the Summer of 1890.[42] To even so liberal a prelate as John Keane, classical political economy was "a system avowedly without heart and without ethics" which "regarded the working men" … as "mere instruments for making money:[43] therefore, classical political economy must go.

A second and related reason for the hostility or at any rate suspicion of the market economy evinced by papal social thought is a failure to appreciate what was actually happening to the European economy. Though the French Revolution and its aftermath interrupted an economic development well under way before 1789, real income per head in France grew at an average of 1.4 per cent per annum from 1800 to 1900 (compared with 1.7 per cent for Germany and only 1.2 per cent for Britain), meaning that the command over goods and services was doubling every fifty years.[44] Yet Villeneuve and the "Catholic School" in general followed Sismondi (a protestant) in attacking industrialization, claiming that "the accumulation and concentration of capital [and] the universal use of machinery would indefinitely multiply the number of the poor."[45] Villeneuve's immediate and pressing concern was relief of the thousands of paupers under his jurisdiction as Prefect of the department of Nord. Lacking an analytical framework within which to investigate the relation between capital accumulation, technical progress and economic growth, he was deceived by appearances into mistaking a *correlation* between industrialization and pauperism (which he investigated extensively and comparatively) for a *causal relation.*

Sixty years later per capita real income had more than doubled in most of Western Europe and North America. But influential bishops – Ketteler (Mainz), Manning (Westminster), Gibbons (Baltimore) and others – were understandably exercised by the alienation of the industrial working classes from traditional *mores* and pressed their concerns upon Leo.[46] Though there was "no moral indictment of capitalism" in *Rerum Novarum,* yet it was "sharply criticized."[47] The benefits of economic growth and the possibility of progressive redistribution under competitive product and factor markets were ignored. Attention was focused instead upon "the enormous fortunes of some few individuals and the utter poverty of the masses:" "rapacious usury" was condemned and a minimum wage recommended.[48]

The third, and I believe the most fundamental reason for papal hostility to "economic liberalism" is not economic but philosophical. As part of its ideological grand strategy in the nineteenth century, the Church of Rome formally committed itself to a particular school of philosophical thought which happens to be extremely difficult, if not impossible, to reconcile with the methodological individualism of economic theory.

During the urbane and peaceable age of the Enlightenment, the Roman clergy differed hardly at all from other men of learning and cultivation.[49] There was widespread interest, especially among Jesuits and Benedictines, in Locke, Leibniz, Kant and other contemporary philosophers. All this was rudely shattered by the French Revolution. From the first, as we have seen, the papacy set its face against "the philosophy of this age which leads to death." There are numerous references to philosophy in the encyclicals from *Charitas* (1791) to *Quanta Cura* (1864),[50] which are with one exception hostile. The exception is *Singulari Quidem* (1856) which for the first time since the Revolution indicates that philosophy might form part of the education of the clergy. Not until Leo XIII is there a permanent change of attitude.

For since the early 1830s there had been a gradual revival of Scholasticism in general and of Thomism in particular, "beginning as an ideological discovery by Catholic philosophers confronted with contemporary problems."[51] Though Thomism met with considerable opposition at first – for all philosophy was deeply suspect – it was strongly advocated from 1850 by the Jesuit publication *Civiltà Cattolica*, and pursued with increasing scholarship and rigor in Italy, France, Belgium and Germany. By 1853 its ideological potency was clearly demonstrated by the *Philosophia Christiana* of G. Sanseverino, which sought to refute not only Hume, Reid, Kant and Schelling but also the ultramontane but heterodox "liberals," Lammenais and Gioberti.[52] Archbishop Emmanuel von Ketteler based his influential *Die Arbeiterfrage und das Christentum* (1864) upon Thomistic theory,[53] and Archbishop Joachim Pecci of Perugia (1846–77) promoted the Scholastic revival in his province.[54]

When the Archbishop of Perugia became Leo XIII he brought with him thirty years of experience in fighting "liberalism." He is supposed to have been the "real instigator" of the *Syllabus* and unlike his predecessor Pio Nono was both scholarly and intellectual.[55] Within two years of his election he issued the *Aeterni Patris* (1879) which made Thomism the official philosophy of the Church of Rome.[56] That encyclical noted that "domestic and civil society ... would certainly enjoy a far more peaceful and secure existence" if "the works of Thomas Aquinas" were taught in schools and universities – in particular "the teachings of Thomas on the true meaning of liberty, which at this time is running into license."[57] *Immortale Dei* (1885) used Scholastic ideas to attack "that liberty which begets a contempt of the most sacred laws of God, and casts off the obedience due to lawful authority."[58] Three years later Leo issued *Libertas* (1888), the most carefully worked out and certainly the most important document of his pontificate: the key to a right understanding of *Rerum Novarum* and all subsequent "social catholicism."

According to the doctrine of *Libertas*, all rational beings have a *natural* liberty: "the faculty of choosing means fitted for the ends proposed." But if the *will* seeks what is contrary to *reason* "it abuses its freedom ... and corrupts its very essence." This is because reason can know the *law* of one's being, the *natural law*, a special case of the *eternal law* of God – and disobedience of that law is automatically punished by relegation to a lower (sub-rational) state of being and therefore loss of liberty. Human positive law does for political society what natural law does for the individual. But it must be for the good of individuals and hence in accordance with natural law.[59] Leo had no difficulty in proceeding on this basis to dispose of the characteristic doctrines of *political* liberalism: sovereignty of the people, democracy, and the so-called "liberties" of religion, speech, the press and teaching.[60] In so doing, however, and possibly without intending it, he also destroyed the intellectual foundation of *economic* liberalism:

> ... the eternal law of God is the sole standard and rule of human liberty, not only in each individual man, but also in the community and civil society. Therefore ... the true liberty of human society does not consist in *every man doing what he pleases, for this would simply end in turmoil and confusion*, and bring on the overthrow of the State ...[61]

Adam Smith's "invisible hand," Whately's "metropolis" are simply ignored. Thomistic philosophy, at least as understood and taught by Leo, would appear to ignore or deny the power of market co-ordination – "through the agency of men, who each think of nothing beyond his own immediate interest" – to create "the wisest means for effecting an object, the vastness of which it would bewilder them even to contemplate."[62]

Leo XIII was the first pontiff since the French Revolution to respond intellectually to its challenge, and the first in modem times to set the teachings of his church within a theoretical framework that was coherent and rigorous when judged by the best standards of its age. It is impossible to overrate his achievement: "With Leo XIII the 'Principles of 1789' at last received baptism at Rome – an adult baptism, received after due confession!"[63] Yet like all human achievement, it had its cost. Part of that cost was an insulation of official doctrine from economic science that persists to this day in the public utterances of the hierarchy.

11.3 Why not socialism?

In view of the hostility to, or at any rate the lack of understanding of, the market economy evinced by *Rerum Novarum*, it might have been expected that socialism would have received a more friendly consideration. "As far back as about 1870," reported Daniel Villey:

> Dupont-White wrote that when a priest is open to modern ideas he favors socialism rather than liberty. The reason is that in socialism he senses some residues of pre-capitalist ideologies, though at the cost of some misunderstandings ... Whether it tends towards a feudal past or a hypothetical collectivist future. Catholic thought appears always to be keen to escape from the present ... [64]

Yet *Rerum Novarum* is nothing if not an anti-socialist tract. Bishop John Keane, first Rector of the Catholic University of America (though subsequently dismissed by Leo), explained why in what is probably the last article by any bishop to appear in a professional journal of economics. A "false individualism:"

> had its birth in English Deism, grew into the system of *laissez-faire*, reached its awful culmination of the Reign of Terror and the Paris commune, and now comes back to the starting point of its vicious circle in the State Socialism so vehemently urged at present.[65]

Upon this construal "socia1ism" – just like capitalism in this respect – is simply another manifestation of the hubristic individualism of the Enlightenment. Leo's encouragement of "associations" is directed against "false individualism" and therefore – according to Claudio Jannet writing in the same journal two years later – "has condemned the principal errors relating to the economic order, and has fixed an impassable gulf between Socialism and the great religious organization over which he presides."[66]

Attractive and plausible as this explanation appears, it leaves some loose ends untied. It hardly accounts for the almost obsessive reiteration of Leo's strictures against socialism, and certainly not for his willingness to modify Thomistic property-rights theory in order to make his point in *Rerum Novarum*. Moreover, it fails to account for the fact that papal teaching in the twentieth century and especially since the 1960s has gone far to bridge the "impassable gulf."

At the very outset of his pontificate Leo condemned the "deadly plague" affecting the body politic of Europe, one of the consequences of which was "the confiscation of property that was once the support of the Church's ministers and of the poor."[67] He returned to the theme in his second encyclical ten months later, *Quod Apostolici Muneris*

(1878), a comprehensive condemnation of "socialists, communists, or nihilists" who "assail the right of property sanctioned by natural law."[68] *Rerum Novarum* is the thirty-eighth of Leo's encyclicals. Anti-socialist themes are to be found in no fewer than thirteen of its predecessors.[69] Some of these references are rather strained, as in *Diuturnum* (1881), where the German Reformation is blamed, or simply comic, as in *Auspicato Concessum* (1882) on St. Francis of Assisi, in which socialism is traced to the Albigensian heresy.

It would seem from his encyclicals that the greatest evil associated by Leo with socialism was its challenge to private property rights and in particular its proposed "regulation of all ecclesiastical property, *starting from the principle that its ownership belongs to the State.*"[70] Since the final loss of the Papal States between 1858 and 1870 and the loss of Rome itself in the latter year, the papacy had become politically impotent. The pope was allowed to use, *but not to own*, the Vatican, St. John Lateran and Castel Gandolfo. Tithes were abolished in Italy in 1888 and in 1890 the Italian parliament nationalized the church's charitable foundations.[71] Many similar invasions of the church's property rights took place in France, Germany, Austria and elsewhere during the last third of the nineteenth century.[72] Though few of these were executed in the name of "socialism" they were invariably advocated and applauded by socialists. It was urgently necessary to identify and denounce the theory which appeared to justify these expropriations. The ostensible and, indeed, the primary concern of *Rerum Novarum* was "the condition of the working classes." But in the circumstances it was inevitable that its dominant theme should be an attack on

> ... the socialists, [who are] working on the poor man's envy of the rich ... striving to do away with private property, and contend[ing] that individual possessions should become the common property of all, *to be administered by the state ...*[73]

"The first and most fundamental principle, therefore, if one would undertake to alleviate the condition of the masses, must be the inviolability of private property."[74]

In a previous article in this *Review*[75] I have shown that the property-rights doctrine of *Rerum Novarum* is actually a conflation of two radically incompatible elements: a traditional, Thomistic theory derived from human rationality, and a pseudo-Lockean version derived from a labor theory of appropriation. Now St. Thomas denied that *individuals* have a natural right to property and followed St. Augustine in attributing to the *state* the right and duty to assign property rights to individuals by human positive law. Useful as Thomism was for Leo in general, therefore, it was wholly unserviceable in defending private property against the temporal power. It was, therefore, necessary to elide that part of St. Thomas's doctrine which would embarrass the ideological defense of the papacy and supplement the remainder with a bowdlerized version of Locke. Thus, right of "outright possession" is acquired "when man ... turned the activity of his mind and the strength of his body towards procuring the fruits of nature," for "by such an act he makes his own that portion of nature's field which he cultivates;" it is not just, therefore, that "the fruits of a man's own sweat and labor should be possessed and enjoyed by anyone else."[76] From this, the encyclical can proceed to the required conclusion (not to be found in Thomas) that the natural right of individuals to own property renders it beyond the "competence" of the state to abolish private property rights.[77]

It has been suggested by de Sousberghe[78] that this confusion was introduced into *Rerum Novarum* by P. Liberatore, who based his first draft of the encyclical on the work of his friend and fellow-Jesuit, Taparelli d'Azeglio. For both the Thomistic and the pseudo-Lockean elements are found in Taparelli's own attempted synthesis of natural law theory. Whether Leo and his drafting committee were simply muddle-headed at this crucial state of the argument, or whether they actually cheated, can never be known. One thing is certain. Though some version of a "natural right" to property is always present in post-Thomistic Scholasticism after Jean Gerson, the labor theory of appropriation was an unquestioned novelty in Roman Catholic doctrine in 1891.

Rerum Novarum is the high-water mark of papal hostility to socialism. After the creation by Mussolini of the Vatican State in 1929, there was a gradual softening of the official policy which continues to this day.[79] By the pontificate of John Paul II, the protestant innovations of *Rerum Novarum* were quietly buried. The "sacred and inviolable" right to private property was denied in *Laborem Exercens* (1981); "Christian tradition has never upheld this right as absolute and untouchable."[80]

I conclude from these facts that the anti-socialist character of *Rerum Novarum* is not so much the sign of a general, theoretical incompatibility of Christianity with socialism, as of the temporary circumstances of the Church of Rome in the nineteenth century. The French Revolution was the most cataclysmic event in the history of the Western Church, far exceeding in its damage to Christianity anything inflicted by the Pornocracy in the tenth century, the Papal Schism in the fourteenth century, or the protestant Reformation in the sixteenth century.

Throughout the nineteenth century, the train of events it set in motion continued to harass and perplex the Christian churches of Europe. For the reasons I have indicated the Roman church was peculiarly vulnerable to the temporal consequences. Only in the twentieth century, and only really after the Second World War, could it again begin to lead a "normal" life.

11.4 Is Adam Smith compatible with St Thomas?

I have argued that the intellectual isolation of Roman Catholic social thought, signalized and to a great extent reinforced by *Rerum Novarum* is not a necessary consequence of Christian theology as understood by the Church of Rome but merely the resultant of two contingent events. The geopolitical location of the papacy rendered it peculiarly vulnerable to the *temporal* consequences of the French Revolution both in the short and in the long run. And the accidental (and quite understandable) failure of Pius VI and VII and their advisers to recognize that the Jacobins had no right to invoke Enlightenment philosophy in support of revolution seriously handicapped the papal response – when it eventually came with Pius IX and Leo XIII – to the *intellectual and spiritual* consequences. The repudiation of socialism arose from the first of these circumstances. There was for most of the century an urgent need for an ideological defense against the repeated attacks by most of the European states upon the property of the church. The adoption of Thomistic philosophy, thereby insulating "social teaching" from economic theory and from any ability to appreciate the virtues of a market economy, arose from the second. It seemed essential to Leo to supply a "philosophy" owing nothing to the Enlightenment within which to reassert Christian theology as against "progress, liberalism and modem civilization."

The first of these is no longer an issue. The Church of Rome has found a stable economic position in the twentieth century and come to terms with it. The attack upon socialism is muted if not abandoned. There seems no compelling reason why the second need continue to be. It is not beyond the ingenuity of philosophers to discover, after all, that Adam Smith is compatible with St. Thomas. Or, failing that, that other philosophical systems will serve as well to formalize the Christian understanding of social phenomena.

12
Market Social Order and Christian Organicism

12.1 Introduction

Centesimus Annus marked the centenary of *Rerum Novarum*,[1] the encyclical commonly regarded as inaugurating modern social teaching in the Church of Rome.[2] In the spirit of its exemplar, it identified "New Things" of today, most obviously "the events which took place near the end of 1989 ... and the radical transformations which followed," variously described as the "collapse ('crisis,' 'fall') of Marxism," "the defeat of so-called 'real socialism'," and "the failure of communism."[3] In explaining this *débâcle* the Pope endeavored to show that it was caused by a serious moral and spiritual error of socialism – "the suppression of private property" – and that this outcome had been predicted by *Rerum Novarum*.[4] As a consequence, *Centesimus Annus* provided a more detailed consideration than any previous papal document of the economic benefits of "the human rights to private initiative, to ownership of property, and to freedom in the economic sector."[5]

These features caused it to be widely acclaimed, especially in the United States, as a belated recognition in Vatican circles of the virtues of capitalism and the essential compatibility of a market social order with Christian anthropology: "a ringing endorsement of the market economy. ... Capitalism is the economic corollary of the Christian understanding of Man's nature and destiny."[6] The Pope "capture[d] the spirit and essence of the American experiment in political economy."[7] But other readers saw *Centesimus Annus* in a very different light. It "focuses on the practical materialism of market economies, their unbridled search for profit, consumerism, and selfishness without solidarity [and] reflects the Pope's concern that what he calls the virus of Western capitalism now threatens to contaminate the lands of Eastern

Europe."[8] Kenneth Minogue identified one passage as "the purest expression of central planning as the solution to big problems" and judged that "This encyclical resembles a lot of socialist literature in assuming that the basic question is management and distribution."[9]

How could *Centesimus Annus* afford scope for such divergent interpretations of its message? I shall argue that *Centesimus Annus* superimposes a set of propositions appreciative of a market social order upon an underlying body of traditional papal social teaching that is alien to, and fundamentally incompatible with, such an order. For whereas traditional Christian social theory envisages society as an *organism*, Enlightenment social theory – in which political economy is an *explanans* and market order an *explanandum* – views society as an *habitat*. My article will establish the intellectual context of papal social teaching, contrast this with the presuppositions of a market social order, exhibit and explain the conceptual dissonance within *Centesimus Annus*, and identify the social–theoretic and theological agenda the encyclical sets for normative social theory.

12.2 The intellectual context of papal social teaching

Papal Social Teaching as we now understand that term was born in 1891 as a delayed reaction to the political and economic upheavals of the nineteenth century.[10] The presupposition of *Rerum Novarum* and the tradition it both receives and transmits is that human society may be properly thought of as a single organism, the "body politic," carefully distinguished from the merely contingent "state."[11] In *Quadragesimo Anno* of Pius XI, "it will be possible to say in a certain sense even of this body what the Apostle says of the mystical body of Christ: 'The whole body (being closely joined and knit together through every joint of the system according to the functioning in due measure of each single part) derives its increase to the building up of itself in love'."[12] Pauline Christology and ecclesiology supply the controlling image in Christian social theory. As an English political theorist had put it three hundred and sixty years before, "yf al the partys of the cyty wyth love be not knyt togyddur in unyte as membres of one body, ther can be no cyvylyte ... [but] there ys perfayt cyvylyte ... where ... al the partys ... be knyt togyddur in parfayt love & unyte, every one dowing hys offyc & duty."[13]

It is an implication of this organicist view of society that its various members must consciously and deliberately work in harmony for the common good, not to mention that there should be a controlling intelligence, or "head," to rule the body. According to *Rerum Novarum*, for example, "A family, no less than a State, is a ... true society, governed by an authority peculiar to itself, that is to say, by the authority of the father."[14] Both the father's authority and that of political rulers come from God and are contingent upon their own obedience to "the eternal law of God" or to the "natural law." In his earlier encyclical *Libertas* (1888), by contrast, Leo XIII taught that "the highest duty is to respect authority, and obediently to submit to just law" for "it belongs to the perfection of every nature to contain within itself that sphere and grade which the order of nature has assigned to it, namely that the lower should be subject and obedient to the higher."[15]

We should note that there is nothing uniquely papal or even Roman Catholic about this doctrine. It is the standard form of virtually all pre-Enlightenment Christian social thought and is, for example, identical to the political theory of the English Coronation liturgy and the Anglican *Book of Common Prayer*.[16] The key concept in an organicist theory of society is what eighteenth-century authors in Britain such as Samuel Johnson and William Paley recognized as the "principle of subordination."[17] For as Archbishop Secker admonished his ordination candidates in 1769, "Without union there cannot be a sufficient degree either of strength or of beauty: and *without subordination there cannot long be union*. Therefore obey, as the apostle directs, them that have the rule over you."[18]

It has often been remarked that the metaphor of the body politic could be and was used by seventeenth-century authors with no Christian-theological underpinnings.[19] Sir William Petty, for example, began his *Political History of Ireland* by noting Francis Bacon's "judicious parallel ... between the *Body Natural* and the *Body Politick*" and made much use of the latter in his own work.[20] Sir Robert Filmer and Jean Bodin each made the body politic resemble the family, and thus extended patriarchy from the latter to the former – exactly as did Leo XIII in *Rerum Novarum*.[21] At least the first of these might be thought of as modern examples of the Aristotelian organicism that Minogue attributed to *Centesimus* Annus.[22] In my opinion, however, the type of organicism characteristic of papal social teaching, though undoubtedly affected by Aristotle through the influence of scholasticism, is primarily Christian and specifically Pauline.

An instructive example of Christian organicism that seems to owe nothing to Aristotle is Jonathan Swift's famous sermon on mutual subjection (c. 1720). The sermon presents the familiar "comparison which St Paul maketh between the Church of Christ and the natural Body of

Man: for the same Resemblance will hold, not only to Families and Kingdoms, but to the whole Corporation of Mankind." We learn from this sermon, moreover, "the Nature of that Subjection which we all owe to one another. God Almighty hath been pleased to put us in a state, where we have perpetual Occasion of each other's Assistance." And "where there is a mutual Dependence, there must be a mutual Duty, and consequently a mutual Subjection."[23] For Christians mutual subjection is an act of love. There is negligible difference at this point between Swift in 1730 and either Thomas Starkey's *Dialogue* two centuries earlier, or Pius XI's *Quadragesimo Anno* two centuries later.

Now it might be argued that an organicist view of society, implying as it does a necessary controlling oversight by the highest temporal authorities, would be congruent with a collectivist, specifically socialist, economic order. Yet, from *Quanta Cura* (1864) of Pius 1X[24] down to *Centesimus Annus* (1991) papal teaching has been hostile to socialism. In part this was simply because it was important for the Church of Rome to provide an ideological defense against the wholesale plunder of its property by European governments in the nineteenth century.[25] At bottom, however, the popes rejected socialism because they believed it represented a false version of that corporate view of state and society, which both they and socialists maintained against liberals. This position is clarified in *Centesimus Annus:* "the fundamental error of socialism is anthropological in nature. Socialism considers the individual person simply as an element, a molecule within the social organism, so that the good of the individual is completely subordinated to the functioning of the socio-economic mechanism."[26] Both Socialists and John Paul II view society as a whole. But, for socialists, the whole is a machine, and "social engineering" the characteristic mode of governance; whereas for the Pope, the whole is a living body. Socialism, for all its emphasis upon the state, views human beings atomistically, and therefore – as Bishop John Keane pointed out in his 1892 review of *Rerum Novarum* for the *Quarterly Journal of Economics* – the objection to socialism is really the same as the objection to capitalism. A "false ... individualism had its birth in English Deism, grew into the system of *laissez-faire* ... and now comes back to the starting point of its vicious circle in ... State Socialism."[27]

12.3 Presuppositions of a market social order

Thanks to Friedrich von Hayek,[28] it is now commonplace to speak of a "theory of spontaneous order"[29] originating in David Hume's development of certain pregnant insights contained in Mandeville's *Fable of the Bees*. This is because the two most important ideas that Hume learned from or, at any rate, shared with Mandeville are, first, "the narrow Bounds of human Knowledge, and the small Assistance we can have, either from Dissection or Philosophy, or any part of Mathematicks to trace and penetrate into. ... Cause a *priori;*" and, second, that "all human Creatures are sway'd and wholly govern'd by their Passions ... even those who act suitably to their Knowledge, and strictly follow the Dictates of their Reason, are not less compell'd so to do by some Passion or other."[30] It is evident that in a world populated by beings of exiguous knowledge and impotent "reason," social order cannot be the result of human design but must emerge – if at all – as the unintended consequence of human slavery to the "passions."

Though Mandeville had muddied the waters by improperly describing the pursuit of luxury as "private vice" leading to "publick benefit," and though in consequence the *Fable* and its author were denounced from the pulpit and the book placed by the Vatican on the *Index Librorum Prohibitarum*, its central message was quickly assimilated by the more powerful minds of eighteenth-century Britain: Joseph Butler, David Hume, Samuel Johnson, Adam Smith, Josiah Tucker, William Paley, Edmund Burke, Edward Gibbon, and T. R Malthus, among many others. That message is simply that the multifarious activities of any large human society – most notably its economic activities – arise *and can only arise* in a gradual, unplanned, accidental, piecemeal fashion in response to the incentives to individual, self-regarding action created by others' needs, wants, and desires. Mandeville's well-worn but untenable claim that the "wealth" of a commercial society is inversely related to its "virtue" was soon filtered out by Adam Smith's careful description in *Theory of Moral Sentiments*, following Josiah Tucker, not to mention Aristotle and the Stoics, of "self-love" as possibly virtuous rather than certainly vicious.[31] Dependence, even as "mutual subjection," came to be seen as morally less worthy than self-reliance.

It was the achievement of the Scottish Enlightenment, above all of David Hume, to transform and develop Mandeville's ideas into a theory that explained not only the market but all of human social life. In Part II of the *Fable* Mandeville pointed out that "we often ascribe to the Excellency of Man's Genius, and Depth of Penetration, what is in Reality owing to length of Time, and the Experience of many Generations ...".[32] Spontaneous order is far more than a short-run economic phenomenon, a mere matching of consumers' demands with producers' outputs. The entire structure of society, Hume argued at various places in his writings – our laws, morals, arts, science, and religion – is the outcome of a vast evolutionary process in which what has proved to be of service to human societies has survived, and what has proved unserviceable has perished. And not simply the public institutions of society, but even the internalization in individual consciousness of the "artificial virtues" of justice, truthfulness, fidelity, chastity, good manners, and the like, are the result of a kind of evolution.[33]

It is evident that this social theory is drastically different from both the Aristotelian and/or Christian organicism discussed above, and from any form of mechanism, either Newtonian or Socialist. Human society is not an *organism*, nurtured by wise and far-seeing statesmen reliably informed by their reason of a natural law given by God to his creatures. Nor is it a *mechanism*, intelligible to science and subject to operation by skillful and disinterested managers and politicians who can know the social welfare function and seek the common good. John Paul II was wide of the mark in supposing that "the rationalism of the Enlightenment" viewed "human and social reality in a mechanistic way."[34] For Hume, Smith, and their followers, society is rather an ecosystem or *habitat*, in which the multifarious customs, arrangements, and institutions evolve blindly from an unknown past to an unknown and unknowable future. The scope for conscious political direction is modest and chiefly confined to maintaining the rules defining property rights subject to tentative, incremental improvement of those rules. It is this vision of society, at least since *The Wealth of Nations*, that has informed the modern science of political economy, and which, in turn, has supplied our understanding of the nature and possibilities of a market social order.

It is important to understand that this has little or nothing to do with the political preferences of any individual economist. Methodological individualism and the doctrine of unintended consequences lie at the heart of the economic way of thinking. Even Marx took the invisible-hand idea for granted, differing only from Smith, Malthus, and the other classical political economists in certain important details of his analysis of growth and fluctuations in a market economy. Hence, writes Marx, "capitalist production begets, with the inexorability of a law of Nature, its own negation." For "what the bourgeoisie ... produces, above all, are its own grave-diggers." Therefore, the anticipated collapse of capitalism and its transformation into a classless society will occur, we learn in *Capital*,[35] as an unintended consequence of the myriad individual, self-regarding actions of capitalists and workers.

12.4 Contradiction in *Centesimus Annus*

It is obvious that *Centesimus Annus* does indeed contain many passages, which, taken by themselves, read like "a ringing endorsement of the market economy."[36] "The social order will be all the more stable, the more it takes [Original Sin] into account and does not place in opposition personal interest and the interests of society as a whole."[37] Therefore, an important cause of the collapse of communism "was certainly the inefficiency of the economic system," itself "a consequence of the violation of the human rights to private initiative, to ownership of property and to freedom in the economic sector."[38] Consequently, "It would appear that on the level of individual nations and of international relations the free market is the most efficient instrument for utilizing resources and effectively responding to needs."[39] Wherefore, "The Church acknowledges the legitimate role of profit as an indication that a business is functioning well. When a firm makes a profit, this means that productive factors have been properly employed and corresponding human needs have been duly satisfied."[40]

While these affirmations are virtually without precedent in papal social teaching, the most remarkable innovation is the stress laid on private initiative. One year before the appearance of *Centesimus Annus* an influential society of German social economists noted that "the entrepreneur, who is the center-piece of the market economy system, is not given adequate recognition in the social doctrine of the Catholic Church." Until the promulgation of *Sollicitudo Rei Socialis* (1987), indeed, capitalists had been viewed solely as employers: their "social aspect ranged [sic] before the economic."[41] "We believe," Werhahn's editors continued, "the time has come for Catholic social teaching to catch up on this deficit. Perhaps the centenary of *Rerum Novarum* in 1991 may provide an occasion."[42] Obligingly, John Paul expanded his brief mention of "creative initiative" in *Sollicitudo Rei Socialis*[43] into an explicit acknowledgment of the important function of business enterprise:

> Organizing ... a productive effort, planning its duration in time, making sure that it corresponds in a positive way to the demands which it must satisfy and taking the necessary risks – all this too is a source of wealth in today's society. In this way the role of disciplined and creative human work, initiative, and entrepreneurial ability becomes increasingly evident and decisive.[44]

These passages are intelligible only upon the assumption that the creative, profit-seeking activity of individuals (i.e., "personal interest") can bring about a state of affairs, *without the conscious direction of any central authority or "head,"* in which "the interests of society as a whole" are served by the satisfaction of "human needs" through "the most efficient instrument for utilizing resources."

Furthermore, the passages just referred to are also congruent with the theory of spontaneous order and a view of society, or, at any rate, of the economy, as a habitat. However, they are not congruent with any version of organicism. It is important to observe that there is more to this strand of *Centesimus Annus* than merely catching up with the Scottish Enlightenment. Genuine theological "value added" originates from two intellectual currents: property, enterprise, and freedom are (putatively God-given) "rights;" and the unavoidable dependence of social order upon market-coordinated "personal interest" is a consequence of "the wound of Original Sin."[45] The former allows the Pope to urge that the failure of socialism and the success of markets were predictable in light of *Rerum Novarum*. The latter, which is of far greater potential significance for the coherence of Christian social thinking, permits the integration of spontaneous order into the rich texture of Augustinian theodicy: perhaps a market economy, like the state, is God's *remedium peccatorum*.

Nevertheless, the textual material of *Centesimus Annus* – which affirms the market economy – is interspersed throughout a larger body of text that either ignores or denies it. The possible achievements of a market economy are severely qualified. There are "many human needs which find no place in the market."[46] "Those who fail to keep up with the times can easily be marginalized."[47] The market fosters "consumerism" and "artificial consumption contrary to the health and dignity of the human person;"[48] wherefore "educational ... work is urgently needed, including the education of consumers" and "the formation of a strong sense of responsibility among producers ...".[49] To some extent, the warnings against consumerism and artificial consumption must be read as a recognition – absent in some recent pro-market propaganda – that a market social order can flourish only where all or most participants are informed and motivated by a coherent set of ethical principles. However, the cumulative effect of these qualifications would seem to imply that the "state has the duty of watching over the common good and of ensuring that every sector of social life, not excluding the economic one, contributes to achieving that good ...".[50] For example, "society and the state must ensure wage levels adequate for the maintenance of the worker and his family;"[51] the market must be "appropriately controlled by the forces of society and by the state so as to guarantee that the basic needs of the whole of society are satisfied."[52] In short, there must be "a coherent vision of the common good," which "demands a correct understanding of the dignity and the rights of the person."[53] Even more drastically *dirigiste*, consciously planned, global solutions to economic problems are envisaged. Internationalization of the economy "ought to be accompanied by effective international agencies that will oversee *and direct* the economy to the common good."[54] What is called for is a special effort to mobilize resources, which are not lacking for the world as a whole, for the purpose of economic growth and common development, redefining the priorities and hierarchies of values on the basis of which, economic and political choices are made.[55]

Propositions of this kind are intelligible only upon the following assumptions: (1) There exists a collectively optimal course of action in each national economy and in the world economy as a whole; (2) Some individuals in each society are in a position to identify such action; (3) Such individuals are, or could be, in a position of political authority; (4) This authority could be exerted by them with sufficient power to achieve their ends; and (5) Power would actually be used by those in authority to achieve the social optimum (i.e., common good) rather than their own private ends. These assumptions are congruent with the organicism of traditional Christian social theory. They are not congruent with the theory of spontaneous order, which denies (2), questions (3) and (4), and (like John Paul himself in his recognition of "the wound of Original Sin") casts serious doubt on (5).

The contradictions in *Centesimus Annus* are more than simply a matter of mixed metaphors. So ambitious a body of social theory might well entitle its author to employ the metaphor of *body* in one place and *habitat* in another – even that of *machine* for that matter – to capture certain features of the complexity of human society. The teaching of Jesus, for example, makes use of a great variety of similes and metaphors to convey the many-sidedness of "the Kingdom of Heaven." But the Pope makes little explicit use of metaphor in *Centesimus Annus*. His reasoning is discursive and ought to be taken at face value. Nor is it plausible to suppose that the encyclical envisages a "third way" between collectivism and the market, which might account for its mixture of affirmation and rejection with respect to the latter. That possibility was firmly rejected in *Sollicitudo Rei Socialis*.[56] In *Centesimus Annus*, the Pope repeated that "The Church has no models to present;

models that are real and truly effective can only arise within the framework of different historical situations through the efforts of all those who responsibly confront concrete problems in all their social, economic, political, and cultural aspects as these interact with one another."[57] Save only for the single adverb *responsibly*, this statement is pure Humean social theory. The contradictions of *Centesimus Annus* are real and need to be understood.

12.5 A Thomistic logic of Christian organicism

In a previous essay I have reported and explained the intellectual isolation of the Church of Rome between the period of the French Revolution and the last decade of the nineteenth century.[58] All versions of liberalism were rejected, thus leaving papal social teaching securely anchored in the tradition of Christian organicism as described in the first section of this article. With admirable consistency the unfashionable political implications of that organicism – authoritarian government, economic inequality, hierarchy in the church, and patriarchy in the family – were unflinchingly accepted and confidently proclaimed. The *Syllabus of Errors*, promulgated by Pius IX in 1864, had forbidden the faithful to believe that "The Roman Pontiff can and should reconcile himself with progress, liberalism, and modern civilization."[59]

The most lucid and powerful statement of papal doctrine in this period was Leo XIII's encyclical *Libertas Praestantissimus* (1888), which John Paul II cited with approval in *Centesimus Annus*.[60] At the end of the eighteenth century the Church of Rome had repudiated "the philosophy of this age" – meaning the Enlightenment – as the chief cause of the French Revolution and the humiliations inflicted on the papacy thereby. In the fourth decade of the nineteenth century the Roman Catholic Church gradually rediscovered Scholastic, particularly Thomistic, philosophy. The genuinely intellectual Leo strongly encouraged this revival in his tenure as Archbishop of Perugia (1846–1877); and within two years of his election as Pope had promulgated the encyclical *Aeterni Patris* (1879), which made Thomism the official philosophy of the Roman Catholic Church. *Libertas* deployed the new philosophical techniques in a full-scale, frontal assault upon political liberalism.

According to that encyclical, all rational beings have a natural liberty: "the faculty of choosing means fitted for the ends proposed." But, if the will seeks what is contrary to reason, "it abuses its freedom of choice and corrupts its very essence." This abuse occurs because reason can know the law of one's being – the natural law – as a special case of the eternal law of God. Disobedience is automatically punished by relegation to a lower state of being and, therefore, loss of liberty. As Thomas Aquinas demonstrated, "the possibility of sinning is not freedom, but slavery." It follows that "the nature of human liberty ... whether in society or in individuals ... supposes the necessity of obedience to some supreme and eternal law, which is no other than the authority of God commanding good and forbidding evil."[61] Leo had no difficulty in proceeding on this basis to annihilate the trendy doctrines of nineteenth-century liberalism: sovereignty of the people, democracy, and the so-called "liberties" of religion, speech, the press, and teaching.[62] John Paul II summarized Leo's doctrine in a clause of *Centesimus Annus* that "sent shivers" down the back of Milton Friedman: "Obedience to the truth about God and man is the first condition of freedom."[63]

It was precisely the rigor and coherence of Leo's Thomistic analysis of human liberty, deeply influential upon all subsequent popes, including John Paul II, which had as its unintended consequence the subsequent insulation of papal social teaching from political economy and the economic way of thinking. In paragraph 10 of *Libertas* there occurs a passage that perfectly rationalizes Christian organicism, and is fundamentally incompatible with – indeed hostile to – the theory of spontaneous order:

> The eternal law of God is the sole standard and rule of human liberty, not only in each individual man, but also in the community and society which men constitute when united. Therefore, the true liberty of human society does not consist in every man doing what he pleases, for this would simply end in turmoil and confusion, and bring on the overthrow of the State ...[64]

According to this way of thinking, it is the function of "competent authority" in society to frame "human law," consistent with that eternal law that comes "before men live together in society," which will bind "all citizens to work together for the attainment of the common end proposed to the community, and forbidding them to depart from this end ...".[65]

It is evident that there can be no compromise between this way of conceiving human society and that of political economy and modern economics. Nor can there be any possibility of an eclectic combination of the two ideas. Either we accept the assumptions of David Hume and his successors, according to which, human knowledge is and must be too feeble to know the consequences in a large society of each individual's actions; or we accept the assumptions of Leo XIII and Thomas Aquinas, according to which, any "competent authority" can and ought to know enough to be able to propose a "common end" to his or her obedient subjects and to constrain them to pursue it. It is not apparent that the logic of one system is any better or worse than that of the other. They differ drastically in their assumptions about the range and power of human knowledge.

12.6 A difficulty of "Christian social thinking"

The conceptual dissonance I have attempted to identify in *Centesimus Annus* presents in sharp focus a problem that has beset Christian social thinking since the first appearance in the eighteenth century of the theory of spontaneous order. How can Christians, being members of a body in which "al the partys ... be knyt togyddur in parfayt love & unyte," permit and approve a social order in which the production and distribution of goods and services (not to mention the very nature of society itself) is the result not of any human design for the common good but of many private, individual actions motivated chiefly by self-love? Neither David Hume nor Adam Smith was interested in this question, although T. R. Malthus certainly was. His preliminary and less than satisfactory attempt to work out a theodicy of unintended consequences gave rise to a body of Christian Political Economy after 1798 to which William Paley, J. B. Sumner, Edward Copleston, Richard Whately, and Thomas Chalmers made important contributions.[66]

According to this tradition, Original Sin and redemption by Christ imply that human life on earth is a state of "discipline and trial" for eternity. Private property is economically necessary, suited to human nature, and consistent with Scripture. Private property, together with the competition produced by scarcity results in the market economy. *The efficacy of the market in organizing human action for wealth creation is evidence of divine wisdom and mercy in turning human frailty to socially beneficent ends.* Though the market produces some poverty and inequality, these may be regarded, for the most part, as a deliberate contrivance by a benevolent God for bringing out the best in His children and training them for the life to come. The impossibility of achieving social progress by legislation is evidence both of God's design – in the creation of the self-regulating economy – and of the moral and religious obligation of Christians to practice charity and compassion. Moreover, true happiness even in this life is largely independent of wealth and station.[67]

Readers of *Centesimus Annus* will see that certain features of this theology of the market have at last begun to be incorporated into papal social teaching. This can be seen particularly in John Paul II's crucial, quasi-Augustinian insight that the market economy is an application of God's remedy for "the wound of Original Sin."[68] But papal social teaching also continues to insist upon a Christian organicism that was largely neglected by the (Protestant) authors of Christian Political Economy, most of whom accepted an ecclesiology that was sketchy, not to say, defective, even by the standards of their own churches. In my opinion, the Christian organicism of *Centesimus Annus* reflects a view of human nature, both individually and collectively, which, in one way or another, is necessary for a Christian understanding of the relations among God, Christ, and the human race.

12.7 Conclusion

If my analysis in this essay is correct, it would seem to follow that there is an important theological task, not only for John Paul II and his advisors, but for all Christians who wish to affirm the benefits of a market social order. The latter depends on a social theory based on a view of society as a habitat, and which alone can explain and rationalize not only the market economy but also liberal democracy, political pluralism, and a wide variety of "liberation" movements. But Christians also need an ecclesiology based upon the view of the Christian Church as the Body of Christ. This organicist ecclesiology is essential for maintaining a coherent soteriology and for undergirding the indispensable ideas of co-inherence, solidarity, and unity, not only of the People of God but of the whole of human society. In some way, an individualistic social theory and an organicist ecclesiology must be brought into harmony and made to coexist.

13
Establishment Social Thinking

"Establishment" describes a legal structure, a theological idea, and a political culture. Because of the second and third of these, it also serves to label a set of attitudes and expectations that may survive long after the first has been dismantled. Attitudes and expectations as much influence the normative social theory of a religious body, specifically the manner in which it combines theological with social-scientific elements, as do formal theological or constitutional doctrine. Therefore, though the establishment of national churches is now merely vestigial, it is instructive to examine Christian social thinking of the present day, which has its origin in that tradition.

I shall first explain why I have chosen to concentrate on the (Protestant Episcopal) Church of England and other Anglican churches, with only occasional glances at the (Presbyterian) Church of Scotland and the (Lutheran) churches of Scandinavia. Then I shall examine the historical background of present-day Anglican social thinking. Finally, I shall analyse the "establishment" way of combining economics with theology in normative social theory and consider some theoretical objections to that enterprise.

13.1 The Church of England as a paradigm

The medieval Western church was a trans-national state with all the usual apparatus of law and government. The effect of Protestant reformation in the sixteenth century was to nationalize religion and politics in European countries furthest from Rome, the seat of government. The process was most complete in Scandinavia and the British Isles, and its essential nature was most clearly seen in Sweden and in England. For, in these two countries, existing episcopal hierarchy and corresponding diocesan and capitular institutions, canon law, even the liturgy and its ceremonial, were all retained with only small modifications; doctrinal innovation was minimized (and, somewhat questionably, even denied by appeal to the Vincentian canon), the decisive change being purely and simply a relocation of supreme temporal authority over the local church from pope to prince.

Nationalization of the church led to a more intimate coexistence of "church" and "commonwealth" than had been the case before the Reformation. The two were never confused conceptually, for "church" was still seen to be a divine society, transcending space and time, uniting the living and the dead;[1] whereas "commonwealth" was understood in its literal sense as a temporal association of human beings for the common good. Yet, in England at any rate, the two were identified politically, if only contingently. Richard Hooker, the greatest and most authoritative Anglican apologist of the Elizabethan period, acknowledged that "a commonwealth is one way, and a church another way, defined." However (he continues with an example), a "schoolmaster" and a "physician" are defined differently, yet "both may be one man." Therefore, "in such a politic society as consisteth of none but Christians, ...one and the selfsame multitude may in such sort be both." This is not the case "under the dominion of infidels;" nor is it "in those commonwealths where the Bishop of Rome beareth sway." But, "in this realm of England; ... with us one society is both Church and commonwealth;" and "our Church hath dependency upon the chief in our commonwealth ... according to the pattern of God's own elect people."[2]

The Elizabethan and Jacobean union of church and state "in this realm of England" was resisted by dissenters both Protestant and papist. During the civil war of the 1640s, an attempt was made by English and Scottish Presbyterians to impose a "Reformed" (i.e., Calvinist) rather than an Anglican unity upon both kingdoms. Its failure, and the Restoration – of monarchy, episcopate and liturgy – in 1660, produced a four-way divergence in British Christianity. After further vicissitudes, Reformed Protestantism was permanently established by the Scottish Parliament in the Church of Scotland by the end of the seventeenth century. In England (including Wales) and Ireland, Anglican Churches of England and Ireland were re-established by the English and Irish Parliaments. Papist dissenters in Ireland (a large majority) were severely discouraged by legal disabilities. Protestant dissenters in England (a small minority) were tolerated but excluded from the universities and from civil or military office.

The Restoration settlement of 1660 to 1662 ratified a legal and constitutional union of church and state in England, which determined political culture until well into the nineteenth century, and which still affects the thinking – and even more the unexpressed attitudes and assumptions – of many today. The coronation of the sovereign, following a ritual based on the fourteenth-century *Liber Regalis,* symbolizes in the most powerful way, now vastly magnified by television, the temporal headship of monarch over the church.[3] The membership of bishops in the House of Lords, and the attention paid by the media to their public utterances, continually advertises the part played by the church in the legislation, governance, and good order of the state. The historic monopoly of the Church of England over higher education, only finally abandoned in the twentieth century, created a governing class the great majority of which before the 1980s had grown up in the Establishment culture of "public" school and Oxford or Cambridge college.

An important part of that Establishment culture is the intellectual legacy of Anglican social thinking since the Reformation. For, though the hierarchy as a whole seldom if ever attempted to formulate doctrine before the decennial, pan-Anglican Lambeth conferences that began in 1867, those individual clerics whose works were studied at the three universities (Oxford, Cambridge, Dublin) often possessed an authority greater in practice than that of any formal pronouncement. Several of these were influential in developing the new science of political economy in the eighteenth and nineteenth centuries. George Berkeley, Josiah Tucker, William Paley, T. R. Malthus, J. B. Sumner, Edward Copleston, and Richard Whately were all clerics; all save Tucker began their ministries with academic appointments, and all except Malthus became dignitaries. Two (Berkeley and Copleston) were made bishops, and two (Sumner and Whately) were made archbishops. Their pioneering work is the first attempt to combine economics with theology in the construction of normative social theory.

Within the Church of Scotland only Thomas Chalmers (1780–1847) made important contributions to this enterprise. As for Sweden, although there was a notable efflorescence of political economy after 1900 in the work of Cassel, Wicksell, Hecksher, and younger members of the "Stockholm School," none of these had any interest in theology, and the normative aspects of their work were more influenced by secular political ideology. By this time, moreover, the intellectual and cultural hegemony of the Church of Sweden, never so powerful and significant as that of the Church of England, had largely disappeared.

For these reasons, and also because of the world-wide dissemination of Anglican culture in many governing *elites* throughout the former British Empire, I shall take the Church of England as a paradigm of the "Established Protestant Church" for purposes of this inquiry.

13.2 History and the present state of Anglican social thinking

The gradual emergence of political economy during the eighteenth century as a distinct branch of "moral science" not only shed new light on the age-old problems of civil government, but it also raised new theological questions. An Anglican attempt to appropriate the first whilst dealing satisfactorily with the second created a powerful and long-lived tradition of Christian social thinking by no means dead today. A Romantic reaction to political economy in the nineteenth century gave birth to a loosely defined "Christian Socialism," which won the affection of many upper-class Anglicans, including several bishops and at least one prime minister, and remained influential until 1945 and beyond. By that time, however, the seeming success of Keynesian economics had attracted Anglican thinkers once again to the possibility of a fruitful alliance of political economy and Christian theology, and a new orthodoxy began to take shape. We must consider each of these phases in more detail.

"Christian political economy"

In some ways, the great philosopher George Berkeley (1685–1753), Fellow of Trinity College Dublin and later Bishop of Cloyne in the Church of Ireland, is typical of the intellectual tradition I wish to describe. His philosophical work was a consequence of well-recognized professional duty to defend the reasonableness and coherence of Christian belief. His economic writings were born of his experience as chief pastor of a poor, rural diocese in a "Third World" country, and they were consistent with his parliamentary duties as a member of the Irish House of Lords. *The Querist*[4] addresses the problem of economic development upon the assumptions, first, that the attitudes of the population are the most important determinant of wealth-creation; second, that it is the "aim of every wise State ... to encourage industry in its members." Berkeley apprehended "some censure ..., for meddling out of my profession; though to feed the hungry and cloth[e] the naked by promoting an honest industry, will perhaps be deemed no improper employment for a clergyman who still thinks himself a member of the commonwealth."[5] *The Querist* was preceded by *An Essay towards Preventing the Ruin of Great Britain*[6] written in the aftermath of the South Sea Bubble, and *Alciphron*,[7] the second dialogue of which analyses the defects of Bernard Mandeville's economic argument for "vice." It was followed by *A Letter on the Project of a National Bank*.[8]

Berkeley's economic writing displays his usual acumen, yet it is not a fully developed specimen of Christian political economy. First, there is no recognition of the "spontaneous order" that may arise when social institutions such as markets exist to coordinate the private acts of individuals motivated by "self-love." The dialogue against Mandeville repeatedly ignores this possibility. Second, the concept of scarcity, eventually definitive for economics, is absent. And because of this, third, no possibility of theological dissonance between Christianity and the facts of economic life is recognized.

Some awareness of the first of these was evinced by Josiah Tucker (1713–1799), a Whiggish but orthodox Dean of Gloucester whom even Karl Marx later called "an honourable man and a competent political economist."[9] Four years before Adam Smith's *The Theory of Moral Sentiments*,[10] Tucker explained that "the main point to be aimed at, is neither to extinguish or enfeeble self-love, but to give it a direction, that it may promote the public interest by pursuing its own!" If this can be arranged, "every individual (whether he intends it or not) will be promoting the good of his country, and mankind in general, while he is pursuing his own private interest."[11] Tucker's crucial insight seems to have been suggested by the theological work of his friend and patron, the celebrated Bishop Butler.[12] Elaborated by a clear explanation – which Tucker himself neglected to provide – of the way in which competitive markets may supply this "direction," it eventually became the leading theme of Adam Smith's *An Inquiry into the Nature and Causes of the Wealth of Nations*.[13]

The final stage of Anglican social thinking before the clear emergence of political economy may be seen in the work of William Paley (1743–1805), Fellow of Christ's College Cambridge and later Archdeacon, then Chancellor, of Carlisle. Paley's *The Principles of Moral and Political Philosophy*[14] was based on his Cambridge lectures during the decade after 1766 and exhibits more clearly than any other work of the time the educational and intellectual assumptions of the English Establishment. The university exists to form the minds and the moral sensibilities of the next generation of clergymen, magistrates, and legislators. Under the guidance of godly and learned tutors, and subject to a discipline of twice-daily prayer in college chapel, all undergraduates must read Holy Scripture and Divinity, selected Latin and Greek classics, and at least the rudiments of moral and political philosophy. Paley's *Principles,* which was immediately adopted by the University of Cambridge as a compulsory text for all undergraduates, expounds the last of these with brilliant lucidity. Duty to oneself, to one's neighbors, and to God occupies five books. Book 6, which is the last, treats of the "Elements of Political Knowledge" and considers the duty of submission to civil government, continues with material on the constitution, the administration of justice, and the establishment of the Church, and ends with two chapters entitled "Population and Provision" and "War and Military Establishments." The underlying, epistemological assumption at Cambridge in 1785 remained what it had ever been. Both individual conduct and a social order pleasing to God can be known and taught, and their study may properly be called a science.

Because a right understanding of how the economy works is necessary to an informed judgment about social ethics, economic study is a branch of Christian moral science. Paley's chapter entitled "Population and Provision,"[15] which sets out a sophisticated version of Berkeley's economic doctrine that generalizes Mandeville, is obviously acquainted not only with Tucker but also with Hume, Steuart, and possibly Smith, and foreshadows many of Malthus's more important ideas. It begins with the premise that "the final view of all rational politics is, to produce the greatest quantity of happiness;" argues that the level of population is a reliable index of social welfare; and analyses the determinants of population in a two-sector ("provisions" and "luxuries") model of inter-regional trade. Impressed by its technical mastery, Keynes felt that "perhaps, in a sense," Paley (rather than Malthus) "was the first of the Cambridge economists."[16] Yet Paley's results depend on a quasi-Keynesian assumption of demand-led supply. There is no resource scarcity in his model.[17] It remained for his successor, T. R. Malthus (1766–1834) to transform Adam Smith's reassuring study of "the nature and causes of the wealth of nations" into the politically and theologically disturbing "dismal science" of political economy.

It is now generally accepted that by postulating "a world characterized by scarcity and *necessary* inequality of access to resources,"[18] Malthus's anti-perfectibilist argument in the first essay, entitled *Essay on Population* (1798), marks the beginning of "political economy" as that term came to be used in the nineteenth century. Diminishing returns are of the essence. Increasing returns to scale – eventually recognized as a serious analytical embarrassment – are relegated to a special case. It is also widely agreed that modern economics is continuous with classical political economy in this respect at least: "economics" is about coping with scarcity.

Berkeley, Tucker, Paley, and their contemporaries had seen no inconsistency between a moral science of wealth-creation and the Christian theology it was their duty to study and teach. But the dominance of scarcity in human affairs, recognized scientifically for the first time in the work of Malthus, presented an ancient theological problem in a novel and threatening form. The ineluctable laws of nature produce poverty and inequality: natural fecundity in finite space. *Why then does God create a world in which most human beings must live in "misery" or "vice"?* Political economy was attacked from the first as "hostile to religion" and viewed with alarm by Romantics throughout the nineteenth century.

Malthus's own attempt to deal with the problem in the first *Essay* was evidently unsatisfactory, and for the next thirty-five years a distinguished succession of Anglican thinkers constructed an intellectual synthesis of classical political economy with Christian theology that I have labelled "Christian Political Economy" and considered at length in previous work.[19] First Paley in response to Malthus, then Malthus himself in subsequent recensions of the *Essay on Population,* later J. B. Sumner (1780–1862), Edward Copleston (1776–1849), and Richard Whately (1786–1863), gradually hammered out a coherent social doctrine for the guidance of legislators in a Christian society. I have elsewhere summarized that doctrine as follows:

> Poverty and social inequality are the inevitable outcome of scarcity: more particularly of population pressures in a world of limited resources. Because of original sin and redemption by Christ, human life on this earth is to be regarded as a state of "discipline and trial" for eternity. Though poverty and inequality may entail some genuine suffering – to be accounted for by the Fall – they may be regarded, for the most part, as a deliberate "contrivance" by a benevolent God for bringing out the best in His children and so training them for the life to come. The social institutions of private property and marriage are economically necessary (and indeed inevitable), suited to human nature, and consistent with scriptural teaching. A combination of the institution of private property with the competition produced by scarcity, results in the market economy. The efficacy of the latter in organizing human activity for the maximization of wealth is evidence of the divine wisdom and mercy in turning human frailty to socially beneficent ends. The

> impossibility of achieving social progress by legislation is evidence both of "design" – in the creation of the self-regulating economy – and of the moral and religious need of Christians to practice charity and compassion. True happiness in this life is largely independent of wealth and station. But in any case wealth is positively correlated with moral worth, itself the result of faithful Christianity. Universal Christian education is then of the highest practical importance, and an essential feature of the traditional union of church and state.[20]

The particular contribution of the Scotch Presbyterian Thomas Chalmers (1780–1847) was to emphasize the last of these points and to provide an original and ingenious, though flawed, *economic* argument for the establishment of the church.[21]

Christian socialism

The term "Christian socialism," which to Marx and Engels was "but the holy water with which the priest consecrates the heart-burnings of the aristocrat,"[22] was the self-designation first of a short-lived movement in the Church of England from 1848 to 1854,[23] and second of its revival as an ideologically vague alliance of Anglican and other theologians and social activists with "a discontinuous and fragmented history" until well into the twentieth century.[24] Organized in the Guild of St. Matthew (1877–1909), the Christian Social Union (1889–1919) and Church Socialist League (1906–1924), its ideas became highly influential among the Anglican *élite*. By the end of the century, about two-thirds of all Oxford undergraduates belonged to the Christian Social Union, and nearly a third of all episcopal appointments were drawn from its membership.[25]

All commentators are agreed in regarding the English Christian socialists as naive and confused about economic matters.[26] This is so because the intellectual roots of Christian socialism lie in the Romantic revolt against political economy, having its origins in Southey's (1803) vehemently hostile review of Malthus's second *Essay on Population*. Southey was abetted by Coleridge; and Wordsworth, Hazlitt, and other repentant ex-Jacobins-turned-Anglican-Tory added powerful literary support to the attack on political economy in general.[27] Donald Winch has lately identified "the schism or fault-line, separating economists from the self-appointed spokesmen for human beings."[28] By the 1830s, Winch states:

> One of the enduring fault-lines in British cultural debate had now been created, and where Coleridge and Southey had led, Carlyle,

> Ruskin, and their nineteenth and twentieth-century admirers followed, often doing so with the same willful disregard for what their chosen antagonists were actually saying.[29]

The slightly heterodox Anglican theologian F. D. Maurice (1805–1872) played a leading part in the first phase of Christian socialism and was acknowledged by a later generation – B. F. Westcott (1825–1901), Henry Scott Holland (1847–1918) and Charles Gore (1853–1932) – "as the great progenitor of their ideals."[30] That Coleridge was crucial in the formation of Maurice's own theological and political ideas is well known.[31] A full century after Southey's review of Malthus, a leading Christian socialist of the day, W. E. Moll (c. 1857–1932), "decried the false political economy which 'teaches men to say that there are those for whom God has placed no plate at the banquet of life.'"[32] This is a direct allusion to a notorious passage in the second *Essay on Population* (1803),which Malthus had expunged three years later in the edition of 1806.[33]

Economics and religion in the English Establishment

Hostility to political economy, and identification of the Establishment *élite* with the "self-appointed spokesmen for human beings," developed only gradually during the second half of the nineteenth century. And, though many traces still linger, the process has equally gradually reversed itself in the second half of the twentieth century.

Of the younger originators of Christian political economy, Edward Copleston (Bishop of Llandaff) died in 1849. But J. B. Sumner (Archbishop of Canterbury) lived until 1862, and Richard Whately (Archbishop of Dublin) until 1863. It is not to be supposed that two such intellectually and politically powerful prelates would have abandoned or even seriously modified their views in the face of the short-lived and obscure flurry raised by Ludlow, Kingsley, and Maurice from 1848 to 1854. Moreover, as E. R. Norman has argued, "by the end of the 1830s ... the most influential of the church leaders were all soaked in the attitudes of Political Economy."[34] What seems to have loosened the hold of political economy on the next generation of Establishment men was a combination of circumstances unique to that time and place.

Parliament had long been willing to enact piece-meal reform as occasional exceptions to *laissez-faire* theory. Country gentlemen, lords, and bishops had no warm love for the new class of industrial capitalists and could usually be persuaded to legislate government intervention when presented with flagrant examples of injustice or misery. During the Hungry Forties, such "exceptions" came thick and fast. Population doubled in the first half of the century, most of the increase occurring in London and the new industrial cities.

A combination of low wages with over-crowding, jerry-building, and a total disregard for private or public sanitation led to the cholera epidemics of the thirties and forties. The ruling class was compelled to attend. "To maintain the traditional patterns of English life" the new cities "must have drains, lavatories, paved roads, houses, policemen, nurses, schools, parks, cemeteries and churches."[35] A stream of legislation was generated, all of it extending the economic responsibility of the state. Meanwhile, the ability of government to meet new demands had been greatly enlarged by the same combination of social, cultural, technical, and material factors that was transforming the private sector. A quarter-century of world war had made government a more powerful and efficient instrument for achieving social goals; and that fact alone suggested new possibilities of state action to the reformers.[36] Norman has shown that bishops such as Wilberforce, Thirlwall, and even Sumner himself, supported Factory Acts and public health legislation whilst continuing to profess their belief in *laissez faire*.

The incipient "fault-line" between economists and most of the educated upper class was widened in the latter part of the century by a secularization of British intellectual culture after the 1860s[37] and the appearance of economics as a distinct profession after 1890. The mid-century conflict between Christian Socialism and political economy had been a debate within the Establishment *élite*. By the beginning of the new century, it had become to a large extent a schism between professional economists and professional ecclesiastics. This may be seen in miniature in the contrast between Alfred Marshall (1842–1924) and R. H. Tawney. Marshall had been destined for the Church and went up to Cambridge in 1861 intending to prepare for Holy Orders. But "after a quick struggle religious beliefs dropped away" and he became an agnostic.[38] Marshall defined the profession of economics in Britain, and his influence was world-wide. His younger colleague Tawney (1880–1962) remained a faithful layman, becoming a leading Christian Socialist, intimate of bishops and influential in church councils.[39] But Tawney believed that capitalism and Christianity are fundamentally incompatible; and though he was a professor at the London School of Economics, he remained on the outer edges of the profession, deeply distrustful of its development. "There is no such thing as a science of economics," Tawney confided to his *Commonplace Book* in 1913, "and

Marshall's talk as to the need for social problems to be studied by 'the same order of mind which tests the stability of a battleship in bad weather' is twaddle."[40]

It was, however, Tawney's lifelong friend William Temple (1881–1944) who opened the door for a *rapprôchement* between economics and religion in the Church of England. Though as a young man Temple had declared that "the alternative stands before us – socialism or heresy,"[41] his later pastoral and political experience as bishop and archbishop gradually convinced him that reality is more complex. Temple's *Christianity and Social Order*,[42] written shortly before he became Archbishop of Canterbury, has been described as "one of the foundation piers of the Welfare State."[43] In the foreword to the 1976 edition, the (Conservative) Prime Minister, Edward Heath, testified to Temple's "immense" influence on all "who were seriously concerned with the social, economic and political problems of his day."

Possibly because of the example of Josiah Stamp (1880–1941), an eminent economist, friend of J. M. Keynes, and powerful public figure who reasserted a modern version of Christian political economy,[44] Temple became aware that faithful Christians, though in accord on theological principles, could be deeply divided on political questions because of legitimate disagreement about the "facts of the case." It is therefore:

> of crucial importance that the Church, acting corporately[,] should not commit itself to any particular policy. A policy always depends on technical decisions concerning the actual relations of cause and effect in the political and economic world; about these a Christian as such has no more reliable judgment than an atheist.[45]

Temple's acknowledgment of the autonomy of "technical" studies of "cause and effect in the ... economic world," together with a certain hesitation – resembling Berkeley's – for "meddling out of [his] profession," caused him to submit the proof-sheets to J. M. Keynes's critical scrutiny. Keynes went out of his way to reassure the Archbishop as to "the right of the Church to interfere[46] in what is essentially a branch of ethics;" and also reminded him of his illustrious predecessors (referred to above). "Bishop Berkeley wrote some of the shrewdest essays on these subjects available in his time; ... Archdeacon Paley is of fundamental importance. The Reverend T. R. Malthus was the greatest economist writing in the XVIII century after Adam Smith; ... Archbishop Sumner's early work was on economics."[47]

Archbishop Temple died before the end of the Second World War, and it was left to a much younger Christian Socialist, Denys Munby (1919–1976), to develop the implications of Temple's recognition of the significance of "technical" knowledge for Christian social thinking.[48] Munby, who had been early affected to the Christian Socialism of Canon V. A. Demant and the "Christendom" group, was made aware of Temple's crucial insight by his friend R. H. Preston, the first Anglican theologian ever with a university degree in economics (London School of Economics). After study at Oxford, Munby became a professional economist, returning in 1958 as professorial fellow of Nuffield College and university reader. Over and above his considerable professional output, Munby produced an important series of books and articles beginning with *Christianity and Economic Problems* (1956) and all dealing in various ways with the matter of that work. A recurrent theme in all these was the incompetence and amateurishness of contemporary Anglican social thought, based on the sound "incarnational" theology of Maurice and his successors but stultified by complete ignorance of the economic facts of life. Though wishing to distance himself from the "old-fashioned" (that is to say, conservative, evangelical and individualist) approach of Lord Stamp, Munby had much the same general message for Anglican leaders of his day: "technical competence" is essential; private enterprise and the profit motive are not all bad; the price mechanism is "the only way in which a large-scale economic system can satisfy the needs of a large population;"[49] the state is neither all-wise nor all-good; and the willingness of prominent Anglican thinkers like Demant, Maurice Reckitt, and T. S. Eliot to grasp at absurd nostrums such as "social credit" can only bring ridicule upon Christian social thinking. In *The Idea of a Secular Society and its Significance for Christians*,[50] Munby criticized the romantic escapism of Eliot's *The Idea of a Christian Society*.[51]

Though influential, Munby never occupied a position of power in the Church of England, and he was evidently a serious threat to many of his contemporaries. But the climate of public opinion was beginning to turn in favor of economics, partly in consequence of the immense prestige enjoyed by Lord Keynes in official circles and the seeming triumph of "Keynesian" policies in fostering post-war prosperity. The moderate interventionism authorized by Keynesian macroeconomics was especially congenial to Establishment leaders; and also to those – such as Keynes himself – who shared their intellectual and political culture even though they had abandoned the religious belief which gave it integrity. For the new political economy seemed to offer a remedy for the economic ills of society without any unsettling need of large-scale, fundamental change. Since the mid-1950s, automatic hostility in the Church to a market social order has gradually dwindled, and since the 1980s there have been signs of a renewed appreciation of Christian political economy.[52]

13.3 Economics and theology in Establishment social theory

Establishment social theory

Munby argued that:

> the problems of economics arise because human beings want to follow different ends, and the means available to them are limited. Economizing only arises in a world where things are scarce, and choice has to be made. In heaven no problem of scarcity arises, and in hell no possibility of choice exists; economics is a science dealing with the conditions of human life in this world.[53]

In so doing, he reasserted the fundamental *empirical* assumption of Christian political economy: the dominance of scarcity in human affairs. He also presented a devastating account of "how a major prophet[54] may fall into major errors in approaching so dreary and mundane a topic as Political Economy, when unarmed with the appropriate kit of burglar's tools."[55] And, in so doing, he reasserted the fundamental *methodological* assumption of Christian Political Economy: that of a demarcation between "religious" and "scientific" knowledge, which safeguards the autonomy of each with respect to the other. Because, in my opinion, the second of these assumptions lies at the heart of any serious attempt within established Protestant churches to combine economics and theology in normative social theory, it requires closer attention

In a previous volume in this series[56] I have described how Richard Whately[57] was led to propound this demarcation, built upon the theological work of his former pupil Samuel Hinds,[58] in response to an ideological threat to Christian political economy after the mid-1820s from the avowedly atheist "philosophic radicals." The radicals' claim that political economy was alone sufficient for rational public policy formulation brought the new science into disrepute with many Anglicans, only too willing to side with the Romantic protest against scarcity already launched by Coleridge and the other Lake Poets. Against his

Tory fellow-churchmen, therefore, Whately had to defend political economy from the charge of being "hostile to religion" and to show that it is *necessary* to public policy formulation. And, against Jeremy Bentham and his disciples, he had to show that political economy, though necessary, is not *sufficient,* for public policy should be based upon a coherent set of values that religion must supply and about which political economy can have nothing to say. But policy can be formulated and implemented only by means of the "technical knowledge" that economics affords and about which (as William Temple rediscovered) religion can have nothing to say. So successful was Whately in this enterprise that his version of the case for the necessarily neutral, "value-free" character of economic analysis – developed and popularized by another of his former pupils, Nassau Senior – passed into the orthodox mainstream of modern economics, and underlies the canonical methodology of J. N. Keynes[59] and Lionel Robbins.[60]

We are now in a position to state precisely the assumptions upon which economics and theology are combined in Establishment social theory:

1. It is assumed that there exists a governing class, or *élite,* which has a well-understood duty to formulate and implement public policy in the light of a generally accepted and legally sanctioned Christian belief and doctrine;
2. The clergy of the national church occupies a clearly defined constitutional position in the state and has the duty (among others) of teaching the Christian religion to all classes in society;
3. It is the duty of the clergy to instruct the governing class in those theological principles that bear upon public policy in a Christian society;
4. The governing class (which itself may – and actually does – include bishops and other higher ecclesiastics) must avail itself of the best available "technical" knowledge to formulate and implement policy;
5. "Technical" knowledge about "the actual relations of cause and effect in the political and economic world" is the intellectual property of a community of experts who may, but who need not, be Christian, "for about these a Christian as such has no more reliable judgement than an atheist;"
6. Regardless of their private religious beliefs, it is the professional duty of the experts to supply the best possible analysis of, and advice about, all relevant political and economic factors.

When a Temple (Christian Archbishop) submits *Christianity and Social Order* to a Keynes (atheist economist), it is evident that each perfectly understands and accepts all the foregoing principles, and also – by implication – Whately's demarcation, which guarantees their coherence.

It is of course possible to criticize this doctrine in a number of obvious ways: assumption 1 may now be false; assumption 2 conflicts with present-day political orthodoxy; assumption 3 may be unfeasible in the pluralistic world of today. It is more fruitful, however, and perhaps of wider interest, to criticize it in terms of assumptions 4, 5, and 6, and of the putative demarcation between religious and scientific "knowledge" that lies behind them.

It is apparent that Establishment social thinking, which takes for granted the scientific autonomy of economics, is at odds with those traditions of Christian social thought – such as neo-Calvinism – which are based upon the assumption that all human knowledge, but especially economics, is subject to correction by religious authority.[61] It would therefore be possible to reject Whately's demarcation and to criticize Establishment thought upon the assumption that economics is or ought to be subordinate to theology. In my opinion, however, the real objections to the Establishment position go much deeper. For not only can Christianity have nothing to say about the "facts" of any state of affairs, as Temple recognized, but also, *it may have nothing to offer the economist by way of "value axioms."* And, if economics were truly "scientific," it could supply none of its own.

Two serious objections to Establishment social theory

The first of my two objections was powerfully stated by Frank Knight[62] and has never been answered to my knowledge. Knight argued that the "gospel of love" was the specifically Christian contribution to ethics; that this must necessarily find expression in relations between individuals; and, that the maxims governing the behavior of *individuals* to one another could not serve to construct a *social* ethic, which must be embodied in impersonal *rules*. Knight held that biblical teaching points toward "a society of antinomian anarchism."[63] Because only a few can fill "the role of parasitic saintliness in the real world, ... evil rather than good seems likely to result from any appeal to Christian religious or moral teaching in connection with the problems of social actions."[64]

Where, then, can we look for ethical guidance? Knight gave no answer. Papal Social Teaching (and its Anglican approximations) relied upon the Natural Law: that knowledge of right and wrong that all can have by the use of right reason. Though not uniquely Christian, the use of right reason can be regarded as sanctioned by "revelation" and hence as a Christian activity. This is where we encounter the second fundamental objection, luminously expounded by the Polish philosopher Leszek Kolakowski in his book entitled *Religion*.[65]

"Right reason" means using language according to the rules established for preserving intelligibility in every-day, or "profane," speech and its extension in "scientific discourse." When those rules are followed faithfully, it is impossible to speak about ethical, religious, or aesthetic matters. Hence the Natural Law is a chimera.

This well-known result of analytical philosophy was formerly used by logical positivists to discredit theology and religion. The chief purpose of Kolakowski's book is to defend both religion and theology in the spirit of Wittgenstein's celebrated maxim: "The language is in order as it is." "Profane" language is "appropriate for reacting to and for manipulating our natural environment;" "sacred" language is "for making it intelligible."[66] It is logically inadmissible to criticize the truth-claims of propositions stated in one language by the criteria appropriate to the other, for "the validity of [religious] beliefs are vindicated differently from the truth of empirical assertions. These two areas of our speaking, thinking, feeling, and acting are fundamentally irreducible to the same store of experience."[67] Hence:

> "the traditional arguments of empiricists are cogent if limited to the assertion 'religious beliefs are empirically empty.' The subsequent verdict, ... 'therefore they are meaningless' is not credible at all. There are no transcendentally valid criteria of meaningfulness and no compelling reasons why the meaningful should be equated with the empirical.[68]

The truth-claims of modern science, indeed, are based upon the criterion of the "efficacy of knowledge," which is logically admissible but essentially arbitrary. To accept that criterion "requires an act of faith[,] and therefore the principle *credo ut intelligam* operates over the entire field of knowledge."[69] The cognitive validity of both science and religion therefore depends on the same, essentially circular process. It is impossible to criticize or appraise the truth-claims of the one by the methods appropriate to the other.

So far, all this is only a more radical, modern-dress version of Whately's demarcation. But Kolakowski's analytical method, by taking us deeper, destroys the possibility of a fruitful collaboration between religion and science such as Whately sought to establish, for if economics affords knowledge of any kind, then it must be "scientific" and not "religious" knowledge. And, at this point, a crucial distinction remarked by Kolakowski becomes relevant.

The disjunction between "fact" and "value" is logically possible in profane language; progress has actually been made in recognizing this disjunction; and the disjunction "ought not to be obscured in the language of science."[70] But the fact/value disjunction cannot exist in sacred language. Kolakowski states: "In the realm of sacred speech, whether an assertion has an ostensibly 'descriptive' or 'normative' form, it has the same validity ... for two reasons. First, because the criteria of validity is the same. It is God's explicit word. ... Secondly, because ... the two kinds of knowledge merge into one."[71] The consequences of this for any collaboration between economics and theology are completely destructive.

As a "scientific," non-"religious" enterprise, economics is stuck with profane language and the necessity of the fact/value disjunction. And the only knowledge, which might be generated by economics, is knowledge of "facts." *But theology cannot generate a set of "values" in abstraction from "factual" perceptions of reality.* Hence, the cognitive paradigm that undergirds Establishment social theory is simply impossible. There can be no division of labor between theologians and economists, the former supplying a set of detached "value axioms" for the latter to use in designing social policy. It is even questionable whether any fruitful conversation upon social policy can occur between a theologian and an economist when the two are wearing their proper hats.

The most radical part of Kolakowski's argument is yet to come. Not only is the profane language of science/common-sense incapable of ethical utterance, but also it is incapable of uttering *truth* of any kind. For, "without the absolute subject, the predicate 'true' can never be vindicated."[72] Hence, "Dostoyevski's famous dictum, 'If there is no God, everything is permissible,' is valid not only as a moral rule ... but also as an epistemological principle;"[73] therefore, "either a God or cognitive nihilism, there is nothing in between."[74]

We may summarize the foregoing objections as follows:

1. Christian theology can make no contribution to social ethics because the latter must be based upon impersonal *rules,* about which Christianity is silent, even hostile (Knight);

2. Economics can make no contribution to social ethics because, like all other scientific inquiry, its necessarily profane formulation makes the truth of its propositions unknowable (Kolakowski);
3. Even if neither objection holds, there can be no collaboration between theology *of any religious variety* (supplying "values") and economics (supplying "facts") because the language-games, which generate each, are incompatible and non-intersecting (Kolakowski).

I do not wish to suggest that these objections are conclusive, only to report that to my knowledge they have not yet been satisfactorily answered. It would seem, therefore, that those who wish to combine economics and theology in normative social theory – in any way that does justice to the cognitive claims of each – have some preliminary, ground-clearing work to do.

14
Economics and the Mutation of Political Doctrine

> ... The private interests and prejudices of particular orders of men ... have given occasion to very different theories of political œconomy ... These theories have had a considerable influence, not only upon the opinions of men of learning, but upon the public conduct of princes and sovereign states. *Wealth of Nations*

14.1 Introduction

The term "political economy" seems to have entered modern discourse for the first time in 1611 in a treatise on government by L. de Mayerne-Turquet.[1] Four years later a fellow "Consultant Administrator," Antoyne de Montchrétien, Sieur de Watteville (c.1575–1621), published his *Traicté de l'oeconomie politique* (1615), which though "a mediocre performance and completely lacking in originality,"[2] marks the beginning of an intellectual enterprise that has continued – with some large ups and downs – to this day. The object of that enterprise is to generalize Aristotle's οικονομικη ("economics") to the level of the πολιτεια ("commonwealth" or "state"). For in Aristotle's *Politics*[3] "economics" is to be construed as "the art of household management" where "household" means a more or less self-sufficient, manorial estate. Hence at the outset "political economy" was an attempt to extend the art of estate management to the needs and resources of a modern nation state, of which France was in the 1600s the foremost example.

It was in this way that Sir James Steuart employed the term in his *Inquiry into the Principles of Political Œconomy* (1767). Although Adam Smith rejected the enterprise as futile and harmful, he accepted the usage, but introduced an important analytical distinction between two senses of "political economy." In the *normative* sense understood by Smith's predecessors and contemporaries (i.e., prescriptions for running France like a manorial fief) the term signifies some "system" of public policy designed to "increase the riches and power" of a country.[4] But in the *positive* sense that is now orthodox though often contested, "what is *properly* called Political Economy" is "a branch of the science of a statesman or legislator:" namely "an inquiry," which is in principle disinterested and open-ended, into "the nature and causes of the wealth of nations."[5]

In this chapter I shall be concerned with the positive sense of "political economy:" that is, as a *body of theory* that purports to explain economic phenomena. For whatever else has changed, one element of continuity that runs from the political economy of Montchrétien and before to that of Joseph Stiglitz and beyond is its inescapable dependence upon theory of some kind.

Another element of continuity is the two-way street that runs between economic theory and political thought and action. In one direction, the "private interests and prejudices of particular orders of men" – not to mention the social and economic circumstances of their time and place – "have given occasion to very different theories of political economy." In the other, "[t]hese theories have had a considerable influence, not only upon the opinions of men of learning, but upon the public conduct of princes and sovereign states."[6]

A complete intellectual history of the relation between economic and political ideas would look carefully in both directions: to the "context" of the conversation, as well as to the "text" of its recorded exchanges.[7] However, I have argued elsewhere that a merely "internalist" attention to "text" can be justified for some purposes.[8] In this essay, therefore, I shall look in one direction only. I shall adopt a rigorously internalist approach, averting my eyes from wars, slumps, classes and cultures, and keeping them fixed on the logic of the ongoing economic-theory conversation. My purpose in so doing is to propose the following strong thesis for debate:

> *The "'new' political economies" of the present day differ sharply in their ideological implications from those of 50 years ago.*

Neoclassical orthodoxy provided the intellectual foundations of a collectivist – or at any rate an interventionist and *dirigiste* – political consensus. But the new political economies have destroyed those foundations, and have replaced them with economic theories far more congenial to an earlier, *laissez-faire* consensus born in the European Enlightenment.

The thesis calls for a three-part division of the history of political economy. My story must begin with the "new" political economy of Adam Smith and his successors: a brief reminder of their decisive rejection of the original *œconomie politique* of such as Montchrétien and Steuart, and of their powerful and influential arguments for widespread *laissez-faire*. It must continue with the long and gradual retreat from *laissez-faire* associated with the emergence and development of "neo-classical," marginalist economics after 1870. For as Joan Robinson once said, "neo-classical economics is the economics of socialism." Many of its leading practitioners believed they had discovered a scientific basis for "economic control:" the management of a national economy by enlightened administrators acting in the public interest – a sophisticated recrudescence of seventeenth-century *œconomie politique*. The story must conclude with a summary of those developments in economic theory in the last 50 years that have destroyed the rational basis of "economic control," and that now predispose the educated public to a favorable reconsideration of Adam Smith's "natural system of perfect liberty and justice."[9]

14.2 "The natural system of perfect liberty and justice"

Shortly before his death in 1797 the great Whig statesman Edmund Burke, who once said that *Wealth of Nations* was "probably the most important book ever written,"[10] could boldly declare that "the laws of commerce, which are the laws of nature, [are] consequently the laws of God."[11]

Some conception of "nature" is central to all eighteenth-century social theory.[12] Human nature is part of nature, and is that which explains the structure and functioning of society. Human nature is characterized by dominance of the "passions" over "reason," by ignorance of the present and by inability to know the future. Of the passions that drive human action, "self-love" is the most constant and therefore normally determinative. Because human knowledge is so limited and fallible, the order we perceive in society would seem to be an unintended consequence of private decisions driven by self-love. Somehow or other, the institutions of any viable society must have evolved so as to harness the self-regarding acts of individuals to the common good (at best), or at any rate to prevent them from destroying the social fabric (at worst). Interference with this "spontaneous order"[13] by central authority is at least as likely to do harm as good. *Œconomie politique* is both futile and mischievous. Not even *le roi soleil* himself is well served by it. Yet social viability under *laissez-faire* depends on the willingness of individuals to observe natural justice: to play the game according to the rules even when the referee is not looking.

These were the presuppositions of Anglophone social thought in the eighteenth century, as collected and developed from Mandeville, Bishop Butler and David Hume by historians and philosophers of the Scottish Enlightenment.[14] Economic theorizing in Britain by the end of the eighteenth century took these thoughts for granted, and developed their implications for public policy in a commercial society.

It is sufficient for my purposes to treat *Wealth of Nations* as a paradigm. Smith owed much to Cantillon, Hume, Josiah Tucker and Quesnay; pointedly ignored Steuart's analytically sound but politically unacceptable contributions; was on friendly terms with Turgot;[15] seems to have been unaware, like most other economists then and later, of Condillac's remarkable achievement; and appeared in print before Paley, whose lectures were delivered at Cambridge between 1766 and 1776 but published only in 1785.

The central feature of the pre-classical analytical tradition is a two-sector, "food" and "luxuries" macro-model,[16] with "Malthusian" population adjustment in response to any change in capital stock, the latter – given the relative size of the putatively "unproductive" public sector – determined by the "parsimony" of a capitalist class driven by "the natural effort of every man to better his condition."[17] All market prices "gravitate" to a "natural" level determined by long-period costs of production;[18] in seeking "only his own gain" each capitalist directs his capital to the most profitable outlet, equalizes the profit rate and so "labours to render the annual revenue of society as great as he can;"[19] for any state of the taste for "luxuries," the "natural" wage rate is an increasing function of the rate of capital accumulation;[20] but parametric increases in the taste for "luxuries" causes production and population to grow, and may permanently increase the real wage. In Smith's version of the model, increasing returns to scale in the "luxuries" sector will amplify the effect upon population and production of any increase in the taste for "luxuries."

Because the interest of "labourers" is "strictly connected with the interest of the society," and because "no society can surely be happy, of which the greater part of the members are poor and miserable,"[21] high and rising wages ought to be a prime object of public policy.

Hence economic growth should be maximized and the taste for "luxuries" encouraged. The more parsimonious are capitalists and the smaller the share of government in national expenditure, the higher the rate of economic growth. The "natural" propensity of capitalists to better their condition through parsimony will flourish best when they are left completely free to pursue their own interest. Both domestic and foreign trade will then grow and diversify, and it is particularly through the latter – as Hume[22] had argued powerfully – that the taste for luxuries is developed. Hence both factors that cause high and rising wages, growth and taste, are interconnected with, and optimally related to, the greatest possible freedom of private enterprise and trade, and the smallest possible participation of government in the economy.

In pre-classical economic theory the primary constraint upon growth was capital. Land was generally treated as abundant. But when the concept of diminishing returns in the agricultural sector gradually became clear to the next generation of economists as an implication of Malthus's first *Essay on Population* (1798), land displaced capital as the chief constraint. Because there is no lasting remedy for land shortage, the so-called "classical" political economy of the new century replaced Adam Smith's "chearful" study of growth and development with a Dismal Science of scarcity. Malthus, West, Torrens, Ricardo, Chalmers, M'Culloch and James Mill viewed the turbulent economy of post-1815 England with the aid of a common model of growth and distribution.[23] It was now profits and wages, rather than prices, which "gravitate" to a "very low rate" at which "all accumulation" and population growth are "arrested:" in stationary state wages and profits are minimal, "and almost the whole produce of the country ... will be the property of the owners of land and the receivers of tithes and taxes."[24]

Notwithstanding this drastic change in analytical focus, the policy implications of the "new" political economy of that day remained very much as before. That the welfare of the laboring poor was and ought to be the prime object of public policy was corroborated by the utilitarian principle of the "greatest good for the greatest number" distilled from the ethical theory of Paley and Bentham. That high wages depended in part on rapid growth, and that the latter was fostered by freedom to employ capital and by the smallest possible share of government in the economy remained uncontested. That a growing taste for "luxury" among the laboring class would increase the stationary-state real wage was strongly reinforced by the distribution-theoretic features of the new model, which demonstrated that a parametric increase in the target ("subsistence") real wage not only increases the absolute but also the relative share of labor income by transferring part of the surplus from landlords to capitalists and laborers.[25] A general predilection for *laissez-faire* that Adam Smith shared with the Physiocrats (who invented the slogan) was given a rigorous justification by Ricardo's new theory of comparative advantage: any duty or bounty on a traded product "causes a pernicious distribution of the general funds of society – it bribes a manufacturer to commence or continue in a comparatively less profitable employment."[26]

Perhaps the most striking political implication of the "new political economy" was the attack on the Poor Laws mounted by Malthus and Chalmers, strongly supported by Ricardo and all the Philosophic Radicals. Adam Smith had criticized the Poor Laws because they interfered with the mobility of labor.[27] But the classical objection was more sophisticated. Because the condition of the poor could only be permanently improved if they themselves sought to achieve higher family incomes by deferring marriage (and therefore procreation), it was essential to harness their self-interest in this cause. Institutions became crucially important, and for the first time in the history of our discipline, rational choice – already taken for granted in the analysis of capitalist behavior – was applied to the behavior of the poor. The Elizabethan Poor Law made it rational for the poor to marry early and maintain a low or zero saving rate.

> The comforts and well-being of the poor cannot be permanently secured without some regard on their part, or some effort on the part of the legislature, to regulate the increase of their numbers, and to render less frequent among them early and improvident marriages. The operation of the system of the poor laws has been directly contrary to this.[28]

Savings banks, parochial education and abolition of the Poor Laws would provide the opportunities and incentives to the poor to develop the habits of independence and self-help essential to their "improvement."

By the mid-1820s classical political economy presented a coherent and powerful body of analysis, building on and refining the work of Adam Smith and his contemporaries, the seeming implications of which supplied the rational basis of a non-interventionist, pro-free-trade, anti-Poor-Law consensus that informed political thinking in Britain and America for most of the nineteenth century. Malthus had his doubts about unemployment and "stagnation," and he broke ranks by defending agricultural protection; but in each of these respects he was viewed as a heretic by the profession and his theories ignored for more than a century. John Stuart Mill[29] allowed the "infant industry" exception to free trade pioneered by Alexander Hamilton.[30] Mill also wrote encouragingly about socialism in 1852.[31] But his posthumous second (actually third) thoughts of 1869 were far more cautious.[32] Karl Marx was the last of the classical economists and is commonly thought of as a socialist author. But there is far less advocacy of socialism in *Das Kapital* than there is in the 1852 edition of Mill's *Principles;* and only few and scattered allusions to any possible socialist economy. Marx's model is basically Smithian, with certain key assumptions modified so as to predict the spontaneous *disorder* and eventual collapse of capitalism, brought about by the self-interested acts of individual capitalists – as if led by an invisible hand.

Save for the almost entirely neglected contributions of Thünen (1783–1850), Cournot (1801–1877) and Dupuit (1804–1866), the intellectual or at any rate the social-scientific landscape of the English-speaking world was filled from 1800 to the mid-1870s with developments of, and variations on, Adam Smith's "natural system of perfect liberty and justice."

14.3 The economics of control

What happened next was the belated assimilation of some insights of Condillac, Thünen, Cournot, Dupuit and Gossen during the so-called "marginal revolution;" the application of marginal concepts to welfare theory; a deployment of neo-classical production and welfare concepts in the theory of "market socialism;" a revival of macroeconomics resulting from the so-called Keynesian revolution; and a grand combination of all these in what Samuelson described as "the neo-classical synthesis." The neo-classical synthesis provided a conceptual framework for yet another "new political economy," which gradually came to replace the presupposition of *laissez-faire* among the educated classes of Britain and America by the middle of the twentieth century. For by 1950 it was taken for granted by all – save a tiny minority of ignored and/or derided dissidents – that modern, technical economic theory had supplied the tools for implementing democratic socialism without the need of revolution. The mood of the time is perfectly captured by Abba Lerner's *The Economics of Control*:

> The fundamental aim of socialism is not the abolition of private property but the extension of democracy. This is obscured by

> dogmas of the right and of the left. The benefits of both the capitalist economy and the collectivist economy can be reaped in the controlled economy. The three principal problems to be faced in a controlled economy are employment, monopoly, and the distribution of income. Control must be distinguished from regulation. Liberalism and socialism can be reconciled in welfare economics.[33]

Even those such as Joseph Schumpeter, who disliked the prospect of socialism, feared it was probably inevitable. "Marx's vision was right," for "the capitalist process not only destroys its own institutional framework but it also creates the conditions for another;" "The March into Socialism" has already begun; "Can socialism work? Of course it can."[34]

To the neoclassical building-blocks of scientific socialism we now turn.

Neo-classical production and distribution theory

The classical concept of diminishing returns to a variable factor of production as applied to the fixed factor, land, was well known to Turgot,[35] Steuart[36] and others long before its rediscovery by Malthus, West, Torrens and Ricardo in 1815. But the first to generalize the principle to any or all factors – and to formalize it by means of that differential calculus the theory cries out for – was J. H. von Thünen (1783–1850), the first volume of whose *Isolierte Staat* appeared in 1826; who seems to have invented the term "marginal;"[37] and whom Marshall much later claimed to have "loved above all my other masters."[38] When production takes place under constant returns to scale (CRS), marginal-product factor pricing exactly exhausts the total product, so generalizing the Malthus–Ricardo theory of income distribution. The classical and pre-classical concept of an aggregate "surplus" of production that Malthus, Ricardo and Marx shared with Petty, Quesnay and Smith is thereby shown to be redundant and misleading.

Thünen's "treatment of distribution anticipated the whole of what later came to be known as the marginal productivity theory of distribution and in some respects he improved even on the presentation of John Bates Clark, which came almost fifty years later."[39] Economists reinvented the wheel as usual, and definitive treatments were duly supplied by P. H. Wicksteed,[40] J. B. Clark[41] and Knut Wicksell,[42] assisted by A. W. Flux's[43] demonstration of product-exhaustion by means of Euler's theorem on homogeneous functions.[44]

It is obvious that the concept of a marginal product implies continuous factor substitutability over the relevant range of the production function. Moreover, the requirement of CRS implied by Euler's theorem can only be met, for any firm with a U-shaped long-run average-cost curve, at the minimum point of the latter. But at long-run equilibrium of a perfectly competitive industry each firm produces at that point, and industry production varies in response to any change in demand by the entry or exit of firms. Hence a CRS set of *industry* production functions may be deemed to exist that – for any given set of factor endowments and for any given state of technique – determine the complete range of production possibilities in a perfectly competitive market economy at full, long-period equilibrium for every possible assignment of factors to industries.

It is equally obvious that these new technical developments, though based on a classical insight, diverge sharply from classical economics in two important respects. In the first place, product exhaustion, normal profits and long-period equilibrium imply precisely that "stationary state" that Ricardo and his friends hoped they would never see. Neo-classical production theory is "static" or timeless, abstracting from accumulation, population growth and technical progress. Secondly, the logical requirement of CRS rules out increasing returns to scale that for Adam Smith and Marx, if not for Malthus and Ricardo, were characteristic of an industrializing economy. As always in any science, growth of knowledge is purchased at a cost.

Neo-classical utility and welfare theory

That value is determined by "subjective" utility rather than by "objective" cost is an idea that goes back to the Scholastics.[45] In 1776 Etienne Bonnot, abbé de Condillac (1714–1780), explained value in terms of marginal utility in *Le Commerce et le Gouvernement*.[46] According to a recent Nobel Prize winner, Condillac developed in that work "a general theory of the generation of surpluses, of general economic equilibrium, and of maximal efficiency."[47] However, Condillac's work was totally ignored for a century; the concept of marginal utility was successively rediscovered by W. F. Lloyd,[48] Jules Dupuit,[49] H. H. Gossen[50] and W. S. Jevons,[51] each in seeming ignorance of his predecessors.

Though Malthus had formulated a demand function as early as 1800,[52] "subjective" value theory was resisted for so long because it denied the "hard core" of the classical research program: Ricardo's labor theory of value. The logjam finally broke in the early 1870s with important publications by Jevons,[53] Carl Menger[54] and Leon Walras,[55] a period now canonized in textbooks as the "marginal revolution." In various ways these authors and their successors constructed the components of what eventually became a complete model of the general interdependence of prices and production in a market economy driven by agents who maximize their own satisfaction by equalizing marginal benefits and costs over all possible activities. When agents maximize subject to constraint, demand curves are negatively sloped because "marginal utility" declines; supply curves are positively sloped because "marginal cost" increases.[56] The congruence, indeed the isomorphism, of a marginal analysis of value and demand with a marginal analysis of cost and supply soon became apparent. By 1937 Paul Samuelson was able to demonstrate in his Harvard doctoral dissertation that "the theory of maximizing behavior" is one of "the foundations of economic analysis."[57]

It is obvious (if not tautologous) that diminishing marginal utility must imply gains from bilateral trade; and easy to show that in multilateral market exchange both demanders and suppliers enjoy an access of utility: "consumers' surplus" and "producers' surplus," respectively. It is also obvious in a CRS world that where imperfect competition keeps price above marginal cost, the total surplus is both reduced and redistributed in favor of suppliers. Hence perfectly competitive markets may be conceived to maximize "economic welfare."

Alfred Marshall[58] and many others drew inferences for redistributive taxation and other welfare-policy measures from these deceptively simple[59] propositions. A. C. Pigou[60] analysed externalities and the resulting "market failures" and defined an aggregate welfare maximum as that state in which the "marginal social product" of all resources in all alternative uses is equalized. But all these attempts at normativity eventually foundered on the rock of interpersonal comparisons of utility. It was left to Vilfredo Pareto[61] to formulate the only welfare criterion to survive L. Robbins's[62] influential attack on interpersonal comparisons nearly 40 years later.

Pareto rejected cardinal utility and worked strictly with ordinal utility, which he renamed "ophelimity." He also rejected Marshall's partial equilibrium analysis in favor of the Walrasian model of general interdependence. Making use of the indifference-curve analysis pioneered by F. Y. Edgeworth,[63] Pareto was able to provide a convincing demonstration of the most important theorem of neo-classical economics: in a world of strictly private goods, general equilibrium of a perfectly competitive economy defines a welfare optimum – in the sense that it is impossible to make any individual better off without making at least one other worse off *as these individuals perceive their own welfare*. Because of the insularity and chauvinism of the Anglo-American world, Pareto's results and their significance, and the

Walrasian framework within which they were presented, were neither known nor understood by most English-speaking economists until Robbins, the "market socialists" and J. R. Hicks[64] had digested and republished them in the 1930s.

"Socialist calculation"

"The nomenclature of utility and disutility ... leads one immediately to ask whether a free enterprise system represents such a use of satisfying wants as to insure society the greatest surplus of utility over disutility."[65] Except in Austria, socialist or, at any rate, reformist sympathies were normal among the first two or three generations of neo-classical economists. The new tools of analysis seemed to afford the possibility of a rational approach to social reform. Wicksell was a life-long socialist who exercised an enduring influence over his younger colleagues of the Stockholm school. Pigou's work has been described as "virtually a blue-print for the welfare state."[66] Pareto's *Cours d'Economie Politique* (1896) examined the allocation of resources to be pursued by a socialist state in achieving "the maximum well-being of its citizens." His compatriot Enrico Barone demonstrated the formal equivalence of the welfare optimum under either capitalism or socialism in his famous article "The Ministry of Production in the Collectivist State."[67]

Friedrich von Wieser, Menger's successor at the University of Vienna, had noted as early as 1893 that a socialist economy would require a complete set of "natural values" (which might be approximated by market prices under competitive capitalism with equal wealth endowments); and that the task of calculating these correctly would be formidable.[68] Wieser's most famous pupil, Ludwig von Mises, made this the basis of a frontal attack on socialism launched in 1920,[69] which inaugurated the celebrated "socialist calculation" debates[70] of the inter-war years. Several distinguished neo-classical economists took up the challenge. Fred M. Taylor (1929) chose "The Guidance of Production in the Socialist State" as the subject of his presidential address to the American Economic Association of 1928. Oskar Lange thanked Mises for drawing attention to an important practical consideration and recommended that "[a] statue of Professor Mises ought to occupy an honourable place in the great hall of the ... Central Planning Board of a socialist state."[71] In a series of influential publications from 1929 to 1937, Taylor, Lange and A. P. Lerner established the orthodox, neo-classical theory of socialism.[72] A central planning board can function like Walras's auctioneer, generating a set of resource prices by trial and error. Managers of state enterprises set product prices at marginal cost and adjust the scale of production by reinvesting any surpluses. At equilibrium, the first-order marginal conditions exist for Pareto optimality: the equality of subjective and objective marginal rates of substitution among all goods, products and factors.

It was only in Lerner's articles on the economics of socialism that the conditions of a welfare optimum in the general equilibrium model were fully specified for the first time. Important contributions by A. Bergson[73] and Hicks[74] set the seal of professional approval on the "new welfare economics." Though the collectivist and free-market versions of the theory are formally equivalent, most orthodox, mainstream economists of the 1930s and 1940s preferred the former. This is because the analysis of increasing returns to scale and the consequent breakdown of competition – pioneered by J. A. Hobson,[75] Piero Sraffa[76] and A. A. Young[77] and made canonical by Joan Robinson[78] and E. H. Chamberlin[79] – showed convincingly that there is no theoretical reason to doubt that price will normally exceed marginal cost in a free-market economy; whereas the putative benefits of perfect competition, unobtainable under capitalism can be simulated under socialism. Many would have agreed with Paul Sweezy that "Marxian economics is essentially the economics of capitalism, while 'capitalist' economics is in a very real sense the economics of socialism."[80] Joan Robinson[81] was an astute critic of the Lange–Lerner–Taylor prescription, but as late as 1980 she could write (with probable allusion to Kantorovitch) that:

> there is an important area where neo-classical theory comes into its own. The concept of a production possibility surface and of the distribution of scarce means between alternative uses are the very foundation of economic planning. While the theory of allocation is withering in the free-market world, it is blossoming afresh in socialist economies.[82]

Full employment

For economists of Joan Robinson's generation it was the theoretical inability of market forces to determine output at *any* point on the "production possibility surface" (save by chance), even more than the impossibility of perfect or even "pure" competition, that was decisive in their rejection of the free-market version of general equilibrium.

Say's law is implicit in the Walrasian model. Unemployment can only exist in general disequilibrium. Excess supply in factor markets is matched by excess demand in some other market or markets, such as the money market. But the putative remedy, falling money-prices in factor markets, is unavailable. Walras's *tâtonnement* and Samuelson's dynamic stability analysis that formalizes *tâtonnement*[83] correctly rule out transactions in disequilibrium. This is because any "false trading" would falsify expectations and therefore change agents' evaluations of their assets. And that in turn would cause supply and demand curves to shift before equilibrium was attained. As a matter of mere logic – in economics though not in mechanics – "it is impossible for a system to get into a position of equilibrium, for the very nature of equilibrium is that the system is already in it, and has been in it for a certain length of past time."[84]

When the *General Theory* appeared in 1936, therefore, Keynes's ruthless disregard of the Walrasian model was received by many as a theoretical breakthrough of the first magnitude.[85] By locating his story in the Marshallian short period and implicitly assuming that with massive underemployment of productive factors all supply curves are horizontal, Keynes could use Marshall's quantity adjustment,[86] rather than Walras's operationally useless price adjustment, to analyse the consequence of a change in "effective demand" (a concept resurrected from Malthus) and so leave all absolute *and relative* prices unchanged. Aggregation is therefore logically admissible[87] and "macroeconomics" legitimated. In a closed economy aggregate supply is the same as aggregate income. And if aggregate demand is an increasing, non-homogeneous function of income with non-unitary slope, equality of aggregate supply with aggregate demand will determine a unique level of output and employment at which what Keynes misleadingly called "equilibrium"[88] will exist. If the slope is less than unity, this quasi-equilibrium will be "stable" in an operational sense. For at any state of expectations, quantity adjustment with unchanged prices affects asset values in a manner that reinforces the adjustment.[89]

"Keynes denies that there is an *invisible hand* channelling the self-centred action of each individual to the social optimum. That is the sum and substance of his heresy."[90] But since aggregate demand can be managed or at least manipulated by government through its control over the banking system and the national budget, an *ex post facto* theoretical justification was provided for the unvarnished empiricism of many national governments during the great depression of the 1930s. More importantly for the new political consensus, a theoretical rationale was supplied for the "full-employment" policies announced with much fanfare by the governments of Britain, Canada, Australia and the United States from 1944 to 1946.

The neo-classical synthesis and mathematical economics

Figure 14.1 represents the production possibilities *AB* of a closed economy with two goods, *Y* and *X*, a social indifference curve *CD* that touches *AB* in E, and a possible position of general underemployment, *U*. *AB*, drawn for given factor endowments and a given state of technique, summarizes neo-classical production theory. *CD*, drawn for given asset endowments and a given state of taste, summarizes the preferences of "society," either as revealed by individual market behavior or as discovered by the managers of a collectivist state.

Suppose production takes place at *U* According to neo-classical assumptions this implies general disequilibrium. Market forces must then be able to perform two functions: (1) to move production to some point on *AB*, such as *F;* and (2) to determine that point at *E* where the marginal conditions for Pareto optimality are satisfied. "Right-wing Keynesians" such as Keynes himself rejected (1) but accepted (2). Moreover, they defended economic "individualism" on ethical grounds as "the best safeguard of personal liberty."[91] "Left-wing Keynesians" such as Joan Robinson,[92] Lerner and many others of that generation rejected both (1) and (2). Moreover, they were skeptical about the value of "personal liberty" under capitalism.

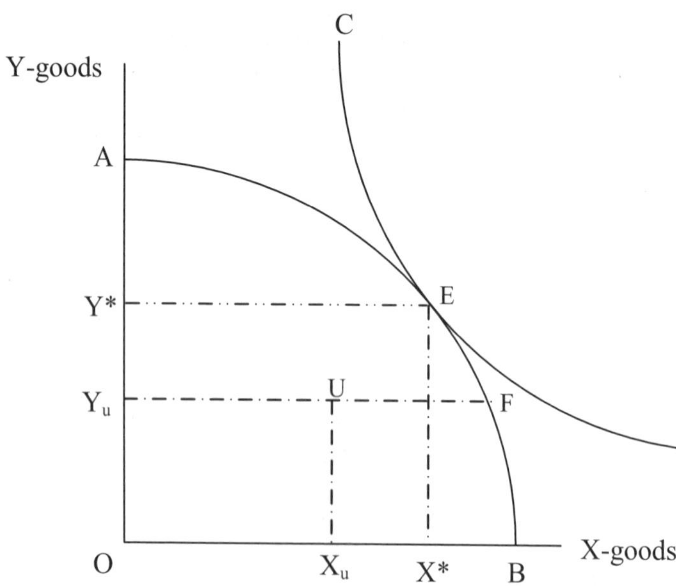

Figure 14.1 The neo-classical synthesis

By far the most influential of all Keynesians after 1946 was Paul Samuelson. With quasi-liturgical symbolism, his world-famous *Economics: An Introductory Analysis*[93] bore on its front cover a version of the "Keynesian cross:" a rising saving function of national income intersecting an horizontal investment function. Samuelson agreed with Keynes that markets might function adequately provided that government stabilize the economy at full employment. But he noted with approval that "nations all over the world [are] moving towards a planned state," disparaged *laissez-faire,* and held that Adam Smith's ideas "have done almost as much harm as good in the past century and a half."[94]

A recent commentator[95] has observed that Samuelson is an American Progressive in the political tradition of Richard T. Ely, J. R. Commons and Thorstein Veblen. But by replacing their theoretically degenerate "institutionalism" with the neo-classical synthesis of Paretian socialism and Keynesian macroeconomics, he "adapted the basic progressive scheme for the scientific management of American society ... to the circumstances of the post-World War II setting."[96] The neo-classical synthesis provided a "left-of-centre Keynesian" model that made the economics of control obvious to the meanest intelligence. Not only in the United States, but in many countries and in many languages millions of undergraduates during the next three decades and more assimilated its underlying message:

> *The economic activity of a human society is a rational system that can be sufficiently comprehended by economic scientists, who may therefore be trusted to design and implement measures to control economic activity so as to achieve politically determined collective goals.*

The "new political economy" of Samuelson and his associates, *mutandis mutatis,* is none other than that French *œconomie politique* that Adam Smith and his followers hoped they had stamped out once and for all.

Powerful and sophisticated mathematical analyses by Samuelson himself and by many others bolstered, for "men of learning," the scientific authority of the economics of control. Frank Ramsey's[97] pioneering work on dynamic optimization set the tone for a large literature on public planning of intertemporal resource allocation through the 1970s and beyond.[98] Influential contributions to mathematical economics were made by Soviet mathematicians during this period.[99] Wassily Leontief's[100] input-output analysis was used as a planning tool in the Soviet Union. John von Neumann's[101] model of "general economic equilibrium" inaugurated two decades of investigation by the best brains in our profession into the existence and stability of equilibrium in Leontief-type multisectoral models, expanding optimally in "time" with Pareto-efficient accumulation.[102]

When the economy is conceived as a machine,[103] in one case indeed as a "sausage machine,"[104] the economics of control looks easy. For most of the century after 1870, there is a rough but detectable correlation between a left-wing political commitment to "social engineering"[105] and mathematical economics. Among the few to dissent from the *dirigiste* consensus were those who rejected or were skeptical of mathematical economics on strictly theoretical grounds, such as Mises, Hayek and Frank Knight.

The "natural system of perfect liberty and justice," being founded on an *ecological* rather than a *mechanical* view of the economy, was banished from polite society for nearly a century.

14.4 "New Political Economies" in the twenty-first century

The economics of control has been subjected in the last 50 years to a rigorous scrutiny from which little of it, if any, has been left standing. Though in Milton Friedman's sense "we are all Keynesians now," the development of macroeconomic theory since the mid-1950s increasingly has tended to cast doubt on its relevance for discretionary demand management. The neo-classical model of general equilibrium, brought to its highest perfection in the work of K. J. Arrow and G. Debreu,[106] is now regarded as useless, whether as a source of welfare-policy prescriptions, or as a rationale either of capitalism or of socialism.[107] Save under dictatorship, the possibility of determining collective goals through the political process currently seems remote. And even if collective goals could be determined democratically – and even if economic tools were available to implement those goals – it is today well understood that agents in the public sector can normally be expected to place their own private goals first, and that these may be in conflict with collective goals.

By undermining the intellectual foundations of a pro-socialist political consensus in these ways, the "new political economies" of the past 20 years or so have produced a detectable ideological effect. Though external factors such as the collapse of the Soviet empire have compelled "men of learning" to re-examine their political assumptions, there can be little doubt that new and very "different theories of political economy" have "had a considerable influence" not only upon them, but upon "public conduct" of the present-day equivalents of "princes and sovereign states." It is no longer fashionable for an economist or a politician to be a socialist.

Macroeconomics

Keynes's novel but somewhat incoherent theorizing was eagerly tidied up by the younger generation.[108] Mathematical models were constructed, the variables of which corresponded with the categories in newly collected national accounts that had been inspired to some extent both by the *General Theory* itself[109] and by Jan Tinbergen's[110] path-breaking econometrics. The most influential of these models was Hicks's[111] elegant compression of Keynes's sprawling masterpiece into the Walrasian framework of *Value and Capital*.[112] It was a small matter thereafter to disaggregate by sector; add equations for the labor market; and generalize for the income-expenditure and monetary effects of interaction with the rest of the world.[113] Given constant-price national accounts, reasonably honest balance of payments estimates, monetary data from the central bank, price indices and official labor force surveys, all relevant magnitudes are in principle observable;[114] and the way opened for the elaboration of large-scale econometric income-expenditure models comprising several hundred equations that proliferated in the 1960s and 1970s.[115]

"Keynesian" models were and are typically fix-price.[116] Therefore, the discovery of the "Phillips Curve"[117] presented a major theoretical and policy challenge to the "new political economy" of the 1950s. Samuelson and Solow[118] "roughly" estimated the Phillips Curve for the United States, transformed it into a relation between unemployment and price inflation, and presented it as a "menu of choice" for policy makers, soon to be widely known as the "trade-off." The trade-off quickly entered the vocabulary of macroeconomics and was canonized in the sixth edition of Samuelson's textbook.[119] It was soon noted that a stable trade-off would destroy economic control in an open economy under fixed exchange rates, for both the world inflation rate and the world unemployment rate must then rule.[120] But a more general, and far more radically subversive, attack on economic control through demand management was launched at that time by E. S. Phelps and M. Friedman.[121]

Milton Friedman was one of the very few leading economists of his generation to have resisted the "disease" that struck down Samuelson and so many others; and had produced important theoretical studies critical of the Keynesian framework.[122] Friedman's presidential address to the American Economic Association of 1967[123] exploited the expectational instability of the Phillips Curve recognized by Samuelson and Solow[124] to show that with adaptive expectations, the long-run trade-off is vertical at what he called the "natural rate of unemployment."[125]

Though the profession eventually rejected Friedman's provocatively Smithian terminology in favor of the uneuphonious "NAIRU," the idea itself immediately became embedded in the core of macroeconomic theory. The fuzzy concept of "involuntary unemployment," always suspect,[126] ceased to be theoretically interesting; more importantly, so did the correlative concept of "full employment," central to the economics of control. The level of employment became instead a topic in the microeconomics of labor markets with costly information.[127] A. A. Alchian[128] showed that temporary unemployment is a matter of rational choice for a worker expecting a better position, and is therefore "productive." It is now customary to analyse the unemployment level in terms of separation and hiring rates of flow, determined exogenously by such factors as productivity, the minimum wage and the extent of unemployment insurance.[129]

Friedman's dramatic result was based upon the *ad hoc* assumption of adaptive expectations, which implies that agents are irrational. But real human beings learn from experience[130] and form their expectations probabilistically on the basis of all available information.[131] Before long, therefore, the next generation of young turks replaced Friedman's rough-and-ready expedient with the theoretically impeccable assumption of rational expectations.[132] Combined with the theoretically appealing but empirically implausible assumption of instantaneous market clearing, this so-called New Classical macroeconomics generates the notorious "policy-ineffectiveness proposition."[133] Not only can discretionary demand management have no permanent or long-run effect, it can have no transitory or short-run effect either. All Keynesian "multipliers" are always and everywhere zero, which is thus the *ne plus ultra* of the anti-Keynesian reaction against the economics of control.

As might be expected, this unwelcome result was vigorously contested by even younger turks, and "New Keynesian" models have proliferated that incorporate rational expectations but reject instantaneous market-clearing, and in a variety of ways purport to justify those wage and price rigidities that enable Keynesian models to produce real effects in the short run.[134] Meanwhile, a "Post-Keynesian" tradition that originated with Joan Robinson and other "left-wing Keynesians" has persisted, united in little except a general predilection for a Sraffian rather than a

Marshallian explanation of relative prices[135] and a dislike of mainstream economic theorizing, from which it has become increasingly cut off. In either the "New Keynesian" or "Post-Keynesian" case it turns out that "virtually every practical theory of macroeconomics" must be based at least in part on "hypotheses other than rationality."[136]

So far as the macroeconomic half of the economics of control is concerned, agreement is possible on one matter only. Consensus is dead.

Information, prices and markets

In the world of the Arrow-Debreu model, every good is a dated, contingent commodity, and all agents have full information about the present and a complete set of futures and insurance markets to protect them against uncertainty. No "decisions" have to be made. The market is a "human-link analogue computer" that grinds out the only information required by the system: a complete set of prices that correspond with the "shadow prices" required for socialist production.[137] Production managers simply find "in the book of blueprints the appropriate page corresponding to current (and future) factor prices"[138] and could be replaced by robots. Therefore, "in Samuelson's classic textbook" and in all its many imitators, the economic problem was summarized in three questions about total production: "What?" "How?" and "For whom?" The important theoretical "question about who decides or how decisions are to be made" was ignored. "While decision makers in the socialist economies paid more attention to these questions, they had neither the information nor the incentives to make good decisions."[139] Most microeconomic theorizing of the past 50 years has therefore been addressed to some of the more interesting questions posed by the Arrow-Debreu model. What are the welfare implications when Pareto optimality requires a shift along the production-possibility frontier, or when some of the marginal conditions are not satisfied, or when some of the markets are missing? How useful is the model when the assumptions of perfect information and perfectly competitive equilibrium are relaxed, as they must be if transactions costs are to be admitted into the analysis? What becomes of the model when, because of informational constraints, it is more efficient for some production to be coordinated by entrepreneurs making real decisions than by relative prices; and how is this "hands-on" coordination affected by the "principal–agent" problem and by "bounded rationality" – limitations on our merely human capacity to receive and process information? What are the implications of the answers to these questions for the relative efficiency of socialist and market economies?

Early discussion of the first of these sets of questions took place in the context of Hecksher-Ohlin trade theory. A change in trade policy (for example) that increases national income might be supposed Pareto-superior if the gainers could compensate the losers and remain better off.[140] But Scitovsky[141] showed that this result could be reversed with a different income distribution, thereby undermining the comforting distinction that originated with J. S. Mill between the "laws of Production," which are scientific, and "those of Distribution," which are at least in part political.[142] Further analysis by Samuelson[143] showed that "we cannot even be sure that group A is better off than group B even if A has collectively more of everything."[144] It was also originally in the context of trade theory that a different question arose for welfare economics: If the conditions of Pareto optimality are violated in some parts of the economy, are they sufficient welfare criteria elsewhere? The "general theory of second best" of R. G. Lipsey and K. Lancaster[145] showed that they were not. A "second best," such as lowering import duties without perfect competition, may be no better than a third-best option. Since the existence of government virtually necessitates second-best policies, the marginal conditions can be no guide for intervention in a mixed economy. A more general formal objection to the usefulness of the Arrow-Debreu model arises from its multidimensional conception of a "commodity," which almost guarantees the impossibility of a complete set of futures and insurance markets; and that, more seriously, must undermine the assumption of perfect competition.[146]

Incompleteness of the market arises from transactions costs, for markets require resources in order to function (even Walras's auctioneer may expect to be paid) and some are too costly to be worthwhile. The existence and significance of transactions costs was pioneered by Ronald Coase,[147] who showed that they can explain why some economic activity is coordinated by firms rather than by markets and, more generally, why economic institutions exist. A "new institutional economics"[148] that developed from these ideas has focused on imperfect information in a variety of ways: asymmetric information; the "principal-agent problem," which is a special case of the former;[149] "strategic intelligence;"[150] and "bounded rationality."[151] Much of this has been brought together by Oliver Williamson[152] who has shown that hierarchical business organization is more efficient than atomistic markets in a world of limited information.

Coase, or at any rate those who developed his ideas in the "law-and-economics" branch of economic theorizing, analysed the efficiency of various institutional arrangements and argued that private property rights are crucial in providing incentives to efficient resource allocation and use. This doctrine and the presupposition in favor of privatization that it justifies have been contested by Joseph Stiglitz[153] who observes – ironically, in view of Coase's own original contribution – that it ignores the pervasiveness of transactions and information costs. In a world of perfect information, public and private ownership are equally efficient. However "neither the model of the market economy nor the [equivalent] model of the market socialist economy provided a good description of the markets they were supposed to be characterizing."[154] The economic (as distinct from the merely political) arguments for privatization rather are based upon the inability of governments to make and keep commitments to reward the efficient and punish the inefficient, the more powerful incentives to innovation and risk-taking in the private sector, and the ultimate sanction of bankruptcy.[155] These arguments, like the "new institutionalism" noted above, take imperfect competition for granted and formalize it as an efficient response to imperfect information.

Friedrich von Hayek was perfectly aware, in his original attack upon socialism, that information is the heart of the matter, and believed that only a regime of private property and free markets could generate the information necessary to coordinate production and distribution in a complex, industrial society.[156] The counter-attack of Taylor, Lange and Lerner was apparently successful because of the suitability of neo-classical price and welfare theory for their model of market socialism. Hayek and Mises did not believe in neo-classical economics. They rejected the mechanistic concept of static equilibrium and preferred to study economic agents in real time, engaging in a continual process of learning, not only about means but also about ends. Therefore, it seemed to them at the time – as it still seems to their neo-Austrian school successors[157] – that it is unhelpful to model market processes formally, and is merely obfuscating to use mathematics. They were thus at a disadvantage in controversy, and the socialists were deemed for a generation and more to have won the high ground. Though Hayek later published important work on knowledge in economic theory[158] and was duly awarded a Nobel Prize, his ideas were long ignored by the profession, and lately he has been criticized by Stiglitz – who agrees with his general conclusions about socialism – for his unwillingness to use the language of his fellow economists. Because "Hayek ... failed to develop formal models of the market process, it is not possible to assess claims concerning the efficiency of that process."[159] Following the lead of George Stigler[160] and George Akerlof,[161] Stiglitz and his collaborators rejected Austrian methodological defeatism[162] and began in the 1970s to construct "a number of such simple models."[163] Their results confirm the uselessness of the Walras–Pareto–Arrow–Debreu model considered above, either as "the economics of socialism" or as a rationale of Adam Smith's "invisible hand." While generally corroborating the Austrian belief in the superiority of free markets over central planning and control, they encourage a pragmatic, piecemeal view of the public sector in a mixed economy. "The socialist-versus-market debate posed the question the wrong way. The correct question is not which mode of production is superior, but what are the comparative advantages of each sector, what should be the appropriate scope of each?"[164]

Collective decision

Even if a well-defined production-possibility hyper-surface existed, and even if national governments possessed the power – through skillful use of macroeconomic analysis and econometric modeling – to determine aggregate production on that hyper-surface, the fourth textbook question posed by Stiglitz would remain: "Who decides, and how?" Without a complete set of perfectly competitive markets, the question applies *a fortiori* to necessarily political determination of the socially optimum point on that hyper-surface.

One of the "presuppositions of Harvey Road ... may perhaps be summarized in the idea that the government of Britain was and would continue to be in the hands of an intellectual aristocracy using the method of persuasion ... Keynes tended till the end to think of the really important decisions being reached by a small group of intelligent people."[165] The high tradition of disinterested duty that distinguished the British civil service in Keynes's day guaranteed that the "intelligent people" would seek the common good. But in countries such as the United States where democracy – not to say populism – was more obtrusive, it was feared by many that monetary and fiscal policy formulated by government for short-term political gain would have an inflationary bias. It was also increasingly recognized that demand management is subject to long lags and therefore likely to be destabilizing. These considerations induced Friedman[166] to recommend that the objective of "full employment" be abandoned, that government eschew discretionary demand management, and that rules be designed and followed to ensure that monetary and fiscal changes act as automatic stabilizers.

Friedman's distrust of "Keynesian" policy was congruent in various ways with a large body of theoretical analysis on the relation between political motivation and macroeconomic instability.[167] It should be noted that these objections to discretionary policy would remain even if New Keynesians should eventually be successful in establishing a generally acceptable theoretical framework for demand management.

But even if discretionary demand management were efficacious and even if democratic governments could be trusted to employ it, the problem would remain of just where to be on the production-possibility hyper-surface. How can we know the socially optimal composition of output? Why should we suppose that agents in the public sector would act so as to achieve it?

Bergson[168] and Samuelson[169] devised a social welfare function as a positive study of "the consequences of various value judgments"[170] and were able to demonstrate that many of the Paretian conditions hold regardless of value judgments. But no unambiguous criteria of social welfare emerged. For any set of factor endowments and individual tastes, there is an infinite number of Pareto-optimum outcomes depending on asset endowments; and explicit value judgments are required in order to rearrange he latter. A different approach was taken by Kenneth Arrow,[171] who analysed voting procedures as a means of determining collective preferences, building on recent psephological work[172] and bringing to its highest development a line of inquiry going back to Lewis Carroll.[173] As is well known, Arrow concluded that "if we exclude the possibility of interpersonal comparisons of utility, then the only method of passing from individual tastes to social preferences which will be satisfactory and which will be defined for a wide range of sets of individual orderings are either imposed or dictatorial."[174] The seeming impossibility of truly collective choice did not deter a tradition of analysis based upon Bowen,[175] according to which, "one can induce the median-voter's preferences for public policies from the policies that actually emerge under majority rule."[176] But H. G. Brennan and L. Lomasky[177] have shown that neither the "median voter" model nor the belief that an invisible hand operates in voting so as to bring about a socially beneficent outcome from narrowly interested voting is tenable. Voting is an "expressive" act, like cheering at a football match; and "fully rational voters will not reliably vote for the political outcomes they prefer." "Democracy is flawed as a mechanism for making social decisions and, on occasion, may produce genuinely disastrous results."[178]

Secure in the selfless devotion of a noble army of upper-class statesmen, bureaucrats and colonial administrators, British economists from Jevons to Joan Robinson – almost all of whom were themselves drawn from that class – ignored the problems of political process and political agency. It was otherwise in less favored lands. Mathematical models of the economics of smuggling, law enforcement and bribery appeared in Italy before the end of the eighteenth century.[179] By the 1880s Italian economists such as Mazzola and Pantaleoni had begun to investigate the public economy within an exchange framework.[180] In Sweden, starting in the early 1900s, Wicksell and Erik Lindahl studied "just taxation" in the theoretical context of politically determined budgeting.[181] In the United States, as in Sweden, what became known as "Public Choice" theory grew out of the study of public finance, with seminal papers by R. A. Musgrave,[182] H. R. Bowen,[183] J. M. Buchanan[184] – who received a Nobel Prize for his many contributions to this field – and Samuelson.[185] The importance of this branch of economic theorizing lies in "the elementary fact that collective or public-sector decisions emerge from a political process rather than from the mind of some benevolent despot."[186] Political process includes the "supply-side" actions of bureaucrats, who may be taken to be rational maximisers pursuing private goals.[187] Bureaucrats, and all others in a political process, compete in "rent-seeking." Public regulation of the private sector is the outcome of competing attempts by powerful interests to "capture" rents,[188] as Adam Smith[189] well understood. Yet another "New Political Economy" has emerged from this tradition in the past 15 years that aims at *explaining* economic policy rather than designing or recommending it, and that rejects the "conventional assumption that policy is determined by maximizing a social welfare function."[190]

Little if anything remains of "the presuppositions of Harvey Road." Neither political procedures nor bureaucrats, it would seem, can do much either to identify collective preferences or – if they could somehow be discovered – to satisfy them.

14.5 The bottom line

Half a century ago an almost unbroken consensus existed among "intelligent people" to the effect that the state can and ought to manage, regulate or control economic activity in the public interest. "Intelligent people" were sustained in this conviction by "state-of-the-art" economic theory, which proved that it was so. Adam Smith[191] had warned that "the duty of superintending the industry of private people, and directing it towards employments most suitable to the interest of the society" was one "for the proper performance of which no human wisdom or knowledge could ever be sufficient." But the "neo-classical synthesis" showed he was wrong and his warning went unheeded.

In the last quarter of the twentieth century this consensus has collapsed. Left-wing governments in New Zealand, Australia and many other countries compete in their zeal for deregulation, privatization and rolling back the welfare state. The Labour Prime Minister of Great Britain calls himself in all sincerity a "Thatcherite." No doubt the reasons for this are complex and many, and include the rapid and unexpected withering away of socialism in the Soviet empire. If any new, "neo-conservative" consensus is emerging it would seem to be based on a superstitious veneration of "market forces" somewhat similar, *mutatis mutandis,* to that blind faith in "economic control" that characterized the old consensus.

Insofar as "neo-conservatism" has any genuine intellectual foundations, however, these are to be sought, in part, in the present state of economic theory. The "new political economies" of the late twentieth and early twenty-first centuries have been powerfully corrosive of those collectivist or at any rate *dirigiste* doctrines taken for granted by Knut Wicksell and Thorstein Veblen, Beatrice Webb and Harold Laski, Fred M. Taylor and Abba Lerner. *Œconomie politique* has been rolled back once again. The onus of proof has been shifted, and now lies with those who wish to assert a more important part for the state in the economic life of the nation.

Notes

1 Political Economy *and* Christian Theology

1. Waterman (1965) p. 123, note 101.
2. Waterman (1965) p. 123, note 98.
3. Demant (1936)
4. A Group of Churchmen (1922)
5. Heath (1976)
6. Munby (1960) p. 157.
7. Waterman (1965) p. 120.
8. Waterman (1965) pp. 116–20.
9. Marx, *A Contribution to the Critique of Political Economy,* transl. (1904) N. I. Stone, p. 11.
10. Marx and Engels, *Manifesto of the Communist Party,* ed. (1959) L. S. Feuer.
11. Laslett (1965).
12. Wolfe (1998) (2003).
13. Keynes (1936) p. 383.
14. Collini (2000) p. 15.
15. Skinner (1969) pp. 45–9.
16. Waterman (2002) pp. 459–62.
17. Nelson (1991) (2001).
18. Raeder (2002).
19. McLean (2003).
20. Keynes (1972) p. 168.
21. Raeder (2002) ch. 4.
22. Haddow (1939) p. 67.
23. Keynes (1972) p. 170.
24. Stephen (1881) vol. ii, p. 456.
25. Willey (1934) (1940) (1949).
26. Barker (1957) p. 13.
27. Laslett (1960) p. 105, note; see McPherson (1962).
    Laslett (1960) pp. 92ff.
28. Skinner (1978) vol. i, pp. ix, 50.
29. Skinner (1978).
30. Young (1998) p. 218.
31. Clark (1994); Haakonssen (1996).
32. Stephen (1881) vol. i, p. 43
33. E.g. Crimmins (1983) (1990); Hole (1989); Waterman (1991a) and Chapter 3 below.
34. E.g. Vance (1985); Norman (1987); Corsi (1988); Levy (2001); Raeder (2002).
35. Faccarello (1999).
36. Waterman (1991c) pp. 113–14, 160, 224; Fontana (1985).
37. Stephen (1881) vol. ii, ch. XI.
38. Winch (1996) p. 421.

39. Feyerabend (1988) p. 21.
40. E.g. Feyerabend (1988) ch. 13.
41. Waterman (1998) pp. 303–4, 312–13
42. Skinner (1969) p. 28.
43. Waterman (2003).
44. Winch (1965) (1978) (1983) (1987) (1996).
45. Winch (1996) p. 6.

2 Why the English "Enlightenment" Was Different

1. Gay (1966) p. 69.
2. Cassirer (1951) p. 134.
3. Cassirer (1951) pp. 135–6.
4. Gilley (1981) p. 104.
5. Hume "Of Superstition and Enthusiasm," p. 46.
6. Hume, "Of Superstition and Enthusiasm," p. 48.
7. Young (1998) p. 218.
8. Jacob (1981) Pocock (1980) (1985).
9. Haakonssen (1996) p. 3.
10. Roche (1998) part II, ch. 11; Clark (1985).
11. See Chapter 3 below.
12. Pocock (1982) p. 86.
13. Johnson (1755) *A Dictionary of the English Language.*
14. Locke (1695) *An Essay Concerning Human Understanding* pt IV, pp. xix, 7. Hereafter cited as Locke, *Human Understanding.*
15. de Cahusac (1969) vol. iv (vol I in 1969 edn).
16. Lough (1968) p. 299.
17. Kafker and Kafker (1988) pp. 95–101.
18. Hume "Of Superstition and Enthusiasm," pp. 47, 48.
19. Knox (1950) pp. 361–5.
20. Knox (1950) ch. XVI.
21. Rack (1989) pp. 183–202; Knox (1950) pp. 521–4.
22. Lavington (1749–51) *Enthusiasm of the Methodists and Papists Compared.*
23. Casaubon (1656) *A Treatise Concerning Human Understanding,* p. xii. Hereafter cited as Casaubon, *A Treatise.*
24. Casaubon, *A Treatise,* pp. xviii–xix.
25. Casaubon, *A Treatise*, pp. 173–5.
26. More (1656) *Enthusiasmus Triumphatus; or, A Brief Discourse of the Nature, Causes, Kinds, and Cure of Enthusiasm.*
27. Cooper (1708) *Characteristics of Men, Manners, Opinions, Times, etc. By the Right Honourable Anthony Earl of Shaftesbury.* Hereafter cited as Cooper, *Characteristics.*
28. Locke *Human Understanding,* pt. IV, pp. xviii, 10, 11.
29. Locke *Human Understanding,* IV, pp. xix, 16.
30. Cooper *Characteristics,* vol. i, p. 39.
31. Cooper *Characteristics,* vol. i, pp. 37–9.
32. Cooper *Characteristics,* vol. ii, p. 173.
33. Cooper *Characteristics,* vol. ii p. 129.

34. Grean (1967) pp. 19–20.
35. Pocock (1985) pp. 532.
36. Boswell (1791), *Boswell's Life of Johnson, together with Boswell's Journal of a Tour to the Hebrides ...* ,6 Vols, vol. vi, p. 269. Ed. (1934) G. B. Hill.
37. Boswell, *Life of Johnson,* vol. v, p. 111. Ed. (1934) G. B. Hill.
38. Pocock (1985) p. 531.
39. Graves (1815) *The First Praelection, delivered as Professor of Divinity, by ... Richard Graves, King's Professor of Divinity in Trinity College Dublin. To which are annexed the Regulations for the Examination, directed by the Statute fixing the Duty of the Professorship of Divinity, and the List of Books recommended to the Students etc.*
40. Holy Bible, *Acts* 25: 19.
41. Jaucourt "Superstition" in *L'Encyclopédie ou Dictionnaire Raisonné des Sciences des Arts et des Métiers, Par une Societé de Gens de Lettres,* eds Diderot and D'Alembert (1751–1772) vol. xv (vol. III in 1969 edn).
42. Kafker and Kafker (1988) pp. 175–7.
43. Voltaire "An Important Study by Lord Bolingbroke, or the Fall of Fanaticism," pp. 297, 298.
44. Hume "Of Superstition and Enthusiasm," p. 47.
45. Collins (1713), *A Discourse of Free-Thinking,* pp. 37, 38, 41.
46. Pocock (1982).
47. Roche (1998) pp. 382–9.
48. Voltaire (1767?), *Philosophical Dictionary,* pp. 11–15, 55–6, 126–7.
49. Whelan (1988).
50. Wilberforce (1797) *A Practical View of the Prevailing Religious System of Professed Christians, in the Higher and Middle Classes of this Country, Contrasted with Real Christianity* p. 287. Hereafter cited as Wilberforce, *A Practical View.*
51. Smith (1776) *An Inquiry into the Nature and Causes of the Wealth of Nations,* 2 vols, vol. ii, p. 809.
52. Holbach (1769), *The System of Nature, or Laws of the Moral and Physical World,* pp. 116–29.
53. Meslier (1915) p. 36.
54. Waterland (1843), *The Works of the Rev. Daniel Waterland, D. D.,* 6 vols, vol. iv, pp. 466–7. Hereafter cited as Waterland, *Works.*
55. Logan (1982) pp. 64 *et passim.*
56. Whelan (1989) pp. 9–10.
57. Whelan (1989) p. 49.
58. Voltaire "An Important Study," p. 207.
59. Meslier (1915) p. 338.
60. Clark (1985).
61. See chapter in this volume.
62. Stephen (1876) *History of English Thought in the Eighteenth Century,* 2 vols, vol. ii, pp. 405–26.
63. Keynes (1972), *Essays in Biography,* vol. x in *The Collected Writings of John Maynard Keynes,* 29 vols, vol. x, p. 368.
64. See Chapter 5 below.
65. Maclaurin (1775), *An Account of Sir Isaac Newton's Philosophical Discoveries,* pp. 3–5, 401–11.
66. See Chapter 5 below.

67. Watson (1785), *A Collection of Theological Tracts*, 6 vols. i, p. xii. Cited hereafter as Watson, *A Collection.*
68. Wilberforce, *A Practical View*, pp. 39–40, 467.
69. Heitzenrater (1984) vol. i, p. 114.
70. Rack (1989) p. 276.
71. Pocock (1980) p. 101.
72. Gregory (1790), *An History of the Christian Church from the Earliest Periods to the Present Time*, 2 vols, vol. ii, p. 299. Cited hereafter as Gregory, *History of the Christian Church.*
73. Gregory, *History of the Christian Church*, vol. ii pp. 298–9.
74. Watson, *A Collection*, vol. i, p. vii.
75. Sumner (1801), *An Essay, Tending to Show that the Prophecies, Now Accomplishing, Are an Evidence of the Truth of the Christian Religion*, p. 30.
76. Pocock (1982) p. 83.
77. Wallace (1746) 1969.
78. Watson, *A Collection*, vol. i, p. xiv.
79. Hey (1796–8) *Lectures in Divinity, Delivered in the University of Cambridge*, 4 vols, vol. iv, pp. 104–5. Cited hereafter as Hey, *Lectures*
80. Hey, *Lectures*, vol. iv, pp. 96, 99.
81. Milner (1794–1809) *The History of the Church of Christ*, 4 vols, vol. iv, p. 917. Cited hereafter as Milner, *History of the Church.*
82. Milner, *History of the Church*, vol. iv, pp. 478, 479.
83. Waterland, *Works*, vol. iv, p. 608.
84. Pocock (1980) p. 106.

3 Theology and Political Doctrine in Church and Dissent

1. Clark (1985) p. 281.
2. Clark (1985) p. 334.
3. Richard Hooker (1594–1662), *Of the Laws of Ecclesiastical Polity, Book VIII*, vol. ii, p. 2, in *The Works of that Learned and judicious Divine Mr Richard Hooker, Containing Eight Books of the laws*, vols. Hereafter cited as Hooker, *Ecclesiastical Polity.*
4. Green (1978) pp. 143–54.
5. Sykes(1959) p. 10.
6. Green (1978) p. 14.
7. Proctor and Frere (1958), *The Book of Common Prayer, with a Rationale of Its Offices* p. 114.
8. Davies (1961) p. 139.
9. Proctor and Frere (1958) p. 54, n. 2.
10. See, e.g. Davies (1961) pp. 213–26.
11. *Book of Common Prayer* (hereafter *BCP*); *Morning Prayer*, Te Deum.
12. *BCP, Articles of Religion* (1562) no. 36.
13. *BCP, Holy Communion*, Common Preface.
14. *BCP, Public Baptism of Infants.*
15. *BCP, Catechism.*
16. *BCP, Public Baptism of Infants.*
17. *BCP, Solemnization of Matrimony*; my italics.

18. Robinson *(1957) passim.*
19. *BCP, Collects*: Holy Saturday, newly composed by John Cosin in 1662.
20. *BCP, Holy Communion*, Prayer of Consecration.
21. Cyprian, *Epist.* 73, 21, cited in Johannes Quasten (1950–60) *Patrology,* 3 vols, vol. ii, p. 373.
22. Athanasius, *De incarn.* 54, cited in Quasten (1950–60) vol. iii, p. 71.
23. Bindley (1925) p. 39.
24. *BCP, At Morning Prayer*, Athanasian Creed "Quicunque vult."
25. Quasten (1950–60) vol. i, pp. 23–7.
26. *BCP*, "Quicunque vult" (n. 24); my italics.
27. Petty (1662–90), *The Economic Writings of Sir William Petty*, p. 129.
28. 24 Henry VIII, xii.1532–3, cited *OED*; my italics.
29. Watson (1785), *A Collection of Theological Tracts*, 6 vols; vol. vi, p. 111; my italics.
30. Burke (1791), *Reflections on the Revolution in France*, pp. 95, 31.
31. *BCP, Solemnization of Matrimony.*
32. *BCP, Consecration of Bishops.*
33. *BCP, Collects.*
34. *BCP, Ordering of Priests.*
35. *BCP, Ordering of Deacons.*
36. *BCP, Consecration of Bishops.*
37. *BCP, Catechism.*
38. Cross (1969) pp. 129, 126.
39. Ratcliff (1936) pp. 51, 86.
40. Bond (1987) ed. *Certain Sermons or Homilies (1547)* and *A Homily against Disobedience and Wilful Rebellion (1570)* p. 169; my italics.
41. Hooker, *Ecclesiastical Polity*, Bk VIII, vol. i, p. 5.
42. Cross (1969) p. 30.
43. Bond (1987) p. 209.
44. Joseph Priestley (1817–32), *The Theological and Miscellaneous Works etc. of Joseph Priestley, LL.D. F.R.S., etc with Notes by the Editor,* J. T. Rutt (ed.), 25 vols, vol. xxii, pp. 354, 357, 351. Hereafter cited as Priestley, *Works*; Richard Price (1776–78) *Two Tracts on Civil Liberty, the War with America, and the Debts and Finances of the Kingdom*, 2 vols, vol. ii p. 42, 8n.(a).
45. Priestley, *Works*, vol. xxii pp. 356–8.
46. Bradley (1990) pp. 10, 16.
47. David Rivers, *Observations on the Political Conduct of the Protestant Dissenters* (London, 1798), cited in the *Anti-Jacobin Review* (1798) p. 626. Hereafter cited as Rivers, *Observations.*
48. Rivers, *Observations* p. 630.
49. Robbins (1961) p. 7.
50. Price, *Tracts*, vol. i, pp. 5–6.
51. Sandoz (ed.) (1991) pp. 1016, 1027.
52. Priestley, *Works*, vol. xxii, p. 11; cf. Price, *Tracts*, vol. ii, p. 13 n.(a).
53. Price, *Tracts*, vol. i pp. 7–8.
54. Priestley, *Works,* vol. xxii, p. 11.
55. Priestley, *Works*, vol. xxii, pp. 341, 335.
56. Price, *Tracts,* vol. i, p. 11.
57. Priestley, *Works*, vol. xxii, pp. 383, 20, 18, 19, 25–6, 13; my italics.

58. Watts (1978) p. 478.
59. Priestley, *Works,* vol. xv, p. 45, vol. xxii, p. 131; see also vol. v, pp. 154, 264, 267, 285, 489, 493.
60. Hooker, *Ecclesiastical Polity*, Bk VIII, v, 8.
61. Priestley, *Works*, vol. xxii, p. 315.
62. Bradley (1990) p. 137 and *passim*; my italics.
63. See, e.g., Priestley, *Works,* vol. xxv pp. 36–8.
64. Price (1789) in Sandoz (ed.) (1991) p. 1015.
65. Priestley, *Works,* vol. v, pp. 10, 480, 233; XXII p. 364, V p. 493.
66. These works, published between 1772 and 1804, may be found Priestley, *Works,* vols. ii, v, vi, vii–x, xv, and xi–xiv, respectively.
67. Priestley, *Works,* vol. v, pp. 105–21; vol. vii pp. 522–3; vol. v, p. 151.
68. Priestley, *Works,* vol. v, p. 434.
69. Burke (1791) pp. 147–8; cited in Priestley, *Works,* vol. xii, p. 191.
70. Priestley, *Works,* vol. xxii, pp. 88–99, 229, 233, etc. and 373–9.
71. Davies (1961), ch. 4.
72. Priestley, *Works,* vol. xv, pp. 325–48.
73. See, e.g., Sandoz (1991) pp. 1022, 1025, 1027; Priestley, *Works,* vol. xv, pp. 356, 381, 443.
74. Godwin (1796) *Enquiry Concerning Political Justice and Its Influence on Morals and Happiness* 2nd edn, 2 vols, vol. ii, p. 340.
75. Waterman (1991c) pp. 72–81.
76. Swift (1744) *On Mutual Subjection.*
77. Sisson (ed.) (1976) *The English Sermon: an Anthology,* 3 vols, vol. ii, pp. 309, 312.
78. Godwin (1798) *Enquiry Concerning Political Justice and Its Influence on Morals and Happiness* 3rd edn, 2 vols, vol. ii, pp. 501, 503.

4 Intellectual Foundations of Tory Doctrine

1. Viner (1972) p. 95.
2. Viner (1972) p. 96.
3. Boswell (1791), *Life of Johnson* p. 346, ed (1953) R. W. Chapman.
4. Lovejoy (1936).
5. Consider this example from a recent review of the "state of the art" of eighteenth-century historiography: "... we study the era in which English and Scottish writers for the first time engaged in fully secular discussion of their society and its destinies, *from which point British intellectual history can begin to be written"* (my italics), in J. G. A. Pocock (1985), *Virtue, Commerce and History: Essays in Political Thought and History, Chiefly in the Eighteenth Century,* p. 33.
6. Watson (1785), *A Collection of Theological Tracts,* vol. vi, p. 111. Hereafter cited as Watson, *Tracts.*
7. Ellis Sandoz (ed.), *Political Sermons of the American Founding Era, 1730–1805,* p. 1021.
8. Sandoz (1991) pp. 1021, 1017, 1018–19; my italics.
9. Sandoz (1991) p. 1016.
10. See, for example, Priestley (1817–32), *The Theological and Miscellaneous Works & c. of Joseph Priestley,* vol. xxv, pp. 35–8. Hereafter cited as Priestley, *Works.*

11. See Haakonssen (1981), *The Science of a Legislator: the Natural Jurisprudence of David Hume and Adam Smith* for a succinct account of Smith's view of subordination.
12. Boswell, *Life of Johnson*, p. 289 (ed.) (1953) R. W. Chapman.
13. Adam Smith (1759), *The Theory of Moral Sentiments*, pp. 114–25, 303–5, 369–71, 306.
14. Adam Smith (1762), *Lectures on Jurisprudence*, pp. 318, 402. Cited hereafter as Smith *Lectures*.
15. Hamowy (1987).
16. Smith, *Lectures*, p. 402
17. Greenleaf (1964), pp. 15–19, 125–7.
18. Jenyns (1790), *A Free Inquiry into the Nature and Origin of Evil*, in *The Works of Soame Jenyns, Esq.* vol. iii. Cited hereafter as Jenyns, *Works*.
19. Boswell *Life of Johnson*, p. 223, ed. (1953) R. W. Chapman.
20. Lovejoy (1936) p. 203.
21. Jenyns, *Works*, vol. iii, p. 45.
22. Alexander Pope, *Poetical Works*, ed. Herbert Davis (1966) London, p. 270; cit. Lovejoy, p. 206.
23. Lovejoy (1936) p. 206.
24. Lovejoy (1936) p. 206.
25. Johnson was, of course, a "moral giant:" but he was "little fitted for abstract speculation. He was an embodiment of sturdy prejudice, or, in other words, of staunch beliefs which had survived their logical justification." See Leslie Stephen (1881), *English Thought in the Eighteenth Century*, vol. ii, p. 206. This is quite typical of the tendentious and inaccurate way in which Stephen dealt with ideas which did not fit his schema of a steady and inexorable progress to the telos of Late Victorian agnosticism. (For other examples see Chapter 5 in this volume, "A Cambridge via media in Late Georgian Anglicanism.") Lovejoy follows suit. In the eighteenth century only Voltaire and Johnson attacked the Great Chain of Being on philosophical grounds. "Dr. Johnson's attack upon the theory ... was, somewhat surprisingly, the more profound and more dialectical." See Lovejoy (1936) pp. 251–3.
26. Johnson (1787), "Review of a Free Inquiry into the Nature and Origin of Evil" in *The Works of Samuel Johnson, LL.D.*, 11 vols, vol. x, pp. 226, 227, 236. Cited hereafter as Johnson, *Works*.
27. Johnson (1755), *A Dictionary of the English Language*
28. Jenyns, *Works*, pp. 47, 48, 49.
29. Piozzi (1786), *Anecdotes of Samuel Johnson* p. 39. Hereafter cited as Piozzi, *Anecdotes*.
30. Boswell, *Life of Johnson*, p. 312, ed. (1953) R. W. Chapman.
31. Johnson *Works*, vol. x pp. 229, 232, 230.
32. Johnson, *Works*, vol. x, p. 236.
33. Johnson, *Works*, x, p. 254; my italics.
34. For a convenient summary of St. Augustine's political theory see Herbert A. Deane (1963), *The Political and Social Ideas of St Augustine*, esp. ch. IV.
35. Viner (1972) p. 94.
36. See Boswell, *Life of Johnson*, pp. 1154–5, ed. (1953) R. W. Chapman, for Johnson's views on the agreement possible between "wise" Whigs and Tories.

37. At least as both Adam Smith and Samuel Johnson seem to have understood that term.
38. Cited in many histories: e.g. Davies (1961) p. 139.
39. For the standard, legal version of the Prayer Book see *The Book of Common Prayer from the Original Manuscript Attached to the Act of Uniformity of 1662, and Now Preserved in the House of Lords* (London, 1892). This is the version, only as modified by Victorian spelling, that remains the official liturgy of the Church of England and may still be purchased. It is the culmination of a process of revision which began when the first Prayer Book of King Edward VI (1549) was drastically altered in 1552. Intervening recensions date from 1559 and 1604. Since 1552 revisions have been slight and conservative. A recension nearer to the original of 1549 was made in 1637 for the Church of Scotland. This formed the basis for the Scottish Episcopalian Prayer Book of 1764 and through that, for the Prayer Book of the Protestant Episcopal Church in the USA (1789, revised, 1892). In each of these many versions – as in those made for various colonial churches in the nineteenth century – the wording of the clause in the Catechism misquoted by Lovejoy remains the same as the 1549 original (see note 23 above).
40. In Chapter 3 of this volume, "Theology and Political Doctrine in Church and Dissent."
41. These citations may be verified by reference to the currently authorized (1662) recension of the Book of Common Prayer at the offices for the Communion, Baptism of Infants, Communion, and Solemnisation of Matrimony respectively.
42. For an illuminating exposition of St. Paul's theological organicism which lies behind all this see Robinson (1957), *The Body: A Study in Pauline Theology*.
43. St. Cyprian of Carthage, in Epist. 73, 21: cited Quasten (ed.), vol. ii, p. 373.
44. Book of Common Prayer, office for Consecration of Bishops.
45. See Figgis (1965), *The Divine Right of Kings*; Acton (1952) *Essays on Church and State*, Woodruff (ed.), especially ch. III; Ullmann (1970), *The Growth of Papal Government in the Middles Ages: A Study in the Ideological Relation of Clerical to Lay Power.*
46. Cited Cross (1969), *The Royal Supremacy in the Elizabethan Church*, p. 129.
47. Ratcliff (1936) p. 51.
48. Ullman (1970).
49. Bond (ed.) (1987), *Certain Sermons or Homilies and A Homily Against Disobedience and Wilful Rebellion*, p. 169; my italics.
50. Hooker (1594–1662), *Of the Laws of Ecclesiastical Polity, Book VIII*, in *The Works of that Learned and Judicious Divine Mr Richard Hooker, Containing Eight Books of the Laws*, 3 vols, vol. ii, pp. 2, 5, 7. Cited hereafter as Hooker, *Polity*.
51. Hooker, *Polity*, vol. ii, p. 2.
52. E.g., Greenleaf (1964), *passim*.
53. Petty (1662–90), *The Economic Writings of Sir William Petty*, p. 12.
54. Filmer (1680), *Patriarcha and Other Political Writings of Sir Robert Filmer*
55. Greenleaf (1964), chs V, VII.
56. *Book of Common Prayer*, Catechism.
57. Swift, *Irish Tracts, 1720–3* and *Sermons*, vol. ix of *The Prose Works of Jonathan Swift*, pp. 143, 144. Cited hereafter as Swift, *Tracts*.

58. Thomas Starkey (1538?), *A Dialogue between Pole and Lupsett,* p. 37.
59. Swift, *Tracts,* p. 145.
60. So says Walter Jackson Bate, *Samuel Johnson* (1984) pp. 191–200 *passim.* Bate, unlike Stephen, has no desire to denigrate Johnson. His purpose is to warn us against anachronism in appreciating Johnson's politics; and to caution us against too ready an acceptance of Boswell's portrait of Johnson as a Tory in Boswell's own "romantic" and "snobbish" mould. Moreover, his brief account of Johnson's understanding of "subordination" (pp. 195–6) is more or less consistent with the overall argument of this article. Nevertheless his well-meant caveat has the cumulative effect of leaving its reader without any strong sense of either the philosophic or the theological components of Johnson's political thought.
61. Boswell, *Life of Johnson,* p. 1155, ed. (1953) R. W. Chapman.
62. Piozzi, *Anecdotes,* p. 42.
63. Bond (1987) p. 209.
64. Boswell, *Life of Johnson,* pp. 1154–5. ed. (1953) R. W. Chapman.
65. Edmund Burke (1791), *Reflections on the Revolution in France and on the Proceedings in Certain Societies in London Relative to that Event,* p. 95: "He who gave our nature to be perfected by our virtue, willed also the necessary means of its perfection – He willed therefore the state – He willed its connexion with the source and original archetype of all perfection." See also p. 31: "Our political system is placed in a just correspondence and symmetry with the order of the world, and with the mode of existence decreed to a permanent body composed of transitory parts; wherein, by the disposition of a stupendous wisdom, moulding together the great mysterious incorporation of the human race ..."
66. Daubeny (1798), *A Guide to the Church, in Several Discourses Addressed to William Wilberforce, Esq., M.P.*
67. Coleridge (1830), *On the Constitution of Church and State According to the Idea of Each,* vol. x in *The Collected Works of Samuel Taylor Coleridge;* W. E. Gladstone (1838) *The State in its Relations with the Church.*
68. Clark (1985)

5 A Cambridge *"Via Media"*

1. Mathers (1985), pp. 255–83.
2. Stephen (1881), *History of English Thought in the Eighteenth Century,* vol. i, pp. 405–26.
3. Stephen (1881)*;* cf. J. H. Overton and F. C. Relton (1906) pp. 257–62, in which the same cast is more cautiously denominated "the most intellectual party then in the Church."
4. Stephen (1881) vol. i, p. 406: "The starting point of the Cambridge school may be illustrated by Bishop Law's 'Considerations on the Theory of Religion,' which was published in 1745, and reached a seventh edition in 1784." See also Clark (1985), pp. 311–13.
5. See Le Mahieu (1976) p. 22; Cole (1987) pp. 1–6.
6. Stephen (1881) vol. i, pp. 425, 420, 380.
7. Le Mahieu (1976) p. 14.

8. No complete list of the membership of the Hyson Club exists. Gilbert Wakefield (1792), *Memoirs of the Life of G. W., B.A., Written by Himself,* pp. 526–32 (hereafter cited as Wakefield, *Memoirs*) and Edmund Paley (ed.) (1825) *The Works of William Paley, D. D.,* vol. i, p. 68 (hereafter cited as Paley, *Works*) mention ten members in all, of whom Waring, Beadon, Vince, Pretyman and Milner are common to both. It is generally supposed that Law, Hey, Jebb, Watson and Frend were members at various times but there is no certain evidence for any but Jebb *(Public Characters for 1802–03,* pp. 506–7). It is clear from his selection of members that Edmund Paley wished to emphasize his father's association with "safe" men such as Beadon (later Master of Jesus then bishop of Gloucester) and Pretyman.
9. It is now acknowledged that we may speak of a specifically English "Enlightenment," which by contrast with the infidelity associated with "Enlightenment" elsewhere, evinced "a determination to maintain orthodox belief against the revival of ancient philosophy." See Pocock, "Clergy and commerce. The conservative enlightenment in England," in *L' Eta dei Lumni: Studi storici sul Settecento Europeo in onore di Franco Venturi,* Naples 1985, I, p. 554 and *passim*; also Jacob (1976), *The Newtonians and the English Revolution,* and (1981) *The Radical Enlightenment: Pantheists, Freemasons and Republicans*. See also Chapter 2 in this volume.
10. Lincoln (1938) p. 22.
11. Clark (1985), ch. v passim.
12. Clark (1985) pp. 311–15
13. Wakefield, *Memoirs,* p. 107.
14. Knight (1971) pp. 61–2, ch. ix, x *passim.*
15. According to Knight (1971) p. 218, Frend eventually refused to believe even in negative roots because their existence could not be proved. William Godwin (1798), *Political Justice,* 3rd edn., vol. ii, pp. 491–3, maintained that in a society without private property, no one would have to work more than half an hour a day to provide the necessaries of life. Needless to say, Paley and Watson were bitterly reviled as turncoats by their former friends among the radicals. See John Gascoigne (1986) "Anglican Latitudinarianism and political radicalism in the late eighteenth century," *History,* lxxi, 36 and n. 74.
16. Le Mahieu (1976) p. 15.
17. Watson (1776), *An Apology for Christianity, in a Series of Letters Addressed to Edward Gibbon, Esquire,* p. 2.
18. *The Anti-Jacobin Review and Magazine,* 1802, 528.
19. Paley, *Works,* vol. iv, pp. 72, 137, 157, 162, 75, 154.
20. Paley, *Works,* vol. iv, pp. 511–12; see Clarke (1974) pp. 86–7.
21. Paley, *Works,* vol. iv. pp. 4, 145.
22. Christie (1984) pp. 160–1.
23. Paley, *Works,* vol. iv, pp. 73–5; cf. Thomas Aquinas *Summae Theologiae* II-II, Q 66, arts. 1, 2.
24. Paley, *Works,* vol. iv, p. 158.
25. Paley, *Works,* vol. iv, p. 80; cf. Deane (1963) pp. 104–6.
26. Paley, *Works,* vol. iv, p. 61.
27. Paley, *Works,* vol. iv, p. 75.
28. Paley, *Works,* vol. iv, p. 25.

29. Paley, *Works,* vol. iv, p. 477; Clarke (1974) p. 84; Ehrman (1983) p. 63.
30. Le Mahieu (1976) pp. 91–114.
31. Clarke (1974) p. 73.
32. Edmund Paley reports that "Mr Paley was of every party, and friends with men of all parties, but never exclusively attached to any:" Paley, *Works,* vol. i, p. 72.
33. Stephen (1881), vol. i, pp. 454–8, 464, is very scathing about Watson. A far more sympathetic account of Watson's growing political conservatism and its motivation, though one which concedes all to Stephen in the matter of latitudinarianism, is found in Norman Sykes (1934) ch. viii.
34. Richard Watson (ed.) (1785), *A Collection of Theological Tracts,* 6 vols. Hereafter cited as Watson, *Tracts.*
35. Le Mahieu (1976) pp. 18–19; Richard Watson (1817), *Anecdotes of the Life of Richard Watson, Bishop Llandaff,* 2 vols. Hereafter cited as Watson, *Anecdotes.*
36. Watson, *Anecdotes,* vol. i. p. 62.
37. Sykes (1934) p. 351.
38. Watson, *Tracts,* vol. i. pp. v, xii.
39. See the annotations to the "Contents" of each volume, especially to Lardner (vol. ii), Brett and Taylor (vol. iii), all the authors in vol. iv, Lardner and MacKnight (vol. v), and Fowler (vol. vi).
40. Watson, *Tracts,* vol. i, pp. xxxii, vii.
41. Watson, *Tracts,* vol. i, p. xvi.
42. Watson, *Tracts,* vol. i, p. xiv.
43. Watson, *Tracts,* vol. i, pp. xiii, x, xii. A closely similar cast of villains appears in Paley's Cambridge lecture notes (1767–76) printed by his son in Paley, *Works,* vol. i, p. 415.
44. Stephen (1881) vol. i, pp. 76–7, 345.
45. Watson, *Tracts,* vol. i. p. xv.
46. Watson, *Tracts,* vol. i. pp. vi, ix, x, xi, xiii, xv, xvi.
47. Paley, *Works,* vol. iv. p. 13.
48. Le Mahieu (1976) pp. 21–3, 114.
49. Paley, *Works,* vol. vii, p. 426; see Cole (1987) p. 4.
50. Paley, *Works,* vol. vii, pp. 175–6, 177, 178, 180, 181, 182.
51. Paley, *Works,* vol. v, p. 368n.
52. Paley, *Works,* vol. v, pp. 65–6.
53. Watson, *Tracts,* vol. vi. p. iii; Paley, *Works,* vol. i, pp. 123–4. *The Clergyman's Companion* was mistakenly attributed to Paley himself and printed by R. Faulder & Son (1808) in *Sermons and Tracts by the Late Rev. William Paley, D. D., Archdeacon of Carlisle, Subdean of Lincoln, etc., etc.*
54. Halévy (1949) p. 392.
55. Hey (1796–98), *Lectures in Divinity Delivered in the University of Cambridge,* 4 vols, vol. ii, p. 2; see also vol. ii, p. 47. Hereafter cited as Hey, *Lectures.*
56. Winstanley (1935) p. 176.
57. Stephen (1881), vol. i, p. 426.
58. Overton and Relton (1906) p. 262.
59. Hey, *Lectures,* vol. ii, pp. 48, 200, 43.
60. Hey, *Lectures,* vol. ii, pp. 93, 94, 97.
61. Hey, *Lectures,* vol. ii, pp. 97, 98, 6. Hey even noted with approval that "the *Copts* in Ægypt have divine service in a language they do not understand:" Hey, *Lectures,* vol. ii, p. 97.

62. Hey, *Lectures,* vol. ii. pp. 99, 102.
63. There is no note 63 in the original.
64. Hey, *Lectures,* vol. ii, p. 101. Hey's belief that there can be progress in understanding of scripture was shared by Edmund Law and it is instructive to contrast the inferences each drew from this. Law concluded that assent to formularies should not be required: Hey that it should be. See Gascoigne (1986) p. 25 and n. 17.
65. Stephen (1881), vol. i. p. 420. Stephen quotes Foster: "Where mystery begins religion ends," and suggests that the sentiment is at least characteristic of the school ["of Paley"] Did Stephen actually read Hey's lectures?
66. *Tracts for the Times. No.* 87, *On Reserve in Communicating Religious Knowledge,* part 4 (London: 1840) pp. 21–2: cf. Hey, *Lectures,* vol. ii, pp. 100–3. See also *Tract* 87, 56, for a view of the dangers of "system" which closely resembles that of the Cambridge school.
67. This was the edition used by Stephen: its pagination differs greatly from the first edition used in this article.
68. The author is grateful to the librarian of Trinity College, Toronto, for access to the original collection of the university.
69. Pretyman-Tomline (1812), *Elements of Christian Theology,* 2 vols, vol. ii, p. 572. Hereafter cited as Pretyman-Tomline, *Elements.*
70. Pretyman-Tomline, *Elements,* vol. i, pp. xv–xvii. The citation of Lardner (a Dissenter) and Macknight (a Scotch Presbyterian) is revealing for it shows that Pretyman agreed with Watson, as against Anglicans who "never read dissenting Divinity," that what matters is not "the quarter from whence the matter was taken, but whether it was good." See Watson, *Tracts,* vol. i. p. xix.
71. Pretyman-Tomline, *Elements,* vol. i, p. xiii.
72. King's scholars proceeded automatically to the BA and to a fellowship by lapse of time, being exempt by statute from university examinations. No King's man read for the Tripos until after the changes of the 1860s.
73. The author is indebted to Dr Michael Halls, modern archivist, for access to the library records.
74. John Overton (1802), *The True Churchman Ascertained: or, An Apology for those of the Regular Clergy of the Establishment who are sometimes called Evangelical Ministers: occasioned by the publications of Drs Paley, Hey, Croft; Messrs. Daubeny, Ludlam, Polwhele, Fellowes; the Reviewers, etc., etc.,* 2nd edn, pp. 156; 19, 24, 141;129, 245–7.
75. [J. B. Sumner], *Apostolical Preaching Considered, in an Examination of St. Paul's Epistles,* London 1815; hereafter cited as Sumner, *Apostolical.* See J. H. Newman, *Apologia Pro Vita Sua,* new impression, 1929, pp. 8–9.
76. Sumner, *Apostolical,* pp. v, 29, 58–60, 61–2; cf. Watson, *Tracts,* vol. i, pp. xvi–xvii.
77. J. B. Sumner (1802), *An Essay Tending to Show that the Prophecies, Now Accomplishing, Are an Evidence of the Truth of the Christian Religion,* pp. 9, 30–47. Like "Paley and his School" Sumner believed, as Watson put it, that the thesis *Pontifiex Romanorum est ille Antichristus quem futuram Scriptura praedixit* is a "primary pillar of the reformed faith:" see Watson, *Tracts,* vol. v, p. vii.
78. Sumner, *Apostolical,* p. 179.

79. Sumner, *Apostolical*, pp. 250, 251, 252, 242.
80. Sumner, *Apostolical*, pp. 24, 33, 69, 247.
81. Sumner (1816), *A Treatise on the Records of the Creation: with particular reference to Jewish history, and the consistency of the principle of population with the wisdom and goodness of the deity*, 2 vols, vol. i, pp. xii, 22; vol. ii, 252, 255, 262, 296–7, 362–2, 372–2.
82. Sumner (1824), *The Evidence of Christianity Derived from Its Nature and Reception*, pp. 162, 192, 205, 329–30, 407.
83. Carus (ed.) (1847), *Memoirs of the Life of the Rev. Charles Simeon, M. A., Late Fellow of King's College, and Minister of Trinity Church, Cambridge, With a Selection of his Writings and Correspondence*, p. 178.
84. The author is indebted to Dr William Parry, the librarian, for access to the library records.
85. [Edward Copleston] (1819), *A Second Letter to the Right Hon. Robert Peel, M.P. for the University of Oxford, on the Causes of Pauperism and on the Poor Laws, By one of his Constituents*, pp. 23–4.
86. Though Whately was dissatisfied with Paley's moral philosophy, his hatred of "party spirit," his willingness to affirm the orthodox formularies combined with his detestation of any persecution of heterodoxy, his powerful attack on Humean skepticism, his devotion to logic and his Whiggish political conservatism mark him as one born out of his time, accidentally transposed from Cambridge of the 1780s.
87. Stephen (1881), vol. i, p. 380.

6 *Wealth of Nations* as Theology

1. Stigler (1982) p. 108.
2. Heyne (1976)
3. E.g. Walsh (1961).
4. McCloskey (1983, 1985).
5. McCloskey (1985) pp. 4–9.
6. Sagoff (1997) pp. 968, 972, 980.
7. Nelson (1991)
8. Nelson (2001)
9. Tillich (1967)
10. Smith, Adam (1776), *An Enquiry into the Nature and Causes of the Wealth of Nations* (hereafter abbreviated in the text as *WN*). Republished, eds (1976) R. R. Campbell, A. S. Skinner and W. B. Todd.
11. E.g. Viner (1958); Smith (1759); Raphael (1985); Teichgraeber (1986); Nicholls (1992); Minowitz (1993); Fitzgibbons (1995); Winch (1996).
12. Brown (1994) p. 13.
13. Smith, *WN*.
14. Smith, *WN* Book IV, ch. ix, para. 28.
15. Glahe (ed.) (1993) pp. 345–6.
16. Smith (1759), *The Theory of Moral Sentiments*.
17. See Chapter 7 below.
18. Boswell (1791), *Boswell's Life of Johnson, Together with Boswell's Journal of a Tour to the Hebrides …*, 6 vols, vol. ii, p. 323. ed. (1934) George Birkbeck Hill.

19. Ricardo (1820?), *The Works and Correspondence of David Ricardo,* vol. vii, pp. 212–3.
20. Leibniz (1710), *Theodicy: Essays on the Goodness of God, the Freedom of Man and the Origin of Evil.*
21. Williams (1927).
22. Augustine of Hippo, Saint (413–26), *De Civitate Dei contra Paganos,* iv, p. 2. Hereafter cited as Augustine, *De Civitate.*
23. Augustine, *De Civitate,* ii, p. 21.
24. Waterman, *Revolution, Economics and Religion,* pp. 76–7.
25. Heimann (1978).
26. Pownall (1776), *Letter to Adam Smith, being an examination of several points of doctrine laid down in his inquiry, into the nature and causes of the wealth of nations,* pp 1, 23, 48.
27. Newton (1756), *Four letters from Sir Isaac Newton to Dr Bentley containing some arguments in proof of a deity,* p. 1.
28. Maclaurin (1775), *An account of Sir Isaac Newton's philosophical discoveries,* 3rd edn, p. 4.
29. E.g. Minowitz (1993) p. 141 and *passim.*
30. Minowitz (1993), ch. 7.
31. Stewart (1793), *Account of the Life and Writings of Adam Smith, LL.D.,* p. 274.
32. Ross (1995) p. 109.
33. See Chapter 2 in this volume.

7 The Sudden Separation of Political Economy

1. Coats (1993) vol. ii.
2. E.g. Kleiman (1987)
3. The sections on "Diminishing Returns, Misery and Vice," "Ideological Crisis in the 1820s" and "Whately's Demarcation" are summaries of my previous work in these areas and I must apologize for repeating them here. I have done so because they are important elements in the larger story I now wish to tell, a story which was not apparent to me when I began my research more than a decade ago.
4. E.g. Gay (1966–69).
5. Chadwick (1975) p. 9.
6. E.g. Cassirer (1951) pp. 7–12; Berlin (1956) p. 14.
7. Gilley (1981).
8. Ross (1995) p. 100.
9. Maclaurin (1775) *An Account of Sir Isaac Newton's Philosophical Discoveries,* 3rd edn., pp. 3, 5. Hereafter cited as Maclaurin, *An Account.*
10. Maclaurin, *An Account,* p. 401.
11. Maclaurin, *An Account,* pp. 410–11.
12. Maclaurin, *An Account,* p. 411.
13. E.g. Butler (1726), *Fifteen Sermons Preached at the Rolls Chapel* and *A Dissertation upon the Nature of Virtue,* Sermon xv. Hereafter cited as Butler, *Fifteen Sermons.*
14. Pocock (1985) p. 537 *passim;* see also Gascoigne (1989).
15. Jacob (1981).

16. Berkeley (1735–37), *The Querist,* vol. vi in *The Works of George Berkeley,* 9 vols, pp. 87–184. Hereafter cited as Berkeley, *Works.*
17. Berkeley, *Works,* vol. vi, pp. 105, 103.
18. Mandeville (1714–28), *The Fable of the Bees: Or, Private Vices, Public Benefits,* 2 vols. Hereafter cited as Mandeville, *Bees.*
19. Mandeville, *Bees,* F. B. Kaye (ed.) "Introduction," vol. i.
20. Butler, *Fifteen Sermons.*
21. Waterman (1997), "Recycling Old Ideas: Economics Among the Humanities", *Research in the History of Economic Thought and Methodology,* 15: 237–49. pp. 240–1.
22. E.g. Butler, *Fifteen Sermons,* p. 166.
23. Butler, *Fifteen Sermons,* pp. 32, 36, 37–8, 67.
24. Butler, *Fifteen Sermons,* pp. 164–202.
25. Hayek (1978) p. 263.
26. Hume (1739–40), *A Treatise of Human Nature.* Hereafter cited as Hume, *A Treatise.*
27. Hume, *Political Essays,* Essay 14. Hereafter cited as Hume, *Political Essays.*
28. Hume, *A Treatise,* p. xxi.
29. Hayek (1978) p. 264.
30. E.g Hamowy (1987).
31. Shelton (1981) pp. 126–32.
32. Marx, *Capital* (1867–1895), vol. i, p. 711 n. 2. Transl. (1954) S. Moore, E. Aveling, F. Engels.
33. Tucker (1753–65), *The Collected Works of Josiah Tucker,* vol. iv. Hereafter cited as Tucker, *Works*
34. Tucker, *Works,* vol. iii, p. 58.
35. Tucker, *Works,* vol. iii, p. 48.
36. Shelton (1981) p. 165.
37. Tawney (1925) p. 192.
38. Tawney (1925) p. 10.
39. However, I have lately discovered in the archives of Sidney Sussex College, Cambridge a bound set of lecture notes on "Moral and Political Philosophy" by a former Fellow of the College, the Revd John Hey (1734–1815), later the first Norrisian Professor of Revealed Theology. From internal evidence the lectures appear to date from the early 1770s, though frequently revised until the mid-1780s. Three or four of these lectures deal with what we should now recognize as price theory, and arise in the context of Christian casuistry and the doctrine of the "just price."
40. Mizuta (1996).
41. Smith (1759), *The Theory of Moral Sentiments.*
42. Smith (1776), *An Inquiry into the Nature and Causes of the Wealth of Nations.*
43. Ross (1995) p. 118.
44. E.g. Raphael and Macfie (1976) pp. 19–20.
45. Ross (1995) p. 109.
46. Ross (1995) p. 382.
47. Butler, *Fifteen Sermons,* Sermon XV.
48. Mandeville, *Bees,* vol. ii, p. 104.
49. Minowitz (1993) pp 139, 40.

50. "Providence" is mentioned twice in the Introduction to the *Essay on, Trade,* and at one other point "Liberty of Conscience" is explicated as "Every Man is permitted to worship GOD in the Way he thinks the *right* and *true*" (Tucker, *Works,* vol. ii, p. xi, p. 33). Aside from these there is no other trace of theological language; less indeed than in *WN*.
51. Minowitz (1993).
52. See Chapter 6 in this volume.
53. Ross (1995) p. 340.
54. See Ross (1995) pp 59, 118.
55. E.g. Samuelson (1978).
56. However, I have recently argued that Smith's account of the relation between the profit rate and capital accumulation rules out any "canonical" interpretation of *WN* that incorporates diminishing returns into its growth theory. Waterman (1999), "Hollander on the 'Canonical Classical Growth Model': A Comment," *Journal of the History of Economic Thought,* 21, 3: 311–13.
57. Eltis (1984) ch. 3.
58. E.g. Negishi (1993).
59. Plato, *The Republic,* pp. 158–9. Translation (1954) P. Shorey.
60. Schumpeter (1954) pp. 251, 254, 257; see also Stangeland (1904).
61. Steuart (1767), *An Inquiry into the Principles of Political Oeconomy,* 2 vols, vol. i, p. 37. Hereafter cited as Steuart, *Inquiry.*
62. Steuart, *Inquiry,* vol. i, pp. 130–1.
63. Schumpeter (1954) pp. 259–60.
64. Turgot (1768), "Observations sur le mémoire de M. de Saint-Péravy en faveur de l'impôt indirect," in *Écrits Économiques.*
65. Cantillon (1755), *Essai sur la Nature du Commerce en Général,* p. 82.
66. Schumpeter (1954) p. 257.
67. Wallace (1761), *Various Prospects of Mankind, Nature and Providence.*
68. Malthus (1798), *An Essay on the Principle of Population as it Affects the future Improvement of Society with Remarks upon the Speculations of Mr Godwin, M. Condorcet, and Other Writers,* p. 142.
69. Tucker, *Works,* vol. i, pp. 127, 128.
70. Steuart, *An Inquiry,* vol. i, p. 116.
71. Paley (1785), *The Principles of Moral and Political Philosophy,* in W. Paley (1825), *The Works of William Paley, DD,* ed. Edmund Paley, 7 vols, vol. iv. Hereafter cited as Paley, *Works.*
72. Keynes (1933) vol. x p. 79.
73. Waterman (1996) "Why William Paley was 'The First of the Cambridge Economists'", *Cambridge Journal of Economics* 20: 673–86. pp. 674–6.
74. See Chapter 5 in this volume.
75. Waterman (1996) "Why William Paley was 'The First of the Cambridge Economists.'"
76. Waterman (1996), "Why William Paley was 'The First of the Cambridge Economists,'" p. 681; Winch (1996) pp. 370–1.
77. Cited in Keynes (1971) p. 83.
78. Paley, *Works,* vol iv, p. 480.
79. Malthus (1798), *Essay on Population,* pp. 143–4.
80. Malthus (1798), *Essay on Population,* pp. 15–16.

81. Waterman (1983), "Malthus as a theologian: the 'First Essay' and the relation between political economy and Christan theology," in *Malthus: Past and Present,* eds J. Dupâquier, A. Fauve-Chamoux and J. Grebenik.
82. Waterman (1991c), *Revolution, Economics and Religion: Christian Political Economy*, pp. 126–35, 160–70.
83. Malthus (1817), *Essay on Population*; see Malthus (1803), *Essay on Population*, ed. (1989) P. James, variora edn with variora of 1806, 1807, 1817, 1826, 2 vols, vol. ii, p. 250.
84. Stigler (1952) Lloyd (1969) Waterman (1987) "On the Malthusian theory of long swings," *Canadian Journal of Economics* 20, 2: 257–70.
85. Hollander (1997) pp. 27–39.
86. Waterman (1991c), *Revolution, Economics and Religion: Christian Political Economy*, p. 112.
87. Paley, *Works*, vol. iv, pp. 159–71.
88. Malthus (1803), *Essay on Population*, in ed. (1989) P. James, 2 vols, vol. ii, p. 127.
89. Winch (1996) pp. 402, 418.
90. James (1979) pp. 116–21; Pyle (1994).
91. Anon. (1807), *A Summons of Awakening, or, the Evil Tendency and Danger of Speculative Philosophy*, p. 123.
92. Toynbee cited in Winch (1996) p. 6.
93. Cited in Jones (1968) p. 437.
94. Boswell (1934) vol. ii p. 323, ed. (1934) G. B. Hill.
95. O'Brien (1993) p. 144, n. 1.
96. Burke, *The Writings and Speeches of Edmund Burke,* eds (1981–97) P. K. Langford *et al.*, 9 vols, vol. ix, p. 125.
97. *Eclectic Review*, January 1832, p. 9.
98. Carey (1856), *Principles of Political Economy*, vol. iii, p. iv.
99. Waterman (1991c), *Revolution, Economics and Religion,* pp. 160–76.
100. Waterman (1991c), *Revolution, Economics and Religion,* ch. 5.
101. Waterman (1991c), *Revolution, Economics and Religion,* p. 202.
102. Whately, "Oxford Lectures on Political Economy," *Edinburgh Review* (September 1828) pp. 171, 172.
103. Waterman (1991c), *Revolution, Economics and Religion,* pp. 180–6, 204–6.
104. Whately, E. J. (ed.) (1866), *Life and Correspondence of Richard Whately, D. D., late Archbishop of Dublin*, 2 vols, vol. i, 66–7. Hereafter cited as Whately, *Life and Correspondence*.
105. Whately (1832), *Introductory Lectures in Political Economy*. Hereafter cited as Whately, *Introductory Lectures*.
106. Whately, *Introductory Lectures*, p. vi.
107. See Chapter 8 in this volume.
108. Hinds (1831), *An Inquiry into the Proofs, Nature and Extent of Inspiration and into the Authority of Scripture*, pp. 5, 7, 150–1.
109. Whately, *Introductory Lectures*, pp. 29, 30, 31.
110. Whately, *Introductory Lectures*, pp. 28–9.
111. Whately, *Introductory Lectures*, p. 32; my italics.
112. Corsi (1987).
113. Senior (1852), *Four Introductory Lectures on Political Economy, Delivered before the University of Oxford.*

114. Keynes (1891).
115. Robbins (1932).
116. Whately, *Life and Correspondence,* vol. i, pp. 66–7.
117. Samuelson (1958) pp. 37–8.
118. Whately, *Introductory Lectures,* pp. 93–4, 96.
119. *Westminster Review,* January 1832, p. 10.
120. Winch (1993).
121. James (1979) p. 167.
122. The expression "the spontaneous order of nature" first occurs in J. S. Mill's posthumous essay "On Nature" in "Three Essays in Religion," vol. x of *Collected Works of John Stuart Mill (*1963–91) 33 vols.
123. Darwin (1859), *The Origin of Species by Means of Natural Selection, or, The Preservation of Favoured Races in the Struggle for Life.*
124. Keynes (1933), *Essays in Biography,* vol. x, in eds E. Johnson and D. Moggridge (1971), *The Collected Writings of John Maynard Keynes,* 29 vols, p. 168.
125. E.g. Block, Brennan and Elzinga (1985), Block and Hexham (1986).
126. E.g. Brennan and Waterman (1994); Dean and Waterman (1998); see Waterman (1987) "Economists on the relation between political economy and Christian theology: A preliminary survey," *International Journal of Social Economics* 14, 6: 46–68.
127. Villeneuve-Bargemont (1834), *Economie Politique Chrétienne, ou Recherches sur la Nature et les Causes du Paupérisme, en France et en Europe, et sur les Moyens de le Soulager et de le Prévenir,* vol. i, p. 20; my italics.
128. *Dublin Review,* July 1837, p. 175.
129. See Chapter 11 in this volume.
130. E.g. Vickers (1975) (1976), Tiemstra (1990).

8 Methodology of Classical Political Economy

1. *Edinburgh Review,* October 1837, p. 77.
2. Ekelund and Hébert (1990) pp. 74–5.
3. Hollander (1987) p. 38.
4. Keynes (1930) pp. 34–5.
5. Robbins (1952) p. 24.
6. Robbins (1952) p. 147.
7. Lipsey, Purvis, and Steiner (1988) p. 16.
8. E.g., Katouzian (1980) McCloskey (1986).
9. Waterman (1983), "The ideological alliance of political economy and Christian theology, 1798–1833," *Journal of Ecclesiastical History* 34: 231–44.
10. Malthus (1798), An Essay on the Principle of Population. Hereafter cited as Malthus (1798) *Essay on Population.*
11. Godwin (1798), *Enquiry Concerning Political Justice and Its Influence on Morals and Happiness,* 3rd edn. Hereafter cited as Godwin (1798), *Political Justice.*
12. Paley (1800), *Natural Theology,* vol. v in *The Works of William Paley, D. D.,* ed. (1825) Edmund Paley, 7 vols. Hereafter cited as Paley, *Works.*

13. Sumner (1816), *A Treatise on the Records of Creation; with Particular Reference to the Jewish History, and the Consistency of the Principle of Population with the Wisdom and Goodness of the Deity*, 2 vols. Hereafter cited as Sumner, *Treatise on Creation.*
14. Malthus (1803) (1817), *Essay on Population,* 2nd and 6th edns.
15. Copleston (1819), *A Letter to the Right Hon. Robert Peel, MP for the University of Oxford, on the Pernicious Effect of a Variable standard of Value, especially as it regards the Condition of the Lower Orders and the Poor Laws*. Hereafter cited as Copleston, *Letter to Robert Peel.*
16. Waterman (1991c), *Revolution, Economics and Religion.*
17. Halévy (1952) p. 266.
18. Ricardo (1817), *On the Principles of Political Economy,* vol. i in *The Works and Correspondence of David Ricardo,* 11 vols, hereafter cited as Ricardo, *Works.*
19. Hazlitt (1928) p. 183.
20. *Eclectic Review*, January 1832, p. 9.
21. Wordsworth, *Poetical Works,* ed.(1950) E. de Selincourt, pp. 354–5.
22. Whately (1826), *Elements of Logic, Comprising the Substance of the Article in the Encyclopedia Metropolitana ...*, p. 22. Hereafter cited as Whately, *Elements of Logic.*
23. Whately (1866), *Life and Correspondence of Richard Whately, D. D., Late Archbishop of Dublin,* 2 vols, vol. i, pp. 47–8. Hereafter cited as Whately, *Life and Correspondence.*
24. Senior (1827), *An Introductory Lecture in Political Economy, delivered before the University of Oxford on the 6th day of December, 1826*, p. 7. Hereafter cited as Senior, *Introductory Lecture in Political Economy.*
25. Senior, *Introductory Lecture in Political Economy,* pp. 8, 10.
26. Senior, *Introductory Lecture in Political Economy,* pp. 7, 11.
27. Senior, *Introductory Lecture in Political Economy,* pp. 11, 12, 13, 16.
28. Whately (1828), *Oxford lectures on Political Economy,* p. 171. Hereafter cited as Whately, *Oxford Lectures.*
29. Whately, *Oxford Lectures,* p. 173.
30. Whately, *Oxford Lectures,* pp. 171, 172.
31. Whately, *Oxford Lectures,* p. 31.
32. Whately, *Life and Correspondence*, vol. i, pp. 66–7.
33. Whately (1832), *Introductory Lectures on Political Economy,* p. vi. Hereafter cited as Whately, *Introductory Lectures.*
34. Waterman (1991c), *Revolution, Economics and Religion,* pp. 66–70.
35. Whately, *Oxford lectures,* p. 172.
36. Hinds (1831), *An Inquiry into the Proofs, Nature and Extent of Inspiration, and into the Authority of the Scripture*, pp. 5, 7. Hereafter cited as Hinds, *Inquiry into the Proofs.*
37. Hinds, *Inquiry into the Proofs,* pp. 150–1.
38. Hinds, *Inquiry into the Proofs,* p. 152.
39. Hinds, *Inquiry into the Proofs,* p. 52.
40. Whately, *Introductory Lectures,* pp. 29, 30, 31.
41. Whately, *Introductory Lectures,* pp. 28–9.
42. Whately, *Introductory Lectures,* p. 32, my italics.
43. Whately, *Introductory Lectures,* p. 43.

44. Stewart (1854–60), *The Collected Works of Dugald Stewart, Esq., etc.*, 11 vols, vol. iii. Hereafter cited as Stewart, *Works*.
45. Stewart, *Works*, vol. iii.
46. Pryme (1823), *An Introductory Lecture and Syllabus to a Course Delivered in the University of Cambridge on the Principles of Political Economy*, p. vii, cit. in Waterman (1991c) *Revolution, Economics and Religion*, p. 160.
47. Corsi (1987) p. 94.
48. Jeffrey (1810), "Stewart's philosophical essays," *Edinburgh Review*, 7: pp. 167–211.
49. Stewart, *Works*, vol. iii, pp. 322, 327, 329, 328, 329.
50. Stewart, *Works*, vol. iii, pp. 331, 333.
51. Stewart, *Works*, vol. iii, p. 332.
52. Stewart, *Works*, vol. iii, p. 331.
53. Copleston, *Letter to Robert Peel*, pp. 61–2.
54. Whately, *Introductory Lectures*, pp. 6, 7.
55. Whately, *Introductory Lectures*, pp. 18, 19, 21–2.
56. Whately, *Introductory Lectures*, p. 68.
57. Whately, *Introductory Lectures*, p. 31.
58. Whately, *Introductory Lectures*, pp. 237–8.
59. Whately, *Introductory Lectures*, p. 69; see also p. 239.
60. Whately, *Introductory Lectures*, p. 79.
61. Whately, *Introductory Lectures*, p. 225.
62. Whately, *Introductory Lectures*, p. 91.
63. Whately, *Introductory Lectures*, pp. 97, 61, 145, 146.
64. Whately, *Introductory Lectures*, p. 99.
65. Whately, *Introductory Lectures*, pp. 93–4, 96.
66. Whately, *Life and Correspondence*, vol i, p. 67.
67. Samuelson (1958) pp. 37–8.
68. *Westminster Review*, January 1832, p. 10.
69. Whately, *Introductory Lectures*, pp. 23, 228.
70. Whately, *Introductory Lectures*, p. 240.
71. Whately, *Elements of Logic*.
72. Whately, *Introductory Lectures*, p. 49.
73. Whately, *Introductory Lectures*, pp. 49, 50.
74. Whately, *Introductory Lectures*, p. 59.
75. Paley (1802), *Works*, vol. v
76. Whately, *Introductory Lectures*, p. 103.
77. Whately, *Introductory Lectures*, pp. 147–8.
78. Waterman (1991c), *Revolution, Economics, and Religion*, pp. 213–15.
79. Senior (1852), *Four Introductory Lectures on Political Economy, delivered before the University of Oxford*. Hereafter cited as Senior, *Four Introductory Lectures*.
80. Senior, *Four Introductory Lectures*, Lecture II
81. Senior, *Four Introductory Lectures*, Lecture III
82. Mill (1874), *Essays on Some Unsettled Questions of Political Economy*.
83. Senior, *Four Introductory Lectures*, Lecture IV.
84. McCulloch (1843), *Principles of Political Economy*, new edn, pp. vi–x.
85. Senior (1836), *Outline of the Science of Political Economy*, p. 3.
86. Senior, *Four Introductory Lectures*, pp. 23–5.
87. Bowley (1937) pp. 54, 65.

9 Peasants, Population and Progress

1. Macfarlane (1979), ch. 1.
2. Macfarlane (1979) p. 165.
3. Snooks (1990); McDonald and Snooks (1986).
4. Macfarlane (1979) p. 5.
5. E.g. Tawney (1912); Polanyi (1957).
6. Macfarlane (1979) p. 189.
7. Macfarlane (1979) p. 199.
8. Macfarlane (1979) p. 185.
9. Johnson (1755), *A Dictionary of the English Language.*
10. Malthus (1798), *An Essay on the Principle of Population,* pp. 73, 133. Hereafter cited as Malthus (1798), *Population.*
11. Malthus (1799), *The Travel Diaries of Thomas Robert Malthus,* pp. 48–9, 59–60, 63–5. Hereaftaer cited as Malthus, *Travel.*
12. Malthus, *Travel,* p. 64.
13. Malthus, *Travel,* pp. 118, 131, 202, 278, 282, 244.
14. Chalmers (1832), *On Political Economy in Connexion with the Moral State and Moral Prospects of Society,* pp. 2–11. Hereafter cited as Chalmers, *Political Economy*
15. Chalmers, *Political Economy* p. 461.
16. E.g. Hunt (1992) pp. 87–102.
17. *Book of Common Prayer,* Ordering of Deacons.
18. James (1979); Pullen (1987); Waterman (1991c), *Revolution, Economics and Religion,* pp. 83–7.
19. Laslett (1965) pp. 54–5.
20. Marshall (1985); Martin (1972), ch. 2.
21. Woodforde (1924–31), *The Diary of a Country Parson.*
22. Ferguson (1956).
23. Brown (1982) p. 23.
24. Brown (1982) *passim.*
25. Marx (1867–95), *Capital,* pp 149, 548, 574, 654–5, 679–81. Trans (1930) E. and C. Paul.
26. Hunt (1992) pp. 80, 87, 88, 93 and ch. 4 *passim.*
27. Waterman (1991c), *Revolution, Economics and Religion,* chs 2, 3 and *passim.*
28. Malthus (1798), *Population* pp. 95–9.
29. Malthus (1798), *Population* p. 290.
30. Malthus (1798), *Population* p. 95.
31. Malthus (1803), *Essay on Population,* ii p. 203. ed. (1989a) Patricia James, 2 vols. *The Version Published in 1803, with the Variora of 1806, 1807, 1817 and 1826.* Hereafter cited as Malthus (1803), *Population.*
32. Chalmers, *Political Economy,* pp. 54–5; Waterman "The 'Canonical Classical Model' in 1808 as viewed from 1825: Thomas Chalmers on the National Resources," *History of Political Economy,* 23, p. 233 and Appendix.
33. Samuelson (1978).
34. Chalmers (1808), *An Enquiry in the Nature and Stability of Natural Resources;* Hereafter cited as Chalmers, *Natural Resources;* Waterman (1991), "The 'Canonical Classical Model' in 1808 as viewed from 1825: Thomas Chalmers on the National Resources."

35. Malthus (1820), *Principles of Political Economy, Variorum Edition,* ed. (1989a), John Pullen, 2 vols.
36. Eltis (1980); Costabile and Rowthorn (1985).
37. Chalmers (1826), *The Christian and Civic Economy of Large Towns,* vol. iii Hereafter cited as Chalmers, *Christian and Civic Economy.*
38. Waterman (1992) "Analysis and Ideology in Malthus's *Essay on Population," Australian Economic Papers,* 31: 203–17.
39. Malthus (1815), *An Inquiry into the Nature and Progress of Rent, and the Principles by which it is Regulated*; West (1815), *Essay on the Application of Capital to Law: with Observation Shewing the Impolicy of any Great Restriction on the Importation of Corn ... by a Fellow of University College*; Ricardo (1815), *An Essay on the Influence of a Low Price of Corn on the Profit of Stock ...*; Torrens (1815), *An Essay on the External Corn Trade.*
40. Waterman (1987), "On the Malthusian theory of long swings," *Canadian Journal of Economics,* 20: 257–70; Waterman (1988), "Hume, Malthus and the stability of equilibrium," *History of Political Economy,* 20: 85–94; Waterman (1991), "The "Canonical Classical Model" in 1808 as viewed from 1825; Thomas Chalmers on the National Resources," *History of Political Economy* 23; Waterman (1992), "Analysis and Ideology in Malthus"s *Essay on Population," Australian Economic Papers,* 31: 203–17, 221–41.
41. Samuelson (1978)
42. Stigler (1952); Lloyd (1969).
43. Malthus (1798), *Population,* p. 83.
44. Malthus (1798), *Population,* pp. 205, 284.
45. Chalmers, *Natural Resources,* p. 258.
46. Waterman (1991c), "The "Canonical Classical Model" in 1808 as viewed from 1825: Thomas Chalmers on the National Resources."
47. Chalmers, *Political Economy,* p. iii; my italics.
48. Chalmers, *Poltical Economy,* pp. 420, 422.
49. Malthus (1822), *Letter to Chalmers.*
50. James (1979) pp. 126–36, 171–5, 449–54.
51. Malthus (1798), *Population,* p 50.
52. Malthus (1798), *Population,* p. 35; see also pp. 92, 310.
53. Malthus (1798), *Population,* p. 77.
54. Malthus (1798), *Population,* pp. 84–5.
55. Malthus (1798), *Population,* pp. 92, 93.
56. Malthus (1803), *Population,* vol. ii, pp. 137–47.
57. Malthus (1803), *Population,* vol. i, p. 287.
58. Chalmers, *Political Economy,* ch. xiv.
59. Chalmers, *Political Economy,* p. 383; Malthus (1803), *Population,* vol. I, pp. 345, 346–7.
60. Chalmers, *National Resources.*
61. Malthus (1827), *Letter to Chalmers.*
62. Malthus (1803), *Population,* vol. ii, pp. 137–47.
63. Malthus (1803), *Population,* vol ii, pp. 156–3; Chalmers, *On Political Economy,* ch. xiv, Brown (1982), *passim.*
64. Lloyd (1833), *Two Lectures on Checks to Population, delivered before the University of Oxford, in Michaelmas Term 1832 ...,* p. 482; Hollander (1986) p. 231, n. 40; Waterman (1991a) pp. 142–3.

65. Houlbrooke (1984) p. 62.
66. Houlbrooke (1984) ch. 4; Hajnal (1965) pp. 101–143; Wrigley and Scholfield (1982) pp. 257–65, 423–44.
67. Malthus (1803), *Population,* vol. i, p. 226.
68. Malthus (1803), *Population,* vol. ii, p. 105.
69. Malthus (1803), *Population,* vol. ii, p. 105.
70. Malthus (1803), *Population,* vol. ii, p. 182.
71. Waterman (1991a), *Revolution, Economics and Religion,* pp. 157–9, 169.
72. Malthus (1803), *Population,* vol. ii, pp. 151, 153, 154, 155, 227.
73. Chalmers (1814), *The Influence of Bible Societies on the Temporal Necessities of the Poor*, p. 11. Hereafter cited as Chalmers, *Influence of Bible Societies.*
74. Chalmers, *Influence of Bible Societies*, p. 12.
75. Malthus (1803), *Population,* vol. i, pp. 161, 162; Chalmers, *Influence of Bible Societies*, p. 12; Chalmers, *Political Economy*, p. 424, etc; cp II Thess 3: 10.
76. Chalmers, *Political Economy*, pp. 411, 424.
77. Chalmers, *Political Economy*, pp. 242, 425.
78. Chalmers, *Political Economy*, p. 71.

10 Property Rights in Christian Social Teaching

1. Keynes (1891); Robbins (1932); Hutchison (1938); Knight (1940); Koopmans (1957).
2. Friedman (1953).
3. E.g. Nagel (1963); Samuelson (1963); Boulding (1968); Wong (1973); Morgenstern (1974).
4. Lakatos and Musgrave (eds) (1985) and Lakatos (1978).
5. E.g. Lipsey (1981).
6. E.g Latsis (1976); Blaug (1978).
7. Worland (1977); Varian (1974–5).
8. Hollis and Nell (1975); Lindbeck (1977); Hunt and Schwartz (1972).
9. Vickers (1976); Goudzwaard (1979).
10. Furubotn and Pejovich (1972).
11. Locke (1690), *Two Treatises of Government* II: 25, ed. (1967) P. Laslett. References identify Treatise and paragraph. Hereafter cited as Locke, *Two Treatises.*
12. Locke, *Two Treatises* I, II.
13. Laslett (1967) pp. 45–51.
14. See Laslett (1967), ch. III; Tully (1980) pp. 53–9.
15. Kelly (1977) p. 84.
16. *Genesis*, I: 28.
17. *Genesis*, I: 28.
18. Tully (1980) p. 98.
19. Locke, *Two Treatises* I: 28, 29.
20. Locke, *Two Treatises* I: 86.
21. Locke, *Two Treatises* II: 25.
22. MacPherson (1962) p. 201.
23. MacPherson (1962) p. 201.
24. Locke, *Two Treatises* II: 27.

25. Locke, *Two Treatises* II: 27.
26. Locke, *Two Treatises* II: 27.
27. Locke, *Two Treatises* II: 31.
28. Macpherson (1962) p. 201.
29. Locke, *Two Treatises* II: 123.
30. Locke, *Two Treatises* II: 127–130.
31. Locke, *Two Treatises* II: 120.
32. Tully (1980) pp. 164–5.
33. Hooker (1593–1662), *The Works of that Learned and Judicious Divine, Mr Richard Hooker*, Book I, ch. x, para. 8.
34. Tully (1980) p. 167.
35. Tuck (1979) p. 13.
36. Tuck (1979) p. 16.
37. Tuck (1979) p. 18.
38. Thomas Aquinas, *Summa Theologiae,* vol xxxvii, ed. and trans. (1975a) Gilbey, QQ II–ii, pp. 57–62, 60 vols. Hereafter cited as Aquinas, *Summa*
39. Tuck (1979) pp. 20–4.
40. Tuck (1979) pp. 25–28.
41. Tuck (1979) pp. 26–7.
42. Tuck (1979) p. 29 and *passim.*
43. Tully (1980) p. 116.
44. Locke, *Two Treatises* II: 25.
45. Tully (1980) pp. 38–50, 59–63.
46. Locke, *Two Treatises* II: 27.
47. Locke, *Two Treatises* II: 27.
48. Tully (1980) pp. 104–10.
49. Hundert (1972) p. 16.
50. Soloway (1969) p. 70.
51. Marx, *Capital*, trans. (1930) E. and C. Paul, p. 680n.
52. Chalmers (1833), *On the Power, Wisdom, ... and Goodness of God, as Manifested in the Adaptation of External Nature to the Moral and Intellectual Constitution of Man*, pp. 190, 192.
53. Locke, *Two Treatises* II: 42.
54. Hundert (1972) p. 16.
55. Schumpeter (1954) p. 120; c.f. Laslett (1967) p. 103; Olivecrona (1974) pp. 229–301.
56. Schumpeter (1954) p. 310.
57. Schumpeter (1954) pp. 59–98; Hundert (1972) p. 22.
58. Locke, *Two Treatises* II: 94, see also pp. 95, 124, 127, 131, 134.
59. Viner (1963) p. 554.
60. MacPherson (1962) pp. 220–1.
61. MacPherson (1962) p. 199.
62. MacPherson (1962) pp. 199–220.
63. Viner (1963) p. 558; Hundert (1972) pp. 16–17.
64. Tully (1980) p. 131.
65. Locke, *Two Treatises* II: 28.
66. Laslett (1964) pp. 152–3; Hundert (1972) p. 17; Tully (1980) pp. 136–9.
67. Tully (1980) pp. 131–5.
68. Tully (1980) pp. 98–101.

69. Laslett (1967) p. 69.
70. Clark (1979).
71. Viner (1978) p. 71.
72. Viner (1978) p. 70.
73. Carlen (1981) vol ii.
74. Camp (1969) pp. 1–13.
75. Camp (1969) p. 53.
76. Carlen (1981) vol. ii, 115: 1.
77. All references bracketed below refer to Claudia Carlen's *The Papal Encyclicals 1740–1981* (1981) (5 vols). References to this source identify volume, encyclical number and paragraph.
78. Locke, *Two Treatises* II: 27, 28.
79. Locke, *Two Treatises* I: 88, 89.
80. Tully (1980) pp. 162–3.
81. Tully (1980) pp. 98–100, 163–7, c.f. Hampsher-Monk (1981) pp. 554–5.
82. Viner (1978); Camp (1969); de Sousberghe (1950).
83. Calvez et Perrin (1959) pp. 259–68.
84. Aquinas, *Summa,* vol xxxviii, ed and trans. (1975b) Lefébure, QQ 66, II-ii, pp. 62–89.
85. Aquinas, *Summa Theologiae,* vol. xxxviii, QQ 66.1, p. 65.
86. Aquinas, *Summa Theologiae,* vol. xxxviii, QQ 66.2, p. 69.
87. Aristotle, *Politics,* trans. (1959) H. Rackham, pp. 85–7.
88. de Sousberghe (1950) p. 592, n. 32.
89. de Sousberghe (1950) p. 593.
90. Viner (1978) p. 72.
91. de Sousberghe (1950) p. 593, n. 33.
92. Ryan (1978) p. 60.
93. Tuck (1979) pp. 20–31, 45–57.
94. Tuck (1979) p. 54.
95. Tuck (1979) p. 22.
96. de Sousberghe (1950) p. 593.
97. Jannet (1893) p. 161.
98. Chaigne (1965) p. 156.
99. Piux XI (1931) Quadragesimo Anno.
100. Carlen (1981) vol. iii, 209: 117.
101. Carlen (1981) vol. iii, 209: 45, 49.
102. Carlen (1981) vol. iii, 209: 52, 53.
103. Carlen (1981) vol. iii, 209: 114.
104. Carlen (1981) vol. v, 267: 59–67.
105. Carlen (1981) vol. v, 267: 34.
106. Carlen (1981) vol. v, 267: 19.
107. Carlen (1981) vol. v, 267: 43.
108. Carlen (1981) vol. v, 267: 112.
109. Carlen (1981) vol. v, 267: 109–111.
110. Carlen (1981) vol. v, 267: 121.
111. Carlen (1918) vol. v, 207: 21.
112. Camp (1969) pp. 75–76; Chaigne (1965) p. 162.
113. Flannery (1975) ch. 64.
114. Flannery (1975) pp. 975–6.

115. Flannery (1975) pp. 977–8.
116. Carlen (1981) vol. v, 275.
117. Carlen (1981) vol. v, 275: 22.
118. Carlen (1981) vol. v, 275: 23.
119. Carlen (1981) vol. ii, 115: 15.
120. Carlen (1981) vol. v, 280: 64.
121. Furubotn and Pejovich (1972) p. 1139; my italics.
122. Jacobsen and Lipman (1959) p. 62.
123. Furubotn and Pejovich (1972) p. 1140.
124. Furubotn and Pejovich (1972) p. 1140.
125. Cross and Ekelund (1981) p. 39.
126. Furubotn and Pejovich (1972) p. 1139; my italics.
127. Marshall (1893) p. 389.
128. Furubotn and Pejovich (1972) p. 1157.
129. Carlen (1981) vol. ii, 115: 16.
130. Carlen (1981) vol. iii, 209: 7, 9.
131. Pius XI (1931) p. 3.
132. Carlen (1981) vol. ii, 115: 46.
133. Keane (1891) p. 26
134. Keane (1891) p. 28.
135. Keane (1891) p. 33.
136. Keane (1891) p. 34.
137. Carlen (1981) vol. iii, 209: 41.
138. Carlen (1981) vol. iii, 209: 42.
139. Carlen (1981) vol. v, 280: 5.
140. Vickers (1975) p. 13.

11 The Intellectual Context of *Rerum Novarum*

1. Pius XI (1931), *Quadragesimo Anno.*
2. Camp (1969) pp. 1–13.
3. Hales (1960) chs 1, 2.
4. Carlen (1981), *The Papal Encyclicals,* vol. i, 25: 32, 5 vols. References to this source identify volume number, encyclical number and paragraph.
5. Carlen (1981) vol. i, 26: 15.
6. Hales (1960) p. 129.
7. Hales (1960) pp. 106–10.
8. de Maistre (1793), *Lettres d'un Royaliste Savoisien.*
9. Barruel (1797–98), *Memoires pour server à l'Histoire du Jacobinisme.*
10. Hales (1958), ch. 6; Hales (1960) ch. 18.
11. Carlen (1981) vol. i, 33: 14, 15, 19, 17.
12. *New Catholic Encyclopedia* (1967) xii, 513–14, 607–08.
13. Carlen (1981) vol. i, 63.
14. Denziger (1937) pp. 482–90.
15. Corrigan (1938) p. 295.
16. Price (1798), *A Discourse on the Love of our Country.*
17. Burke (1790), *Reflections on the Revolution in France.*
18. Paine (1791–92), *The Rights of Man: Being an Answer to Mr Burke's Attack on the French Revolution* (2 Parts).

19. Wollstonecraft (1790), *A Vindication of the Rights of Man, in a Letter to the Right Honourable Edmund Burke: Occasioned by his Reflections on the Revolution in France* (2nd edition).
20. Mackintosh (1791), *Vindiciae Gallicae: Defence of the French Revolution.*
21. Godwin (1793, 1796, 1798), *Enquiry Concerning Political Justice and Its Influence on Morals and Happiness* (3rd edn, 1798, corrected).
22. E.g. Dickinson (1977) p. 272; Christie (1984) p. 159.
23. Waterman (1991c), *Revolution, Economics and Religion.*
24. Samuelson (1978)
25. Clark (1985) pp. 393–408.
26. Norman (1976); Waterman (1986) "Christian Political Economy: Malthus to Margaret Thatcher," in *Religion, Economics and Social Thought,* eds Block and Hexham.
27. Cross (1958) p. 206.
28. Villey (1959) pp. 93, 94, 99.
29. Briefs (1983) p. 253.
30. *Collection of Chalmers Papers in New College, Edinburgh,* ref. CHA 4.80.19.
31. Senior (1836) *An Outline of the Science of Political Economy.*
32. *Edinburgh Review* (October 1837) pp. 77, 83.
33. *Collection of Chalmers Papers,* 4, 185, 32.
34. Ring (1935); Tuan (1927) ch. 5, Moon (1921) pp. 20–5.
35. Villeneuve-Bargemont (1834), *Économie Politique Chrétienne, ou Récherche sur la Nature et les Causes du Paupérisme: en France et en Europe, et sur les Moyens de la soulager et de la prévenir,* 3 vols. Hereafter cited as Villeneuve-Bargemont, *Économie Politique Chrétienne.*
36. *Dublin Review* (July 1827) p. 176.
37. *Dublin Review* (July 1837) p. 175.
38. Villeneuve-Bargemont, *Économie Politique Chrétienne,* vol. i, p. 20; my italics.
39. Ring (1935) pp. 211–39.
40. Moody (1961) p. 84, n. 17.
41. Blanqui (1880) pp. 476–9.
42. Soderini (1934) pp. 189–93; Moody (1961) pp. 75–9.
43. Keane (1891) p. 28.
44. Cameron (1989) p. 235.
45. *Dublin Review* (1837) pp. 178–9.
46. Soderini (1934) ch. 17–10.
47. Briefs (1983) pp. 251–2.
48. Carlen (1981) vol. ii, 115: 1, 3, 43–5.
49. Blanning (1981); Beales (1985); Goldie (1981).
50. Carlen (1981) vol. i, 30: 3; 33: 5; 35: 8; 40: 5; 6: 53: 6, 7, 15; 63: 2, 3.
51. *New Catholic Encyclopedia* (1967) xii, 1164.
52. *New Catholic Encyclopedia* (1967) xii, 1166–7.
53. *New Catholic Encyclopedia* (1967) viii, 170.
54. Schmandt (1961) p. 20.
55. Schmandt (1961) pp. 19, 47 n. 6.
56. Carlen (1981) vol. ii, 80: 31 and *passim.*
57. Carlen (1981) vol. ii, 80: 28, 29.
58. Carlen (1981) vol. ii 93: 37.
59. Carlen (1981) vol. ii, 103: 1–9.

60. Carlen (1981) vol. ii, 103: 15–25.
61. Carlen (1981) vol. ii, 103: 10; my italics.
62. Whately (1831), *Introductory Lectures in Political Economy*, p. 62.
63. Hales (1954) p. 273.
64. Villey (1959) p. 99.
65. Keane (1891) p. 40.
66. Jannet (1893) p. 40.
67. *Inscrutabile Dei Consilio* (1878) in Carlen (1981) ii, 78: 2, 3.
68. Carlen (1981) vol. ii, 69: 1.
69. Carlen (1981) vol. ii, 78: 2, 3; 79 *passim;* 84: 23; 86: 2, 3; 87: 22; 88: 14; 90: 1, 2; 91: 8, 22; 93: 23, 24, 31; 97: 34; 103: 16; 108: 8; 112: 14.
70. *Dell'alto del' Apostolico Seggio* (1980) in Carlen (1981) vol. ii, 112: 6; my italics.
71. Hales (1958) chs. 9, 11.
72. Hales (1958) ch. 19.
73. Carlen (1981) vol. ii, 115: 4; my italics.
74. Carlen (1981) vol. ii, 115: 15.
75. See Chapter 10 above.
76. Carlen (1981) vol. ii, 115: 9, 10.
77. Carlen (1981) vol. ii, 115: 4.
78. de Sousberghe (1950)
79. See Chapter 10 above.
80. Carlen (1981) vol. v, 280: 65.

12 Market Social Order and Christian Organicism

1. Carlen, ed. (1981), *The Papal Encyclicals* (5 vol) ii, 115. References to this source identify volume number, encyclical number, and paragraph.
2. John Paul II (1991) nos 1 and 2.
3. John Paul II (1991) nos 12, 23, 26, 27, 35, 42.
4. John Paul II (1991) nos 12 and 24.
5. John Paul II (1991) no. 24.
6. Neuhaus (1991), "Review Essay," *National Review* XLIII, no. 11 Special Supplement, pp. S8–9.
7. Novak (1991), "Review Essay," *National Review* XLIII, no. 11, Special Supplement, pp. S11–12.
8. Poggioli, "Broadcast," National Public Radio, 2 May 1991.
9. *Centesimus Annus,* no. 28; and Minogue (1991), "Review Essay," *National Review* XLIII, no. 11, Special Supplement, pp. S7–8.
10. See Chapter 11 above.
11. Carlen (1981) vol. ii, 115: 33, 32.
12. Carlen (1981) vol. iii, 209: 990.
13. Thomas Starkey (1530), *A Dialogue Between Pole and Lupsett,* p. 37.
14. Carlen (1981) vol. ii, 115: 13.
15. Carlen (1981) vol. ii, 103: 13, 15.
16. See Chapter 3 above.
17. See Chapter 4 above.
18. Watson (1785), *A Collection of Theological Tracts,* 6 vols, vol. vi, p. 111; my italics.

19. Greenleaf (1964) chs V and VII *passim.*
20. Petty (1662–90), *The Economic Writings of Sir William Petty*, ed. (1986) C. Hull, p. 12.
21. Filmer (1680), *Patriarcha and Other Political Writings of Sir Robert Filmer,* ed. (1949) P. Laslett; and Greenleaf (1964) chs V and VII.
22. Minogue (1991).
23. Swift, *The Prose Writings of Jonathan Swift,* 14 vols, vol. ix, ed. (1939–68) H. Davis, pp. 143–4.
24. Carlen (1981) vol. i, 63: 4.
25. See Chapter 11 above.
26. John Paul II (1991) no. 13.
27. Keane (1892), "The Catholic Church and Economics," *Quarterly Journal of Economics* 6, p. 40.
28. See von Hayek (1968), "The Legal and Political Philosophy of David Hume," in *Hume: A Collection of Critical Essays* and (1978) *New Studies in Philosophy, Politics, and the History of Ideas.*
29. Hamowy (1987).
30. Mandeville (1728, 1732), *The Fable of the Bees: or Private Vices, Publick Benefits,* ed. (1988) Kaye, 2 vols, vol. ii, p. 104; vol. i, p. lxxix. Hereafter cited as Mandeville, *Fable of the Bees.*
31. Smith (1759), *The Theory of Moral Sentiments,* republished, ed. (1976) West, pp. 27–88, 440–4, 487–94; and Shelton (1981) pp. 91–2.
32. Mandeville, *The Fable of the Bees,* ii, 142.
33. Haakonssen (1981), pp. 12–35; and (1993) "The Structure of Hume's Political Thought," in *The Cambridge Companion to Hume,* ed. Norton, pp. 188–92; and (1994) (ed.), pp. xxvi–xxviii. David Hume, *Political Essays*
34. John Paul II (1991) no. 13.
35. Marx (1887), *Capital,* 3 vols., ed. Engels, trans. (1954) Moore and Aveling, vol. i, pp. 714–15, ch. XXXII *passim.*
36. Neuhaus (1991)
37. John Paul II (1991) no. 25.
38. John Paul II (1991) no. 24.
39. John Paul II (1991) no. 34.
40. John Paul II (1991) no. 35.
41. Werhahn (1990) pp. 5, 31.
42. Werhahn (1990) p. 5.
43. John Paul II (1988) no. 15.
44. John Paul II (1991) no. 32.
45. John Paul II (1991) no. 25.
46. John Paul II (1991) no. 34.
47. John Paul II (1991) no. 33.
48. John Paul II (1991) no. 36.
49. John Paul II (1991) no. 36.
50. John Paul II (1991) no. 11.
51. John Paul II (1991) no. 15.
52. John Paul II (1991) no. 35.
53. John Paul II (1991) no. 47.
54. John Paul II (1991) no. 58.
55. John Paul II (1991) no. 28.

56. John Paul II (1988) no. 41.
57. John Paul II (1991) no. 4.
58. See Chapter 11 above.
59. Corrigan (1938), p. 295.
60. John Paul II (1991) no. 4.
61. Carlen (1981) vol. ii, 103: 5, 6, 11.
62. Carlen (1981) vol. ii, 103: 15–25.
63. John Paul II (1991) no. 41; and Milton Friedman (1991), "Review Essay," *National Review* XLIII, no. 11, Special Supplement, pp. S3–S4.
64. Carlen (1981) vol. ii, 103: 10.
65. Carlen (1981) vol. ii, 103: 9.
66. See Chapter 7 above.
67. See Chapter 7 above.
68. John Paul II (1991) no. 25.

13 Establishment Social Theory

1. See Chapter 3 above.
2. Hooker (1594–1662), *The Works of that Learned and Judicious Divine, Mr Richard Hooker, with an Account of his Life and Death by Isaac Walton,* 3 vols, eds (1888) Keble, Church and Paget, vol. 7, ch. 1, paras 2, 5, 7.
3. Ratcliff (1936); and see also Chapter 3 above.
4. Berkeley (1735–37), *The Works of George Berkeley, Bishop of Cloyne,* 9 vols, ed. (1950) Luce and Jessop, vol. vi, pp. 87–184. Hereafter cited as Berkeley, *Works*.
5. Berkeley, *Works,* vol. vi, pp. 103, 105.
6. Berkeley, *Works,* vol. vi, pp. 61–85.
7. Berkeley, *Works,* vol. iii.
8. Berkeley, *Works,* vol. vi, pp. 185–7.
9. Marx (1887), *Capital,* 3 vols. ed. Engels, trans. (1954) Moore and Aveling, vol. i, 711 n. 2.
10. Smith (1759), *The Theory of Moral Sentiments,* republished 1976, ed. West.
11. Tucker (1755), *The Elements of Commerce and Theory of Taxes,* 6 vols, Privately printed by J. Tucker in 1755, first published (Schuler) in 1931. Reprinted in Josiah Tucker (1993), *The Collected Works of Josiah Tucker,* pp. 59, 61.
12. Shelton (1981) pp. 14–15, 21.
13. Smith (1776), *An Inquiry into the Nature and Causes of the Wealth of Nations,* republished 1976, ed. West.
14. Paley (1785), *The Principles of Moral and Political Philosophy,* vol. iv of *The Works of William Paley, D. D.,* 7 vols, ed. (1825) E. Paley. Hereafter cited as Paley, *Works.*
15. Paley, *Works,* vol. iv, book 6, ch. 11.
16. Keynes (1933), *Essays in Biography,* vol. x in *The Collected Writings of John Maynard Keynes,* 30 vols, Royal Economic Society (1971–89). Hereafter cited as Keynes, *Collected Writings.*
17. Waterman (1996), "Why William Paley was 'The First of the Cambridge Economist,'" *Cambridge Journal of Economics,* 20: 673–86.

18. Winch (1996) p. 239; Winch's italics.
19. Waterman (1983), "The Ideological Alliance of Political Economy with Christian Theology 1798–1833," *Journal of Ecclesiastical History* 34: 231–44; (1991c), *Revolution, Economics and Religion,* and see also Chapter 9 above.
20. Waterman (1991c), *Revolution, Economics and Religion,* p. 258.
21. Waterman (1991c), *Revolution, Economics and Religion,* pp. 230–40.
22. Kamenka (1983) p. 230.
23. Christensen (1962).
24. Norman (1987) p. 3; Jones (1968).
25. Jones (1968) p. 218; Norman (1987) p. 172.
26. E.g Munby (1956) pp. 92–5; Jones (1968) pp. 444–9; Preston (1983) pp. 31–2; Norman (1987) pp. 8–9.
27. Waterman (1991c), *Revolution, Economics and Religion,* p. 198; Winch (1996) pp. 288–322.
28. Winch (1996) p. 418.
29. Winch (1996) p. 402.
30. Norman (1987) p. 14.
31. E.g. Christensen (1962) pp. 13–16.
32. Jones (1968) p. 437.
33. Malthus (1803), *Essay on Population,* in ed. (1989) James, *T. R. Malthus, An Essay on the Principle of Population, The Version published in 1803, with the Variora of 1806, 1807, 1817, and 1826,* 2 vols, vol. i, p. xiii.
34. Norman (1976) pp. 136–7.
35. Chadwick (1966) vol. i, p. 326; Waterman (1991c), *Revolution, Economics and Religion,* pp. 256–7.
36. Deane (1969) pp. 214–16.
37. Keynes, *Collected Writings,* vol. x, p. 168.
38. Keynes, *Collected Writing,* vol. x, p. 167.
39. Preston (1979) pp. 83–8.
40. Tawney (1972) p. 72.
41. Cit. Preston (1983) p. 27.
42. Temple (1942), *Christianity and the Social Order,* Foreword by the Right Honourable Edward Heath, MBE, MP. Introduction by the Reverend Canon Ronald H. Preston, London 1972.
43. Munby (1960) p. 157.
44. Stamp (1926) (1931) (1936) (1939)
45. Temple (1942) p. 40.
46. Temple's phrase: see Temple (1942) p. 29.
47. Keynes to Temple, 3 December 1941, cited. Iremonger (1948) pp. 438–9.
48. Waterman (1990), "Denys Munby on Economics and Christianity," *Theology,* March/April: 108–16.
49. Munby (1957) (1958).
50. Munby (1963).
51. Eliot (1939).
52. Waterman (1986), "Christian Political Economy: Malthus to Margaret Thatcher," in *Religion, Economics and Social Thought* eds Block and Hexham (Preston, 1991).
53. Munby (1956) pp. 44–5.

54. Demant (1952)
55. Munby (1956) p. 275.
56. Reprinted as Chapter 8 in this volume.
57. Whately (1832), *Introductory Lectures on Political Economy*
58. Hinds (1831), *An Inquiry into the Proofs, Nature and Extent of Inspiration, and into the Authority of Scripture.*
59. Keynes (1891).
60. Robbins (1952).
61. Waterman (1988), "Can 'Economic Policy' be 'Christian?,'" *Review of Social Economy,* 46: 203: 11.
62. Knight (1939), "Ethics and Economic Reform: III Christianity," *Economica 6.* In Knight (1947), *Freedom and Reform: Essays in Economics and Social Philosophy*, pp. 122–53.
63. Knight (1947) p. 138.
64. Knight (1947) p. 124, 139.
65. Kolakowski (1982).
66. Kolakowski (1982) p. 180.
67. Kolakowski (1982) p. 163.
68. Kolakowski (1982) p. 163.
69. Kolakowski (1982) p. 163.
70. Kolakowski (1982) p. 186.
71. Kolakowski (1982) p. 186.
72. Kolakowski (1982) p. 88.
73. Kolakowski (1982) p. 82.
74. Kolakowski (1982) p. 90.

14 Economics and the Mutation of Political Doctrine

1. Groenewegen (1987) pp. 904–6.
2. Schumpeter (1954) p. 168.
3. Aristotle, *The Politics,* trans. (1967) Rackham, pp. 31–2.
4. Smith (1776), *An Inquiry into the Nature and Causes of the Wealth of Nations,* 2 vols, eds (1976) Campbell, Skinner, and Todd, I. xi. n. 1; II. v, xxxii; IV. i. iii. Hereafter cited as Smith, *WN.* References refer to book, chapter, and paragraph.
5. Smith *WN*, IV, intro; IV.ix.38; emphasis added.
6. Smith *WN,* Introduction.
7. Skinner (1969).
8. Waterman (1998) "Reappraisal of 'Malthus the Economist', 1933–97," *History of Political Economy*, 30: pp. 303–4, 312–13.
9. Smith *WN*, IV. vii. 44
10. O'Brien (1993) p. 144, n. 1.
11. Burke (1781–97), *The Writings and Speeches of Edmund Burke,* 12 vols, eds (1981–2000) Langford *et al.*, vol. ix, p. 125.
12. Waterman (1991c), *Revolution, Economics and Religion,* pp. 28–37.
13. Hayek (1960), pp. 160–1; cf Mill (1874), *Three Essays on Religion*, vol. x, p. 381 in *Collected Works of John Stuart Mill*, eds (1969) Robson *et al.*
14. Hamowy (1987).

15. "[T]heir opinions on the most essential points of political economy were the same" in Stewart (1794), *Account of the Life and Writings of Adam Smith LL.D.* in Adam Smith, *Essays on Philosophical Subjects*, p. 304, republished (1980) eds Wightman, Bryce, Ross.
16. Waterman (1996), "Why William Paley was 'the First of the Cambridge Economists,'" *Cambridge Journal of Economics*, 20: 673–86.
17. Smith *WN*, IV.v.43.
18. Smith *WN*, I.vii.12–16.
19. Smith *WN*, IV.ii.9.
20. Smith *WN*, I.viii.43–5.
21. Smith *WN*, I.ix.p. 9; I, viii, 36.
22. Hume (1752), *Political Essays*, essays 13 and 14.
23. Samuelson (1978).
24. Ricardo (1817), *The Principles of Political Economy and Taxation*, pp. 71–2. Republished 1911. Hereafter cited as Ricardo, *Principles*.
25. See Chapter 9 above.
26. Ricardo *Principles*, p. 210.
27. Smith *WN*, I.x.c. 45–9.
28. Ricardo, *Principles*, p. 61.
29. Mill (1848), *Principles of Political Economy with Some of Their Applications to Social Philosophy*, p. 922, ed. (1909) Ashley. Hereafter cited as Mill, *Principles of Political Economy*.
30. Hamilton (1791), *Report on Manufactures*, reprinted 1913.
31. "The restraints of Communism would be freedom in comparison with the present condition of the majority of the human race." In Mill, *Principles of Political Economy*, p. 210.
32. Mill, *Principles of Political Economy*, pp. 984–9.
33. Lerner (1944) p. xi.
34. Schumpeter (1943) pp. 162, 167, 421–31. Republished 1987.
35. Turgot (1768), "Observations sur le mémorie de M. de Saint-Péravy en faveur de l'impôt indirect," in ed. Cazes (1970) *Ecrits Economiques*.
36. Steuart (1767), *An Inquiry into the Principles of Political Economy*, 2 vols, ed. (1966) Skinner.
37. Marshall (1890), *Principles of Economics*, p. viii, 8th edn 1952. Hereafter cited as Marshall, *Principles*.
38. Schumpter (1954) pp. 465.
39. Blaug (1997) p. 308.
40. Wicksteed (1894).
41. Clark (1899).
42. Wicksell (1901).
43. Flux (1894).
44. Enrico Barone pointed out in 1895 that product exhaustion was implied by Walras's (1874) cost-minimization equations with no mention of marginal product, but his submission to the *Economic Journal* reporting this result was rejected by Edgeworth who was then the editor. Blaug (1997) p. 435.
45. Schumpeter (1954) p. 175.
46. Condillac (1776), *Le Commerce et le Gouvernement*, pp. 83–4, trans. (1997) Eltis.
47. Allais (1992) p. 174.

48. Lloyd (1834), *A Lecture on the Notion of Value delivered before the University of Oxford in 1833.*
49. Dupuit (1844), "On the Measurement of the Utility of Public Works," trans. (1952) Barback in *International Economic Papers,* no. 2.
50. Gossen (1854), *Entwickelung der Gesetze des menschliches Handeln.*
51. Jevons (1862) "Brief Account of a General Mathematical Theory of Political Economy," in Jevons (1871), *The Theory of Political Economy,* republished 1957. Hereafter cited as Jevons, *Theory of Political Economy.*
52. Waterman (1998), "Malthus, Mathematics and the Mythology of Coherence," *History of Political Economy* 30, pp. 579–80.
53. Jevons, *Theory of Political Economy.*
54. Menger (1871), *Principles of Economics,* trans. (1950) Dingwall and Hoselitz.
55. Walras (1874), *Elements of Pure Economics,* trans (1954) Jaffé.
56. The assumption that all choice and production sets are convex was later made explicit Debreu (1959).
57. Samuelson (1947).
58. Marshall, *Principles.*
59. See Blaug (1997) pp. 342–9.
60. Pigou (1920).
61. Pareto (1896), *Cours d'Économie Politique* and Pareto (1908) *Manual of Political Economy.*
62. Robbins (1932).
63. Edgeworth (1881), *Mathematical Psychics.*
64. Hicks (1939a).
65. Blaug (1997) p. 286.
66. Backhouse (1985) p. 166.
67. Barone (1908), "The Ministry of Production in the Collectivist State," trans (1935) Hayek.
68. Wieser (1893), *Natural Value,* republished 1971.
69. Mises (1920).
70. O'Neill (1996).
71. Lange (1936) p. 53.
72. Hoff (1938) pp. 392–4.
73. Bergson (1938).
74. Hicks (1939b).
75. Hobson (1909).
76. Piero Sraffa (1926).
77. Young (1928).
78. Robinson (1933)
79. Chamberlin (1933).
80. Sweezy (1935) p. 79.
81. Robinson (1964b).
82. Robinson (1980) p. xvi.
83. Samuelson (1947) Part II.
84. Robinson (1953–54) reprinted 1964(a), p. 120.
85. "The *General Theory* caught most economists under the age of thirty-five with the unexpected virulence of a disease first attacking and decimating an isolated tribe of south sea islanders." Samuelson (1946) reprinted (1966) p. 1517.

86. Marshall, *Principles,* pp. 287–88.
87. Hicks (1939a) p. 33; (1965) p. 78.
88. Eg. Keynes (1936) p. 28.
89. That the logical impossibility of such a quasi-equilibrium when saving is non-zero was formally remedied by R. F. Harrod's "warranted rate of growth" (1939) p. 16. For Joan Robinson (1953) reprinted (1973) p. 263, the necessary instability of warranted growth was yet another nail in the orthodox coffin.
90. Samuelson (1946) reprinted (1966) p. 1523.
91. Keynes (1936) pp. 378–81.
92. Robinson (1953), reprinted (1973) p. 264.
93. Samuelson (1948).
94. Samuelson (1948) pp. 421, 436.
95. Nelson (2001).
96. Nelson (2001) pp. 50, 52.
97. Ramsey (1927)(1928).
98. E.g. Sen (1967); Arrow and Kurz (1970).
99. E.g. Kantorovich (1939) reprinted (1960) (1942) reprinted (1965); Pontryagin *et al.* (1962).
100. Leontief (1936).
101. Neumann (1938) reprinted (1945–46).
102. Hahn and Matthews (1964) pp. 853–88.
103. Kuenne (1968) pp. 5–10.
104. Samuelson and Solow (1953); See Hahn and Matthews (1964) p. 872.
105. E.g. Veblen (1921) republished (1965) pp. 64, 144, 157–8.
106. Debreu (1954) among others.
107. Stiglitz (1994).
108. As Robinson writes, "Your genius wears seven-league boots, and goes striding along, leaving a paper-chase of little mistakes behind him." Robinson (1973) p. 266.
109. Patinkin (1976).
110. Tinbergen (1935).
111. Hicks (1937).
112. Hicks (1939a).
113. Meade (1951); Swan (1960); Mundell (1961).
114. Provided we can assume that we observe variables in the neighborhood of their equilibrium values. Hicks (1965) p. 16.
115. E.g. Klein and Goldberger (1955); Bank of Canada (1976).
116. Hicks (1965) pp. 76–83.
117. Phillips (1958); Lipsey (1960).
118. Samuelson and Solow (1960).
119. Samuelson (1964).
120. Waterman (1966) "Some footnotes to the 'Swan Diagram': or, How Dependent is a Dependent Economy?,"*Economic Record,* 42, 447–64; cf. Swan (1960).
121. Phelps (1967); Friedman (1968).
122. See e.g. Friedman (1956, 1957, 1961).
123. Later this was called "easily the most influential paper on macroeconomics published in the post-war era," Blaug (1997) p. 678.

124. Samuelson and Solow (1960) p. 193.
125. Friedman (1968).
126. Samuelson (1946), reprinted (1966) p. 1520.
127. Phelps (1970)
128. Alchian (1970).
129. Hall (1979); Ramaswani (1983); Darby, Haltiwanger and Plant (1985); Barro and Lucas (1994).
130. Arrow (1962).
131. Muth (1961).
132. Sargent (1973); Sargent and Wallace (1976); Lucas (1975) and so forth.
133. McCafferty (1990) pp. 343–5.
134. E.g. Mankiew and Romer (1991).
135. Harcourt (1987).
136. Arrow (1987) p. 202.
137. Goodwin (1970) pp. 11–12, 24–8.
138. Stiglitz (1994) p. 6.
139. Stiglitz (1994) p. 201.
140. Pareto (1896); Barone (1908) ; Kaldor (1939); Hicks (1939b).
141. Scitovsky (1941).
142. Mill, *Principles of Political Economy,* p. 21.
143. Samuelson (1947) ch. VIII; Samuelson (1950).
144. Boulding (1952).
145. Lipsey and Lancaster (1956).
146. Stiglitz (1994) pp. 16–18, 34, 38.
147. Coase (1937) (1960).
148. Furobotn and Richter (1997).
149. Ross (1973) Sappington (1991).
150. Greenwald and Stiglitz (1986).
151. Simon (1959) (1978).
152. Williamson (1975) (1985) (1986).
153. Stiglitz (1994) ch. 10.
154. Stiglitz (1994) p. 195.
155. Stiglitz (1986) ch. 7; Sappington and Stiglitz (1987).
156. Hayek (1935).
157. Vaughn (1994).
158. Hayek (1937) (1945) (1978).
159. Stiglitz (1994) pp. 24–5.
160. Stigler (1961).
161. Akerlof (1970).
162. "[T]he fact that the real world is more complicated than any model which we might construct does not absolve us of the need for testing our ideas out using simple and understandable models."
163. Stiglitz (1994) p. 26.
164. Stiglitz (1994) p. 247.
165. Harrod (1952) pp. 192–3.
166. Friedman (1948) (1951).
167. E.g. Kalecki (1943); Akerman (1947); Downs (1957); Nordhaus (1975); MacRae (1977) and so forth.
168. Bergson (1938).

169. Samuelson (1947).
170. Saumuelson (1947) p. 220.
171. Arrow (1951).
172. Bowen (1943); Black (1948); Arrow (1950).
173. Black (1958).
174. Arrow (1951) p. 59.
175. Bowen (1943).
176. Brennan and Lomasky (1993) p. 76.
177. Brennan and Lomasky (1993).
178. Brennan and Lomasky (1993) p. 197.
179. Theocharis (1961) pp. 22–7.
180. Buchanan (1975) p. 384.
181. Musgrave and Peacock (1958).
182. Musgrave (1938).
183. Bowen (1943).
184. Buchanan (1949).
185. Samuelson (1954).
186. Buchanan (1975) p. 385.
187. Mises (1944); Tullock (1965); Niskanen (1971) and so forth.
188. Stigler (1971).
189. Smith *WN*, IV.i–viii.
190. Saint-Paul (2000) p. 915.
191. Smith *WN*, IV.ix.51.

References

A Group of Churchmen. *The Return of Christendom: by a Group of Churchmen with an Introduction by Bishop Gore and an Epilogue by G. K. Chesterton* (London: Allen & Unwin, 1922).

Acton, J. E. E. D. Baron. *Essays on Church and State*, ed. D. Woodruff (London: Hollis and Carter 1952), reprinted (New York: Thomas Y. Cowell, 1968).

Akerlof, G. A. "The Market for 'Lemons:' Quality Uncertainty and the Market Mechanism," *Quarterly Journal of Economics* 86 (1970): 488–500.

Akerman, J. "Political Economic Cycles," *Kyklos* 1 (1947) 107–17.

Alchian, A. A. "Pricing and Society," Occasional Paper No. 17 (Westminster: The Institute of Economics Affairs, 1967).

Alchian, A. A. "Information Costs, Pricing, and Resource Unemployment," in *Microeconomic Foundations of Employment and Inflation Theory*, ed. F. S. Phelps (New York: Norton, 1970).

Allais, M. "The General Theory of Surpluses as a Formalization of the Underlying Theoretical Thought of Adam Smith, His Predecessors and His Contemporaries," in *Adam Smith's Legacy*, ed. M. Fry (London: Routledge, 1992).

The Anti-Jacobin Review and Magazine.

Anonymous. *A Summons of Awakening, or, the Evil Tendency and Danger of Speculative Philosophy* (Hawick, 1807).

Aquinas, St Thomas. *Summa Theologiae*, vol. xxxvii ed. and trans. T. Gilby, vol. xxxviii, ed. and trans. Lefébure (London: Eyre and Spottiswoode, 1975).

Aristotle. *Politics*, trans. H. Rackham (London: Heinemann, 1959).

Arrow, K. J. "A Difficulty in the Concept of Social Welfare," *Journal of Political Economy* 56 (1950): 328–46.

Arrow, K. J. *Social Choice and Individual Values* (New York: Wiley, 1951).

Arrow, K. J. "The Economic Implications of Learning by Doing," *Review of Economic Studies* 29 (1962): 155–73.

Arrow, K. J. "Rationality of Self and Others in an Economic System," in *Rational Choice: The Contrast Between Economics* and Psychology, eds R. W. Hogarth and M. W. Reder (Chicago: University of Chicago Press, 1987).

Arrow, K. J. and Debreu, G. "Existence of an Equilibrium for a Competitive Economy," *Econometrica* 22 (1954): 265–90.

Arrow, K. J. and Kurz, M. *Public Investment, the Rate of Return and Optimal Fiscal Policy* (Baltimore MD: Johns Hopkins Press, 1970).

Athanasius, St. *De incarn. 54*, cited in vol. iii of Johannes Quasten, *Patrologia*, 3 vols (Utrecht: Spectrum, 1950–60).

Augustine, St. *De Civitate Dei contra Paganos*, written 413–26, trans. G. E. McCracken (London: Heinemann, 1957).

Backhouse, R. *A History of Modern Economic Analysis* (Oxford: Blackwell, 1985).

Bank of Canada. *The Equations of RDX2*, Technical Report No. 5 (Ottawa, 1976).

Barker, E. *Political Thought in England, 1848 to 1914* (London: Thornton, Butterworth, 1928).

Barker, E. (ed.) *Social and Political Thought in Byzantium, From Justinian I to the Last Paleologus* (Oxford: Clarendon, 1957).

Barone, E. "The Ministry of Production in the Collectivist State," 1st published 1908, trans. 1935 in F. Hayek, ed. *Collectivist Economic Planning* (London: Routledge, 1935).

Barro, R. J. and Lucas, R. F. *Macroeconomics*, 1st Canadian edn (Burr Ridge, IL: Irwin, 1994).

Barruel, A. de. *Mémoires pour servir a l'Histoire du Jacobinisme* (London, 1797–98).

Bastiat, F. *Harmonies Economiques* (Paris: Guillaumin, 1850).

Beales, Derek "Christians and philosophes: The Case of the Austrian Enlightenment," in *History, Society and the Churches*, eds D. Beales and G. Best (Cambridge: Cambridge University Press, 1985).

Bergson, A. "A Reformulation of Certain Aspects of Welfare Economics," *Quarterly Journal of Economics* 52 (1938): 310–34.

Berlin, I. (ed.) *The Age of Enlightenment* (New York: Mentor, 1956).

Berkeley, G. *The Works of George Berkeley 1709–1753*, Nine Volumes, eds A. A. Luce and T. E. Jessop (London: Nelson, 1948–1957).

Bindley, T. H. *The Ecumenical Documents of the Faith*, 3rd edn (London, 1925).

Black, D. "On the Rationale of Group Decision-Making," *Journal of Political Economy* 56: (1948) 23–34.

Black, D. *The Theory of Committees and Elections* (London: Cambridge University Press, 1958).

Blanning, T. C. W. "The Enlightenment in Catholic Germany," in *The Enlightenment in National Context*, eds R. Porter and M. Teich (Cambridge: Cambridge University Press, 1981).

Blanqui, J. A. *History of Political Economy in Europe*, trans. from 4th French edn E. J. Leonard, New York: Putnam, 1880 (Kelley Reprint, 1968).

Blaug, M. *Economic Theory in Retrospect*, 3rd edn (Cambridge: Cambridge University Press, 1978) 5th edn (Cambridge: Cambridge University Press, 1997).

Block, W., Brennan, H. G. and Elzinga, K. (eds) *The Morality of the Market: Proceedings of an International Conference* (Vancouver: Fraser Institute, 1985).

Block, W. and Hexham, I. (eds) *Religion, Economics and Social Thought: Proceedings of an International Conference* (Vancouver: Fraser Institute, 1986).

Book of Common Prayer and Administration of the Sacraments, and Other Rites and Ceremonies of the Church, According to the Use of the Church of England. . . . and the Form and Manner of Making, Ordaining and Consecrating of Bishops, Priests and Deacons (1549, 1552, 1559, 1604, 1662).

Bonar, J. *Malthus and His Work* (London: Allen and Unwin, 1885).

Bond, R. B. (ed.) *Certain Sermons or Homilies (1547) and A Homily Against Disobedience and Wilful Rebellion* (1570) (Toronto: University of Toronto Press, 1987).

Boswell, J. *Boswell's Life of Johnson, together with Boswell's Journal of a Tour to the Hebrides*, 1st edn 1791, republished, Six volumes, eds George Birkbeck Hill and Revd. L. F. Powell (Oxford: Clarendon, 1934–50).

Boswell, J. *Life of Johnson*, 1st edn 1791, republished, ed. R. W. Chapman (Oxford: Oxford University Press, 1953).

Boulding, K. E. "Welfare Economics," in *A Survey of Contemporary Economics*, vol. ii, ed. B. F. Haley (Homewood IL: Irwin, 1952).

Boulding, Kenneth E. "Economics as a Moral Science," *American Economic Review* (1968): 1–12.

Bowen, H. R. "The Interpretation of Voting in the Allocation of Resources," *Quarterly Journal of Economics* 58 (1943): 27–48.

Bowley, M. *Nassau Senior and Classical Economics* (London: Allen and Unwin, 1937).

Bradley, J. E. *Religion, Revolution and English Radicalism: Nonconformity in Eighteenth-Century Politics and Society* (Cambridge: Cambridge University Press, 1990).

Brennan, H. G. and Lomasky, L. *Democracy and Decision: The Pure Theory of Electoral Preferences* (Cambridge: Cambridge University Press, 1993).

Brennan, H. G. and Waterman, A. M. C. (eds) *Economics and Religion: Are They Distinct?* (Boston: Kluwer, 1994).

Briefs, G. A. "Catholic Social Doctrine, Laissez-faire Liberalism, and Social Market Economy," *Review of Social Economy* 41: 3 (1983), 246–58.

Brown, S. J. *Thomas Chalmers and the Godly Commonwealth in Scotland* (Oxford: Oxford University Press, 1982).

Brown, Vivienne *Adam Smith's Discourse: Canonicity, Commerce and Conscience* (London: Routledge, 1994).

Buchanan, J. M. "The Pure Theory of Government Finance: A Suggested Interpretation," *Journal of Political Economy* 57 (1949): 496–505.

Buchanan, J. M. "Public Finance and Public Choice," *National Tax Journal* 28 (1975): 383–94.

Burke, Edmund. *Reflections on the Revolution in France*, 1st edn 1790, republished (Chicago: Gateway, 1955).

Burke, Edmund. *The Writings and Speeches of Edmund Burke, 1759–1797*, Nine vols, eds P. K. Langford *et al.*, (Oxford: Clarendon, 1981–2000).

Butler, J. *Fifteen Sermons Preached at the Rolls Chapel and a Dissertation upon the Nature of Virtue*, 1st edn 1726, republished, ed. W. R. Matthews (London: Bell, 1969).

Calvez, J. Y. & et Perrin, J. *Eglise et société économique: l'enseignement social des Papes de Léon XIII à Pie XII, 1878–1958*, Deuxième Edition (Paris, 1959).

Cameron, Rondo. *A Concise Economic History of the World from Paleolithic Times to the Present* (New York: Oxford University Press, 1989).

Camp. R. *The Papal Ideology of Social Reform* (Leiden, 1969).

Cantillon, R. *Essai sur la Nature du Commerce en Général*, 1st edn 1755, republished ed. and trans. H. Higgs (London: Macmillan for the Royal Economic Society, 1931).

Carey, H. C. *Principles of Political Economy*, 3 vols (Philadelphia: 1837).

Carlen, C. (ed.). *The Papal Encyclicals 1740–1981*, 5 vols (McGrath Publishing Co. 1981).

Carus, W. (ed.) *Memoirs of the Life of the Rev. Charles Simeon, M. A., Late Fellow of King's College, and Minister of Trinity Church, Cambridge, With a Selection of his Writings and Correspondence* (London, 1847).

Casaubon, M. *A Treatise Concerning Enthusiasme*, 2nd edn, London: 1656. Facsimile reproduction of the second edition of 1656, with an Introduction by Paul J. Korshin (Gainsville FL: Scholars' Facsimiles and Reprints, 1970).

Cassirer, E. von. *The Philosophy of the Enlightenment*, 1st German edn 1932, republished, trans. Fritz C. A. Koelln and James P. Pettegrove (Princeton, NJ: Princeton University Press 1951).

Chadwick, W. O. *The Victorian Church*, vol. vii in *An Ecclesiastical History of England*, ed. J. C. Dickinson (London: Black, 1966).
Chadwick, W. O. *The Secularisation of the European Mind in the Nineteenth Century* (Cambridge: Cambridge University Press, 1975).
Chaigne, H. "The Catholic Church and Socialism," *Cross-Currents*, Spring 1965: 151–69.
Chalmers, T. *An Enquiry into the Nature and Stability of National Resources* (Edinburgh: Moir, 1808).
Chalmers, T. *The Influence of Bible Societies on the Temporal Necessities of the Poor* (Cupar: Tullis, 1814).
Chalmers, T. *The Christian and Civic Economy of Large Towns*, 3 vols (Glasgow: Chalmers & Collins, 1821–26).
Chalmers, T. *On Political Economy in Connexion with the Moral State and Moral Prospects of Society* (Glasgow: Collins, 1832).
Chalmers, T. *On the Power, Wisdom, and Goodness of God, as Manifested in the Adaptation of External Nature to the Moral and Intellectual Constitution of Man* (Bridgewater Treatise, London, 1883).
Chamberlin, F. H. *The Theory of Monopolistic Competition: A Reorientation of the Theory of Value* (Cambridge MA: Harvard University Press, 1933).
Christensen, T. *Origin and History of Christian Socialism, 1848–54* (Aarhus: Universitetsforlaget, 1962).
Christie, I. R. *Stress and Stability in Late Eighteenth Century Britain: Reflections on the British Avoidance of Revolution* (Oxford: Oxford University Press, 1984).
Clark, J. B. *The Distribution of Wealth* (New York: Macmillan, 1899).
Clark, J. C. D. *English Society, 1688–1832: Ideology, Social Structure and Political Practice During the Ancien Régime* (Cambridge: Cambridge University Press, 1985).
Clark, J. C. D. *The Language of Liberty, 1660–1832: Political Discourse and Social Dynamics in the Anglo-American World* (Cambridge: Cambridge University Press, 1994).
Clark, L. M. G. "Women and Locke: Who Owns the Apples in the Garden of Eden?" in Clark, Lorenne M. G. and Lange, L. (eds) *The Sexism of Social and Political Theory* (Toronto, University of Toronto Press, 1979).
Clark, Samuel *Scripture-Doctrine of the Trinity*, 3 vols (London, 1712).
Clarke, M. L. *Paley, Evidences for the Man* (Toronto: University of Toronto Press, 1974).
Coats, A. W. B. *The Sociology and Professionalisation of Economics: British and American Essays*, 2 vols (London: Routledge, 1993).
Coase, R. H. "The Nature of the Firm," *Economica* 4 (1937): 386–405.
Coase, R. H. "The Problem of Social Cost," *Journal of Law and Economics* 3 (1960): 1–44.
Cole, G. "A Note on Paley and his School – Was Sir Leslie Stephen Mistaken?," *Tyndale Bulletin* 37 (1987): 1–6.
Coleridge, S. T. *On the Constitution of Church and State According to the Idea of Each*, 1st edn 1830, republished, vol. x in *The Collected Works of Samuel Taylor Coleridge*, eds K. Coburn, G. Whalley and H. J. Jackson (Princeton NJ: Princeton University Press, 1976).
Collection of Chalmers Papers in New College, Edinburgh.

Collini, S. "General Introduction" to *Economy, Polity and Society: British Intellectual History 1750–1950*, eds. S. Collini, R. Whatmore and B. Young (Cambridge: Cambridge University Press, 2000).

Collins, A. *A Discourse of Free-Thinking*, Faksimile-Neudruck der Erstausgabe London 1713 mit deutschem Paralleltext. Herausgegeben, übersetz und eingeleitet von Günter Gawlick, mit einem Geleitwort von Julius Ebbinghaus. (Stuttgart – Bad Cannstatt: Frommann, 1965).

Cooper. A. A. *Characteristics of Men, Manners, Opinions, Times, etc. By the Right Honourable Anthony Earl of Shaftesbury*, 2 vols, republished, ed. John M. Robertson, (Gloucester MA: Peter Smith, 1963).

[Copleston, E.] *A Letter to the Right Hon. Robert Peel, MP for the University of Oxford, on the Pernicious Effect of a Variable standard of Value, especially as it regards the Condition of the Lower Orders and the Poor Laws* (Oxford, 1819).

[Copleston, E.] *A Second Letter to the Right Hon. Robert Peel, M. P. for the University of Oxford, on the Causes of Pauperism and on the Poor Laws, By one of his Constituents* (Oxford, 1819).

Corrigan, R. S. J. *The Catholic Church and the Nineteenth Century* (Milwaukee: Bruce, 1938).

Corsi, P. "The Heritage of Dugald Stewart: Oxford Philosophy and the Method of Political Economy," *Nuncius*, 2 (1987): 89–143.

Corsi, P. *Science and Religion: Baden Powell and the Anglican Debate* (New York: Cambridge University Press, 1988).

Costabile, L. and Rowthorn, B. "Malthus"s Theory of Wages and Growth", *Economic Journal*, 95 (1985): 418–37.

Crimmins, J. E. "John Brown and the Theological Tradition of Utilitarian Ethics," *History of Political Thought 4* (1983): 523–50.

Crimmins, J. E. (ed.) *Religion and the Development of Political Thought: Thomas Hobbes to J. S. Mill* (London: Routledge, 1990).

Cross, C. *The Royal Supremacy in the Elizabethan Church* (London: Allen and Unwin, 1969).

Cross, F. L. (ed.) *The Oxford Dictionary of the Christian Church* (London: Oxford University Press, 1958).

Cross, M. L. and Ekelund, R. B. "A. T. Hadley: The American Invention of Property Rights and Public Goods," *Review of Social Economy* 39 (1981): 37–50.

Cyprian, St. Epist 73, 21, cited in Johannes Quasten, vol. ii, *Patrology*, 3 vols (Utrecht, 1950–60).

Darby, M. R., Haltiwanger, J. C., and Plant, M. W. "Unemployment Rate Dynamics and Persistent Unemployment Under Rational Expectations," *American Economic Review* 75 (1985): 614–37.

Darwin, C. *The Origin of Species by Means of Natural-Selection, or, The Preservation of Favoured Races in the Struggle for Life*, 1st edn 1859, this edition 1899, with editions and corrections from the sixth and last English edition (New York: Appleton, 1899).

Daubeny, C. *A Guide to the Church, in Several Discourses Addressed to William Wilberforce, Esq., M.P.* (London, 1798).

Davies, H. *Worship and Theology in England: From Watts and Wesley to Maurice, 1690–1850*, vol. iii (Princeton: Princeton University Press, 1961).

Dean, J. M. and Waterman, A. M. C. (eds) *Religion and Economics: Normative Social Theory* (Boston: Kluwer, 1998).

Deane, H. A. *The Political and Social Ideas of St Augustine* (New York: Columbia University Press, 1963).
Deane, P. *The First Industrial Revolution* (Cambridge: Cambridge University Press, 1969).
Debreu, G. *The Theory of Value* (New York: Wiley, 1959).
Demant, V. A. *Christian Polity* (London: Faber and Faber, 1936)
Demant, V. A. *Religion and the Decline of Capitalism* (London: Faber and Faber, 1952).
Denziger, H. *Enchiridion Symbolorum: Definitionum et Declarationum* (Friburgi: Herder, 1937).
Dickinson, H. T. *Liberty and Property: Political Ideology in Eighteenth Century Britain* (New York: Holmes and Meier, 1977).
Dorfman, R. "Modernizing Böhm-Bawerk's Theory of Interest," *Journal of the History of Economic Thought* 23 (2001): 37–54.
Downs, A. *An Economic Theory of Democracy* (New York: Harper, 1957).
Dublin Review.
Dunn, J. *The Political Thought of John Locke: an Historical Account of the Argument of the "Two Treatises of Government"* (Cambridge: Cambridge University Press, 1969).
Dupuit, J. "On the Measurement of the Utility of Public Works," 1st edn 1844, transl. R. H. Barback from *Annales des Ponts et Chaussées*, in *International Economic Papers*, No. 2. (London: Macmillan, 1952).
Eclectic Review.
Edgeworth, F. Y. *Mathematical Psychics* (London: Macmillan, 1881).
Edinburgh Review.
Ehrman, J. *The Younger Pitt, the Reluctant Transition* (Stanford: Stanford University Press, 1983).
Ekelund, R. and Hébert, R. *A History of Economic Theory and Method*, 3rd edn (New York: McGraw-Hill, 1990).
Eliot, T. S. *The Idea of a Christian Society* (London: Faber and Faber, 1939).
Eltis, W. A. "Malthus's Theory of Effective Demand and Growth", *Oxford Economic Papers* 32 (1980): 19–56.
Eltis, W. A. *The Classical Theory of Economic Growth* (London: Macmillan, 1984).
L'Encyclopédie ou Dictionnaire Raisonné des Sciences des Arts et des Métiers, Par une Societé de Gens de Lettres. Mis en ordre & publié par M. Diderot & . . . par M. D'Alembert. (Paris: 1751–72). Facsimile in 3 vols. (Readex Microprint Corp: New York: 1969).
Faccarello, G. *The Foundations of Laissez-faire: The Economics of Pierre de Boisguilbert* (London: Routledge, 1999).
Ferguson, T. "Poor, Welfare and Social Services", in M. R. McLarty (ed) *A Source Book and History of Administrative Law in Scotland* (Edinburgh: Hedge, 1956).
Feyerabend, P. *Against Method* (London: Verso, 1988).
Filmer, R. *Patriarcha and Other Political Writings of Sir Robert Filmer*, 1st edn 1680, ed. P. Laslett (Oxford: Blackwell, 1949).
Figgis, J. N. *Studies of Political Thought from Gerson to Grotius, 1414–1625.* (Cambridge: Cambridge University Press, 1907).
Figgis, J. N. *The Political Aspect of St Augustine's "City of God"* (London: Longmans, 1921).
Figgis, J. N. *The Divine Right of Kings*, 1st edn 1922 (Cambridge: Cambridge University Press, 1965).

Fitzgibbons, A. *Adam Smith's System of Liberty, Wealth and Virtue* (Oxford: Clarendon, 1995)

Flannery, A. (ed.) *Vatican Council II. The Conciliar and Post-Conciliar Documents.* (Wilmington, 1975).

Flux, A. W. "Review of Wicksteed (1894)," reprinted in *Precursors in Mathematical Economics*, eds W. A. Baumol and S. M. Goldfeld (London: London School of Economics, 1968).

Fontana, B. *Rethinking the Politics of a Commercial Society: The Edinburgh Review 1802–1832* (Cambridge: Cambridge University Press, 1985).

Forster, B. A. "Optimal Energy Use in a Polluted Environment," *Journal of Environmental Economics and Management* (1980) 321–33.

Friedman, M. "A Monetary and Fiscal Framework for Economic Stability," *American Economic Review* 37 (1948): 245–64.

Friedman, M. "Les Effets d'une Politique de Plein Emploi sur la Stabilité Economique: Analyse Formelle," *Economie Appliquée* 4 (1951): 441–56; (tr.) in Friedman, *Essays in Positive Economics* (Chicago: University of Chicago Press, 1953).

Friedman, M. "The Methodology of Positive Economics," in Milton Friedman, *Essays in Positive Economics* (Chicago: University of Chicago Press, 1953).

Friedman, M. *Essays in Positive Economics* (Chicago: University of Chicago Press, 1953).

Friedman, M. "The Quantity Theory of Money: A Restatement," in *Studies in the Quantity Theory of Money*, ed. M. Friedman (Chicago: University of Chicago Press, 1956).

Friedman, M. *A Theory of the Consumption Function* (Princeton, NJ: Princeton University Press, 1957).

Friedman, M. "The Lag in Effect of Monetary Policy," *Journal of Political Economy* 69 (1961): 193–238.

Friedman, M. "The Role of Monetary Policy," *American Economic Review* 68 (1968): 1–17.

Furubotn, E. C. and Pejovich, S. "Property Rights and Economic Theory: A Survey of Recent Literature," *Journal of Economic Literature* (1972): 1137–62.

Furobotn, E. C. and Richter, R. *Institutions and Economic Theory. The Contribution of the New Institutional Economics* (Ann Arbor: University of Michigan Press, 1997).

Gascoigne, J. "Anglican Latitudinarianism and political radicalism in the late eighteenth century," *History* 71 (1986).

Gascoigne, J. *Cambridge in the Age of the Enlightenment* (Cambridge: Cambridge University Press, 1989).

Gargan, E. T. (ed.) *Leo XIII and the Modern World* (New York: Sheed and Ward, 1961).

Gay, P. *The Enlightenment: An Interpretation*, 2, vols, (New York: Knopf, 1966–69).

Gilley, S. "Christianity and Enlightenment: An Historical Survey", *History of European Ideas* 1 (1981): 103–21.

Gladstone, W. E. *The State in its Relations with the Church* (London, 1838).

Glahe, F. R. (ed) Adam Smith's "An Inquiry into the Nature and Causes of the Wealth of Nations": *A Concordance* (Lanham, MD: Rowman & Littlefield, 1993).

Glass, D. V. and Eversley, D. E. C. (eds) *Population in History: Essays in Historical Demography*, (London: Arnold, 1965).

Godwin, W. *Enquiry Concerning Political Justice and its Influence on Morals and Happiness* 2 vols, 2nd edn (London: 1796), 3rd edn (Robinson: London, 1798).

Godwin, William. *Enquiring Concerning Political Justice and us Influence on Morals and Happiness*, photographic facsimile of 3rd edn (1798) with variora from first (1793) and second (1796) editions, ed. F.E.L. Prestley (Toronto: University of Toronto Press, 1946).

Goldie, M. "The Scottish Catholic Enlightenment." *Journal of British Studies* 30 (1991): 20–62.

Goodwin, R. M. *Elementary Economics from the Higher Standpoint* (Cambridge: Cambridge University Press, 1970).

Gossen, H. H. *Entwickelung der Gesetze des menschlichen Verkehrs, und der daraus fleissenden Regeln für menschliches Handeln* (Braunschweig: Vieweg, 1854).

Goschen, G. J. "Ethics and Economics," *Economic Journal* (1893): 377–90.

Goudzwaard, B. et. al. "Rough Sketch of a Joint Publication, Meant as an Introduction to an Alternative Type of Economics" (mimeo), (Amsterdam, 1979).

Graves, R. *The First Praelection, delivered as Professor of Divinity, by ... Richard Graves, King's Professor of Divinity in Trinity College Dublin. To which are annexed the Regulations for the Examination, directed by the Statute fixing the Duty of the Professorship of Divinity, and the List of Books recommended to the Students etc. Ist edn 1815* (Dublin: R. Milliken, 1820).

Grean, Stanley. *Shaftesbury's Philosophy of Religion and Ethics: a Study in Enthusiasm* (Athens OH: Ohio University Press, 1967).

Green, I. M. *The Re-Establishment of the Church of England 1660–1663* (Oxford, 1978).

Greenleaf, W. H. *Order, Empiricism and Politics: Two Traditions of English Political Thought, 1500–1700* (London: Oxford University Press, 1964).

Greenwald, B. and Stiglitz, J. F. "Externalities in Economies with Imperfect Information and Incomplete Markets," *Quarterly Journal of Economics* 101 (1986): 229–64.

Gregory, G. *An History of the Christian Church from the Earliest Periods to the Present Time*, 2 vols, (London: Kearsly, 1790).

Groenewegen, P. " 'Political Economy' and 'Economics,' " in *The New Palgrave: A Dictionary of Economics*, 4 vols, eds J. Eatwell, M. Milgate and P. Newman (London: Macmillan, 1987).

Haakonssen, K. *The Science of a Legislator: the Natural Jurisprudence of David Hume and Adam Smith* (Cambridge: Cambridge University Press, 1981).

Haakonssen, K. "The Structure of Hume's Political Thought" in *The Cambridge Companion to Hume*, ed. D. F. Norton (Cambridge: Cambridge University Press, 1994).

Haakonssen, K. "David Hume" in *Political Essays*, ed. K. Haakonssen (Cambridge: Cambridge University Press, 1994).

Haakonssen, K. (ed.) *Enlightenment and Religion: Rational Dissent in Eighteenth-Century Britain* (Cambridge: Cambridge University Press, 1996).

Haddow, A. *Political Science in American Colleges and Universities, 1636–1900* (New York: Appleton-Century, 1939).

Hahn, F. H. and Hollis, M. (eds) *Philosophy and Economic Theory*, (Oxford, Oxford University Press, 1979).

Hahn, F. H. and Matthews, R. C. O. "The Theory of Economic Growth: A Survey," *Economic Journal* 74 (1960): 779–902.

Hajnal, J. "European marriage patterns in perspective" in D. V. Glass, and D. E. C. Eversley, (eds) *Population in History: Essays in* Historical Demography, (London: Arnold, 1965).

Hales, E. E. Y. *Pio Nono. A Study in European Politics and Religion in the Nineteenth Century* (London: Eyre and Spottiswood, 1954).

Hales, E. E. Y. *The Catholic Church in the Modern World, A Survery from the French Revolution to the Present* (New York: Hanover House, 1958).

Hales, E. E. Y. *Revolution and Papacy, 1769–1846* (New York: Doubleday, 1960).

Halévy, E. *The Growth of Philosophical Radicalism*, Ist edn 1938, new edn with corrections, trans. Mary Morris (London: Faber 1952).

Halévy, E. *A History of the English People in the Nineteenth Century*, vol. i, *England in 1815*, trans. E. I. Watkin and D. A. Barker (London: E. Benn, 1949).

Hall, R. F. "A Theory of the Natural Unemployment Rate and the Duration of Unemployment," *Journal of Monetary Economics* 5 (1979): 153–170.

Hallis, Martin and Neil, Edward J. *Rational Economic Man: A Philosophical Critique Neo-Classical Economics* (Cambridge, 1975).

Hamilton, A. *Report on Manufactures*, Ist edn 1791, Reprinted (Washington: Government Printer's Office, 1913).

Hamowy, R. *The Scottish Enlightenment and the Theory of Spontaneous Order*, published for the Journal of the History of Philosophy (Carbondale IL: Southern Illinois University Press, 1987).

Hampsher-Monk, I. "Review of [Tully]" in *History of Political Thought* (1981): 554–59.

Harrod, R. F. "An Essay in Dynamic Theory," *Economic Journal* 49 (1939): 14–33.

Harrod, R. F. *The Life of John Maynard Keynes* (London: Macmillan, 1952).

Hayek, F. A. "Economics and Knowledge," *Economica* 4 (1937): 33–54.

Hayek, F. A. "The Use of Knowledge in Society," *American Economic Review* 35 (1945): 519–30.

Hayek, F. A. *The Constitution of Liberty* (Chicago: University of Chicago Press, 1960).

Hayek, F. A. "The Legal and Political Philosophy of David Hume" in *Hume: A Collection of Critical Essays*, ed. V. C. Chappell (Notre Dame: University of Notre Dame Press, 1968).

Hayek, F. A. "Competition as a Discovery Procedure," in *New Studies in Philosophy, Politics, Economics and the History of Ideas* (London: Routledge and Kegan Paul, 1978).

Hazlitt, W. *The Plain Speaker*, first published 1826, excepts from Table Talk, 1821, 1822 (London: Dent, 1928).

Heimann, P M. "Voluntarism and Immanence: Conceptions of Nature in Eighteenth Century Thought," *Journal of the History of Ideas* 39 (1978): 271–84.

Heitzenrater, R. P. *The Elusive Mr. Wesley, John Wesley His Own Biographer*, 2 vols. (Nashville: Abingdon, 1984).

Hey, J. *Lectures in Divinity*, Delivered in the University of Cambridge, 4 vols. (Cambridge: Cambridge University Press, 1796–8).

Heyne, P *The Economic Way of Thinking*, 2nd edn (Chicago: Science Research Associates, 1976).

Hicks, J. R. "Mr Keynes and the Classics: A Suggested Interpretation," *Econometrica* 5 (1937): 147–59.

Hicks, J. R. *Value and Capital* (Oxford: Oxford University Press, 1939a).

Hicks, J. R. "Foundations of Welfare Economics," *Economic Journal* 49 (1936): 696–712.

Hicks, J. R. *Capital and Growth* (Oxford: Oxford University Press, 1965).

Hilton, B. *The Age of Atonement: The Influence of Evangelicalism on Social and Economic Thought, 1785–1865* (Oxford: Clarendon, 1988).

Hinds, S. *An Inquiry into the Proofs, Nature and Extent of Inspiration, and into the Authority of Scripture* (Oxford: Parker, 1831).

Hobson, J. A. *The Industrial System* (London: Longmans, Green, 1909).

Hoff, T. J. B. *Economic Calculation in the Socialist Society*, 1st edn 1938, trans. M. A., Michael (London: Hodge, 1949).

Holbach, P. H. T., Baron d'. *The System of Nature, or Laws of the Moral and Physical World, with Notes by Diderot,* 1st edn 1770, republished trans., H. D. Robinson, 1868, reprinted (New York: Franklin, 1970).

Hole, R. *Pulpits, and Public Order in England, 1760–1832* (Cambridge: Cambridge University Press, 1989).

Hollander, S. "On Malthus's Population Principle and Social Reform," *History of Political Economy*, 18 (1986): 187–235.

Hollander, S. *Classical Economics* (Oxford: Blackwell, 1987).

Hollander, S. *The Economics of Thomas Robert Malthus* (Toronto: University of Toronto Press, 1997).

Hollis, M. and Nell, E. J. *Rational Economic Man: A Philosophical Critique of Neo-Classical Economics* (Cambridge: Cambridge University Press, 1975).

Hooker, R. *Of the Laws of Ecclesiastical Polity (1594–1662),* Book Eight in *The Works of that Learned and Judicious Divine Mr Richard Hooker, Containing Eight Books of the Laws*, 3 vols, ed. J. Keble, 7th edn rev. R. W. Church and F. Paget (Oxford: Clarendon Press, 1888).

Houlbrooke, R. A. *The English Family 1450–1700* (London: Longman, 1984).

Hume, D. *Political Essays*, 1st edn 1752, republished, ed. K. Haakonssen (Cambridge: Cambridge University Press, 1994).

Hume, D. "Of Superstition and Enthusiasm", in K. Haakonssen (ed.), *Political Essays* (Cambridge: Cambridge University Press, 1994).

Hume, D. *A Treatise of Human Nature*, 1st edn 1739–40, republished, ed. L. A. Selby-Bigge (Oxford: Clarendon Press, 1888).

Hundert, E. J. "The Making of Homo Faber: John Locke between Ideology and History," *Journal of the History of Ideas* (1972): 3–22.

Hunt, E. K. and Schwartz, J. G. (eds) *A Critique of Economic Theory: Selected Readings* (Harmondsworth Penguin, 1972).

Hunt, E. K. *History of Economic Thought: A Critical Perspective*, 2nd edn, (New York: Harper Collins, 1992).

Hutchison, T. W. *The Significance and Basic Postulates of Economic Theory* (London: Macmillan, 1938).

Iremonger, F. A. *William Temple Archbishop of Canterbury. His Life and Letters* (London: Oxford University Press, 1948).

Jacob, M. C. *The Newtonians and the English Revolution* (Ithaca: Cornell University Press, 1976).

Jacob, M. C. *The Radical Enlightenment: Pantheists, Freemasons and Republicans* (London: Allen and Unwin, 1981).

Jacobsen, G. A. and Lipman, M. H. *Political Science*, College Outline Series, (New York, 1959).

James, P. *Population Malthus: His Life and Times,* (London: Routledge and Kegan Paul, 1979).

Jannet, C. "French Catholics and the Social Question," *Quarterly Journal of Economics* 7 (1893): 137–61.

[Jeffrey, F.] "Stewart's Philosophical Essays," *Edinburgh Review* 7 (1810): 167–211.

Jenyns, S. *The Works of Soame Jenyns, Esq.,* 4 vols (London, 1790).

Jevons, W. S. "Brief Account of a General Mathematical Theory of Political Economy," 1st published 1862, in W. S. Jevons, *The Theory of Political Economy,* 1st edn 1871, (New York: Kelley and Millman, 1957).

Johnson, S. *A Dictionary of the English Language,* 1st edn 1755, Facsimile edn, 1977 (London: Times Books, 1977).

Johnson, S. *The Works of Samuel Johnson, LL. D.,* 14 vols (London: Buckland 1787–88).

John Paul II, *Encyclical Letter Centesimus Annus* (1 May 1991).

Jones, P. d'A. *The Christian Socialist Revival: Religion, Class and Social Conscience in Late-Victorian England* (Princeton: Princeton University Press, 1968).

Kafker, F. A. and Kafker, S. L. *The Encyclopedists as Individuals: A Biographical Dictionary of the Authors of the Encyclopédie.* Studies on Voltaire and the Eighteenth Century 257, (Oxford: The Voltaire Foundation, 1988).

Kaldor, N. "Welfare Propositions of Economics and Interpersonal Comparisons of Utility," *Economic Journal* 49 (1939): 549–52.

Kalecki, M. "Political Aspects of Full Employment," *Political Quarterly* 14 (1943): 322–30.

Kantorovich, L. V. "Mathematical Methods of Organizing and Planning Production," 1st published 1939, trans. 1960, Management Science 6 (1960): 363–422.

Kantorovich, L. V. *The Best Use of Economic Resources,* 1st published 1942, trans. P. K. Knightsfield, ed. G. Morton (Oxford: Pergamon, (1965).

Kamenka, E. (ed). *The Portable Karl Marx* (New York: Viking, 1983).

Katouzian, H. *Ideology and Method in Economics* (New York: New York University Press, 1980).

Kaye, F. B. "Introduction" in Bernard Mandeville, *The Fable of the Bees,* vol. i, xvii–cxlvi, facsimile reprint (Indianapolis: Liberty Press, 1988).

Keane, J. J. "The Catholic Church and Economics," *Quarterly Journal of Economics* (1891): 25–46.

Kelly, P. "Locke and Filmer: Was Laslett so wrong after all?" *The Locke Newsletter* (1977): 77–88.

Keynes, J. M. *The General Theory of Employment, Interest and Money* (London: Macmillan, 1936).

Keynes, J. M. *A Treatise on Money,* 1st edn 1930, vols v and vi in *The Collected Writings of John Maynard Keynes,* 29 vols, eds. E. Johnson and D. Moggeridge (London: Macmillan, 1971–82).

Keynes, J. M. *Essays in Biography,* vol x in *The Collected Writings of John Maynard Keynes,* 29 vols, eds. E. Johnson and D. Moggeridge (London: Macmillan, 1971–82).

Keynes, J. N. *The Scope and Method of Political Economy* (London: Macmillan, 1891).

Kolakowski, L. *Religion* (Oxford: Oxford University Press, 1982).

Kleiman, E. "The 'Just Price' in Talmudic literature," *History of Political Economy*, 19 (1987): 23–44.
Klein, L. R. and Goldberger, A. *An Econometric Model of the United States, 1929–52* (Amsterdam: North-Holland, 1955).
Knight, F. H. "Ethics and Economic Reform: III Christianity," *Economica* 6 (1939): 398–422 in *Freedom and Reform: Essays in Economics and Social Philosophy* (New York: Harper, 1947).
Knight, F. H. " 'What Is Truth?', in Economics," *Journal of Political Economy* 48 (1940): 1–32.
Knight, F. *University Rebel: The Life of William Frend 1757–1841* (London: Gollancz, 1971).
Knox, R. A. *Enthusiasm: A Chapter in the History of Religion with Special Reference to the XVII and XVIII Centuries* (Oxford: Oxford University Press, 1950).
Koopmans, T. C. "The Construction of Economic Knowledge," in T. C. Koopman, *Three Essays on the State of Economic Science* (New York, McGraw-Hill, 1957).
Kuenne, R. F. *Microeconomic Theory of the Market: A General Equilibrium Approach* (New York: Macmillan, 1968).
Lakatos, I. and Musgrave, A. (eds), *Criticism and the Growth of Knowledge* (Cambridge: Cambridge University Press, 1970).
Lakatos, I. *The Methodology of Scientific Research Programmes*, in *Philosophical Papers*, vol. i, eds. Worrall and Currie, (Cambridge: Cambridge University Press, 1978).
Lange, O. "On the Economic Theory of Socialism," *Review of Economic Studies* 4 (1936): 123–42.
Laslett, P. "Market Society and Political Theory," *Historical Journal* 7 (1964): 150–4.
Laslett, P (ed.) "Introduction" to *John Locke, Two Treatises of Government*, 1st edn (Cambridge: Cambridge University Press, 1960) 2nd edn (London: Cambridge University Press, 1967).
Laslett, P. *The World We Have Lost*, (London: Methuen, 1965).
Latsis, S. (ed.), *Method and Appraisal in Economics* (Cambridge: Cambridge University Press, 1976).
Lavington, G. *Enthusiasm of the Methodists and Papists Compared*, 3 vols, 1st edn 1749–51, republished, R. Polwhele (ed.), (London: J.& P. Knapton, 1820).
Leibniz, G. W. *Theodicy: Essays on the Goodness of God, the Freedom of Man and the Origin of Evil*, 1st edn 1710, trans. E. M. Huggard, ed. Austin Farrar (London: Routledge, 1951).
Leontief, W. "Quantiative Input–Output Relations in the Economic System of the United States," *Review of Economics and Statistics* 18 (1936): 105–25.
Lerner, A. P. *The Economics of Control: Principles of Welfare Economics* (New York: Macmillan, 1944).
Levy, D. *How the Dismal Got Its Name: Classical Economics and the Ur-Text of Racial Politics* (Ann Arbor: University of Michigan Press, 2001).
Lindbeck, A. *The Political Economy of the New Left*, 2nd edn (New York, 1977).
Lincoln, A. *Some Political and Social Ideas of English Dissent*, 1763–1800 (Cambridge: Cambridge University Press, 1938).
Lipsey, R. G. "The Relation Between Unemployment and the Rate of Change of Money Wage Rates in the United Kingdom, 1862–1957: A Further Analysis," *Economica* 27 (1960): 13–1.

Lipsey, R. G. "The Understanding and Control of Inflation: Is there a Crisis in Macro-economics?," *Canadian Journal of Economics* 14 (1981): 545–76.
Lipsey, R. G. and Lancaster, K. "The General Theory of Second Best," *Review of Economic Studies* 24 (1956): 11–32.
Lipsey, R. G., Purvis, D. D. and Steiner, P. O. *Economics*, 6th edn (New York: Harper and Row, 1988).
Lloyd, P. J. "Elementary Geometric/Arithmetic Series and Early Production Theory," *Journal of Political Economy* 77 (1969): 23–44.
Lloyd, W. F. *A Lecture on the Notion of Value Delivered Before the University of Oxford in 1833* (Oxford: Oxford University Press, 1834).
Locke, J. *A Letter Concerning Toleration*, 1st edn 1689, republished, Latin and English texts rev. and ed. Mario Monturi (The Hague: Nijhof, 1963).
Locke, J. *An Essay Concerning Human Understanding*, 1st edn 1690, ed. Peter H. Nidditch (Oxford: Oxford University Press, 1975).
Locke, J. *The Reasonableness of Christianity as Delivered in the Scripture*, 1st edn 1695, ed. George W. Ewing (Chicago: Regnery, 1965).
Locke, J. *Two Treatises of Government*, 1st edn 1690, ed. and with an introduction and critical apparatus by P. Laslett (London, Cambridge University Press, 1967).
Logan, J. F. "Superstition, Impiety, and an Enlightened Social Order: The Theological Politics of the Abbé Mably," in *The Secular Mind: Transformations of Faith in Modern Europe*, ed. W. W. Wagar (New York: Holmes & Meier, 1982).
Lough, J. *Essays on the Encyclopédie of Diderot and D'Alembert* (London: Oxford University Press, 1968).
Lovejoy, A. O. *The Great Chain of Being. A Study of the History of an Idea* (Cambridge, MA: Harvard University Press, 1936).
Lucas, R. F. "An Equilibrium Model of the Business Cycle," *Journal of Political Economy* 83 (1975): 1113–44.
Macfarlane, A. *The Origins of English Individualism: The Family, Property and Social Transition* (Cambridge: Cambridge University Press, 1979).
Mackintosh, James. *Vindiciae Gallicae: Defence of the French Revolution* (London: Robinson, 1791).
Maclaurin, C. *An Account of Sir Isaac Newton's Philosophical Discoveries*, 3rd edn. (London: Nourse, 1775).
MacPherson, C. B. *The Political Theory of Possessive Individualism: Hobbes to Locke* (Oxford: Clarendon Press, 1962).
MacRae, C. D. "A Political Model of the Business Cycle," *Journal of Political Economy* 85 (1977): 239–63.
Le Mahieu, D. L. *The Mind of William Paley* (Lincoln, Nebraska: 1976).
[Maistre, Joseph Marie de]. *Lettres d'un Royaliste Savoisien* (Lausanne, 1793).
[Malthus, T. R.] *An Essay on the Principle of Population as It Affects the Future Improvement of Society* . . . (London: Johnson, 1798).
Malthus, T. R. *An Essay on the Principle of Population as It Affects the Future Improvement of Society, with Remarks Upon the Speculations of Mr Godwin, M. Condorcet, and Other Writers*, fascimile reprint for the Royal Economic Society of the 1798 1st edn, (London: Macmillan, 1966).
Malthus, T. R. *An Essay on the Principle of Population*, 2 vols, 1803 edn, with variora of 1806, 1807, 1817, 1826, ed. Patricia James, (Cambridge: University Press, 1989).

Malthus, T. R. *An Essay on the Nature and Progress of Rent*, 1st edn 1815, vol. vii in *The Works of Thomas Robert Malthus*, 8 vols, (London: Pickering, 1986).

Malthus, T. R. *An Essay on the Principle of Population, or, A View of Its Past and Present Effects on Human Happiness, with an Inquiry into Our Present Prospects Respecting the Future Removal or Mitigation of the Evils which It Occasions*, published 1817, considered 5th edn, (London: Hunter, 1817).

Malthus, T. R. Letter to Chalmers, 21 July 1822, Ms. CHA 4.21.51 in New College, Edinburgh.

Malthus, T. R. Letter to Chalmers, 18 January 1827, Ms. CHA 4.80.9 in New College, Edinburgh.

Malthus, T. R. Letter to Chalmers, 6 March 1832, Ms. CHA 4.185.32 in New College, Edinburgh.

Malthus, T. R. *The Travel Diaries of Thomas Robert Malthus*, ed. Patricia James (Cambridge: Cambridge University Press, 1966).

Malthus, T. R. *Principles of Political Economy*, Variorum Edition, 2 vols. ed. John Pullen (Cambridge: Cambridge University Press, 1989).

Mandeville, B. *The Fable of the Bees: Or, Private Vices, Public Benefits, 1714–1728*, 2 vols, republished, ed. F. B. Kaye, (Oxford: Clarendon Press 1924) facsimile reprint by Liberty Fund, (Indianapolis: Liberty Press, 1988).

Mankiw, N. G. and Romer, D. (eds) (1991). *New Keynesian Economics*, 2 vols. (Cambridge MA: MIT Press, 1991).

Marshall, A. *Principles of Economics*, 1st edn 1890 (8th edn, London: Macmillan, 1952).

Marshall, A. "Response" to [Goschen], Economic Journal (1893): 387–90.

Marshall, J. D. *The Old Poor Law, 1795–1834*, 2nd edn, *Studies in Economic and Social History* (London: Macmillan, 1985).

Martin, E. W. (ed.). *Comparative Development in Social Welfare* (London: Allen & Unwin, 1972).

Marx, K. *A Contribution to the Critique of Political Economy*, 1st edn 1859, republished, trans. N. I. Stone (Chicago: Charles H. Kerr and Company, 1904).

Marx, Karl. *Capital (1867–1895)* republished, trans. E. and C. Paul (Everyman: London, 1930).

Marx, K. *Capital (1867–1895)* 3 vols, ed. F. Engels, republished, trans. E. Aveling and S. Moore, (Moscow: Progress, 1954).

Marx, K. and Engels, F. "Manifesto of the Communist Party," 1st edn 1848, republished in *Basic Writings on Politics and Philosophy: Karl Marx and Friedrich Engels*, ed. L. S. Feuer (New York: Anchor Books, 1959).

Mathers, F. C. "Georgian churchmanship reconsidered: some variations in Anglican public worship 1714–1830," *The Journal of Ecclesiastical History* 36 (1985).

Mayerne-Turquet, L. de *La Monarchie Aristodemocratique; ou le Gouvernement compose et meslé des troisformes de légitimes républiques* (Paris: 1611).

McCafferty, S. *Macroeconomic Theory* (New York: Harper & Row, 1990).

McCloskey, D. N. "The Rhetoric of Economics," *Journal of Economic Literature*, 21: (1983).

McCloskey, D. N. *The Rhetoric of Economics* (Madison: University of Wisconsin Press, 1985).

McCulloch, J. R. *Principles of Political Economy*, new edn (Edinburgh: Tait, 1843).

McDonald, J. and Snooks, G. D. *Domesday Economy: A New Approach to Anglo-Norman History* (Oxford: Oxford University Press, 1986).
McLarty, M. R. (ed.) *A Source Book and History of Administrative Law in Scotland* (Edinburgh: Hedge, 1956).
McLean, M. "Did Paley Ignore Hume on the Argument from Design?," in *Faith, Reason and Economics: Essays in Honour of Anthony Waterman*, ed. Derek Hum (Winnipeg: St John's College Press, 2003).
Meade, J. E. *The Balance of Payments, Mathematical Supplement* (London: Oxford University Press, 1951).
Menger, C. *Principles of Economics*, 1st published 1871, trans. J. Dingwall and B. F. Hoselitz (Glencoe IL: Free Press, 1950).
Meslier, J. *Superstition in All Ages or Common Sense*, transl. Anna Koop (New York: Eckler, 1915).
Mill, J. S. *Principles of Political Economy with Some of Their Applications to Social Philosophy*, 1st edn 1848, republished, ed. W. J. Ashley (Longmans, Green and Co: London, 1909).
Mill, J. S. *Essays on Some Unsettled Questions of Political Economy*, 2nd edn (London, Longmans, Green, 1874).
Mill, J. S. "Nature" in *Three Essays on Religion*, 1st edn 1874, vol. x of *Collected Works of John Stuart Mill*, 33 vols (Toronto: University of Toronto Press, 1963–91).
Milner, J. *The History of the Church of Christ*, new edn, Eour vols. (London: Cadell, 1834).
Minogue, K. "Review Essay," *National Review* XLIII, no. 11, Special Supplement (1991).
Minowitz, P. *Profits, Priests and Princes: Adam Smith's Emancipation of Economics from Politics and Religion* (Stanford: Stanford University Press, 1993).
Mises, L. von. "Economic Calculation in the Socialist Commonwealth," 1st published 1920, tr. in F. A. Hayek (ed.) *Collectivist Economic Planning* (London: Routledge, 1935).
Mises, L. von. *Bureaucracy* (New Haven: Yale University Press, 1944).
Mizuta, H. *Adam Smith's Library: A Catalogue* (Oxford: Clarendon, 1996).
Montchrétien, A. de. *Traité de l'économic politique*, 1st edn 1615, republished ed. T. Funck-Brentano (Paris: Plon, 1889).
More, H. *Enthusiasmus Triumphatus; or, A Brief Discourse of the Nature, Causes, Kinds, and Cure of Enthusiasm* (London, 1712).
Moody, J. N. "Leo XIII and the Social Crisis," in Gargan (ed.), *Leo XIII and the Modern World*, ch. 3, (New York: Sheed and Ward, 1961).
Moon, P. T. *The Labor Problem and the Social Catholic Movement in France* (New York: Macmillan, 1921).
Morgenstern, O. "Descriptive, Predictive and Normative Theory," *Kyklos*, (1974): 699–713.
Morishima, M. *Ricardo's Economics: A General Equilibrium Theory of Distribution and Growth* (Cambridge: Cambridge University Press, 1989).
Munby, D. L. *Christianity and Economic Problems* (London: Macmillan, 1956).
Munby, D. L. "The Disordered Economic Thinking of the National Church," *Theology* 60 (1957): 92–9.
Munby, D. L. "The Importance of Technical Competence," in *Essays in Anglican Self-Criticism*, ed. David M. Paton (London: SCM, 1958).

Munby, D. L. *God and the Rich Society: A Study of Christianity in a World of Abundance* (Oxford: Oxford University Press, 1960).
Munby, D. L. *The Idea of a Secular Society and Its Significance for Christians* (Oxford: Oxford University Press, 1963).
Mundell, R. A. "The International Disequilibrium System," *Kyklos* 14 (1960): 154–172.
Musgrave, R. A. "The Voluntary Exchange Theory of Public Economy," *Quarterly Journal of Economics* 53 (1938): 213–37.
Musgrave, R. A. and Peacock, A. T. (eds). *Classics in the Theory of Public Finance* (New York: St Martin's, 1958).
Muth, J. F. "Rational Expectations and the Theory of Price Movements," *Econometrica* 29 (1961): 315–35.
Nagel, E. "Assumptions in Economic Theory," *American Economic Review, Papers and Proceedings* (1963): 211–19.
Negishi, T. *History of Economic Theory* (Amsterdam: North Holland, 1989).
Negishi, T. "A Smithian Growth Model and Malthus's Optimum Propensity to Save," *European Journal of the History of Economic Thought* 1: 115–27.
Nelson, R. A. *Reaching for Heaven on Earth: The Theological Meaning of Economics* (Savage MD: Rowman & Littlefield, 1991).
Nelson, R. A. *Economics as Religion: From Samuelson to Chicago and Beyond* (Philadelphia: Pennsylvania State University Press, 2001).
Neuhaus, R. J. "Review Essay," *National Review* XLIII, no. 11, Special Supplement (1991): S8–9.
Neumann, J. von "A Model of General Economic Equilibrium," 1st published 1938, trans. G. Morgenstern, *Review of Economic Studies* 13 (1945–46): 1–9.
New Catholic Encyclopedia, 17 vols, (New York: McGraw-Hill, 1967)
Newman, J. H. *Apologia Pro Vita Sua*, new impression, (London: Longmans, Green, 1929)
Newton, I. *Four Letters from Sir Isaac Newton to Dr Bentley Containing Some Arguments in Proof of a Deity* (London: Dodsley, 1756).
Nicholls, D. "The Invisible Hand: Providence and the Market," in *Values of the Enterprise Culture*, eds Paul Heelas and Paul Morris (London: Routledge, 1992).
Niskanen, W. A. *Bureaucracy and Representative Government* (Chicago: Aldine-Atherton, 1971).
Norman, E. R. *Church and Society in England, 1770–1970* (Oxford: Clarendon, 1976).
Norman, E. R. *The Victorian Christain Socialists* (Cambridge: Cambridge University Press, 1987).
Novak, M. "Review Essay," *National Review* XLIII, no. 11, Special Supplement (1991): S7–8.
O'Brien, C. C. *The Great Melody* (London: Minerva, 1993).
Olivecrona, K, "Appropriation in the State of Nature: Locke on the Origin of Property," *Journal of the History of Ideas* (1974): 211–30.
O'Neill, J. "Who Won the Socialist Calculation Debate?," *History of Political Thought* 17 (1996): 431–442.
Overton, J. H. and Relton, F. C. *The English Church from the Accession of George I to the End of the Eighteenth Century* (London: Macmillan, 1906).
Overton, J. *The True Churchman Ascertained: or, An Apology for Those of the Regular Clergy of the Establishment Who Are Sometimes Called Evangelical*

Ministers: Occasioned by the Publications of Drs Paley, Hey, Croft; Messrs. Daubeny, Ludlam, Polwhele, Fellowes; the Reviewers, etc., etc., 2nd edn, (York: 1802).

Paine, T. *The Rights of Man: Being an Answer to Mr. Burke's Attack on the French Revolution* (Part i. London: Johnson, 1791, Part ii London: Jordan, 1792).

Paley, W. *The Principles of Moral and Political Philosophy* (London: Faulder, 1785).

Paley, W. *The Principles of Moral and Political Philosophy*, 1st edn 1785, vol. iv, in *The Works of William Paley*, D. D., Seven vols, ed. E. Paley (London: 1825).

Paley, W. *Horae Paulina* (London: Faulder, 1790).

Paley, W. *Evidences of Christianity* (London: Faulder, 1794).

Paley, W. *Natural Theology* (London: Wilks and Taylor, 1802).

Paley, W. *Natural Theology*, 1st edn 1802, vol v, in *The Works of William Paley D. D.* 7 vols, ed. E. Paley (London: 1825).

Pareto, V. *Cours d'Economie Politique* (Geneva: Libraire de l'Université, 1896).

Pareto, V. *Manual of Political Economy*, 1st edn 1908, trans. A. S. Schwier, eds. A. S. Schwier and A. N. Page (New York: Kelley, 1971).

Patinkin, D. "Keynes and Econometrics: On the Interaction Between the Macroeconomic Revolution of the Interwar Period," *Econometrica* 44 (1976): 1091–1123.

Petty, W. *The Economic Writings of Sir William Petty, 1662–1690*, reprinted, ed. C. Hull (Fairfield NJ: A. M. Kelley, 1986).

Phelps, F. S. "Phillips Curves, Expectations of Inflation and Optimal Unemployment over Time," *Economica* 34 (1967): 254–81.

Phelps, F. S. (ed.), *Microeconomic Foundations of Employment and Inflation Theory* (New York: Norton, 1970).

Phillips, A. W. "The Relation between Unemployment and the Rate of Change of Money Wages in the United Kingdom, 1861–1957," *Economica* 25 (1958): 283–99.

Pigou, A. C. *The Economics of Welfare* (London: Macmillan, 1920).

Pontryagin, L. S., Boltyanskii, V. G., Gamkrelidze, R. V. and Mishchenko, F. F. *The Mathematical Theory of Optimal Processes*, trans. K. N. Tririgoff. (New York: Interscience, 1962).

Piozzi, H. L. *Anecdotes of Samuel Johnson*, 1st edn 1786, republished (Gloucester, 1984).

Plato, *The Republic*, trans. P. Shorey (London: Heinemann, 1953).

Pius XI, Pope, "Quadragesimo Anno: On Reconstructing the Social Order and Perfecting It Conformably to the Precepts of the Gospel" in *Commemoration of the Fortieth Anniversary of the Encyclical "Rerum Novarum"* (Catholic Truth Society: Oxford 1931).

Pocock, J. G. A. "Post-Puritan England the Problem of the Enlightenment," in *Culture and Politics from Puritanism to the Enlightenment*, ed. Perez Zagorin (Berkeley: University of California Press, 1980).

Pocock, J. G. A. "Superstition and Enthusiasm in Gibbon's History of Religion," *Eighteenth-Century Life* 8 (1982): 81–94.

Pocock, J. G. A. "Clergy and Commerce: The Conservative Enlightenment in England," in *L 'Eta dei Lumi: Studi Storici sul Settecento Europeo in Onore Franco Venturi*, eds. R. Ajello *et al.*, 2 vols, (Naples: Juvene, 1985).

Pocock, J. G. A. *Virtue, Commerce and History: Essays in Political Thought and History, Chiefly in the Eighteenth Century* (Cambridge: University Press, 1985).

Poggioli, S. "Broadcast," National Public Radio, 2 May 1991.
Polanyi, K. *The Great Transformation* (Boston: Beacon, 1957).
Pownall, T. *Letter to Adam Smith Being an Examination of Several Points of Doctrine Laid Down in His Inquiry, into the Nature and Causes of the Wealth of Nations* (London, 1776).
Preston, R. H. *Religion and the Persistence of Capitalism: The Maurice Lectures for 1977 and other Studies in Christianity and Social Change* (London: SCM, 1979).
Preston, R. H. *Church and Society in the Late Twentieth Century: The Economic and Political Task* (London: SCM, 1983).
Pretyman-Tomline, G. *Elements of Christian Theology*, 9th edn, 2 vols (London, 1812).
Price, R. *A Discourse on the Love of our Country*, 1st edn 1789 (Boston: Powars, 1790).
Price, R. *Two Tracts on Civil Liberty, the War with America, and the Debts and Finances of the Kingdom . . . Containing i, Observations on the Nature of Civil Liberty . . . (1776), and ii, Additional Observations . . .* (1777). (London, 1778).
Priestley, J. *The Theological and Miscellaneous Works etc. of Joseph Priestley, LL.D. F.R.S., etc with Notes by the Editor*, ed. J. T. Rutt, 25 vols, first printed 1817–32, reprinted (New York: Kraus Reprint, 1972).
Proctor, F. and Frere, W. H. *The Book of Common Prayer, with a Rationale of Its Offices* (London, 1958).
Pryme, G. *An Introductory Lecture and Syllabus to a Course Delivered in the University of Cambridge on the Principles of Political Economy* (Cambridge, Cambridge University Press, 1823).
Public Characters of 1802–1803, vol. v, in *British Public Characters 1799–1810*, 10 vols (London: 1803).
Pullen, J. A. "Some new information on the Rev. T. R. Malthus", *History of Political Economy*, 19 (1987): 127–40.
Pyle, A. (ed.). *Population: Contemporary Responses to Thomas Malthus* (Bristol: Thoemmes Press, 1994).
Quasten, J. (ed.). *(1950–1960) Patrologia*, 3 vols, (Utrecht: Spectrum, 1950–60).
Rack, H. D. *Reasonable Enthusiast: John Wesley and the Rise of Methodism* (London: Epworth, 1989).
Raeder, L. C. *John Stuart Mill and the Religion of Humanity* (Columbia: University of Missouri Press, 2002).
Ramaswami, C. "Equilibrium Unemployment and the Efficient Job-Finding Rate," *Journal of Labor Economics* 1 (1983): 171–96.
Ramsey, F. P. "A Contribution to the Theory of Taxation," *Economic Journal* 37 (1927): 47–61.
Ramsey, F. P. "A Mathematical Theory of Saving," *Economic Journal* 38 (1928): 543–59.
Raphael, D. D. and Macfie, A. L. "Introduction," in A. Smith, *The Theory of Moral Sentiments*, eds. D. D. Raphael and A. L. Macfie (Oxford: Clarendon, 1976).
Raphael, D. D. *Adam Smith*. (Oxford: Oxford University Press, 1985).
Ratcliff, E. C. *The English Coronation Service* (London: Skeffington and Son, SPCK, 1936).
Ricardo, D. *An Essay on the Influence of a Low Price of Corn on the Profit of Stock ...* 1st edn 1815, reprinted in Vol. iv of *The Works and Correspondence of David Ricardo*, 11 vols, ed. P. Sraffa (Cambridge: Cambridge University Press 1951–73).

Ricardo, D. *On the Principles of Political Economy and Taxation*, 1st edn 1817, republished (London: Dent, 1911).
Ricardo, D. *On the Principles of Political Economy, and Taxation*, 1st edn 1817, reprinted in vol. i of *The Works and Correspondence of David Ricardo*, 11 vols, ed. P. Sraffa, (Cambridge: University Press, 1951–73).
Ring, Mary-Ignatius SND *Villeneuve-Bargemont: Precursor of Modern Social Catholicism, 1788–1850* (Milwaukee: Bruce, 1935).
Rivers, D. *Observations on the Political Conduct of the Protestant Dissenters* (London, 1798), cited in the *Anti-Jacobin Review* (1798).
Robbins, C. *The Eighteenth-Century Commonwealthman* (Cambridge MA: Harvard University Press, 1961).
Robbins, L. *An Essay on the Nature and Significance of Economic Science* (London: Macmillan, 1st edn1932, 2nd edn 1952, (London: Macmillan, 1932, 1952).
Robinson, J. *The Economics of Imperfect Competition* (London: Macmillan, 1933).
Robinson, J. *On Re-reading Marx*, 1st edn 1953, Cambridge: Students' Bookshop, reprinted in vol. iv of J. Robinson, *Collected Economic* Papers (Oxford: Blackwell, 1973).
Robinson, J. "The Production Function and the Theory of Capital," *Review of Economic Studies* 21 (1953): 81–106, reprinted in vol. ii of J. Robinson, *Collected Economic Papers* (Oxford: Blackwell, 1964).
Robinson, J. "Consumer's Sovereignty in a Planned Economy," in *Essays in Honour of Oskar Lange* (1964), reprinted in vol. iii of J. Robinson, *Collected Economic Papers*, (Oxford, Blackwell, 1965).
Robinson, J. *Collected Economic Papers*, Five vols (Oxford: Blackwell, 1951–80).
Robinson, J. "Introduction" to *Classical and Neoclassical Theories of General Equilibrium: Historical Origins and Mathematical Structure*, eds, Vivian Walsh and Harvey Gram (New York: Oxford University Press, 1980).
Robinson J. A. T. *The Body: A Study in Pauline Theology* (London: SCM Press, 1952)
Roche, D. *France in the Enlightenment*, 1st French edn 1993, transl. Arthur Goldhammer, (Cambridge MA: Harvard University Press, 1998).
Ross, I. S. *The Life of Adam Smith* (Oxford: Clarendon, 1995).
Ross, S. A. "The Economic Theory of Agency: The Principals Problems," *American Economic Review 63* (1973): 134–9.
Ryan, M. "Christian Social Teaching and Canadian Society", mimeo, St. Peter's Seminary (London ON, 1978).
Sagoff, M. "Muddle or Muddle Through? Takings Jurisprudence meets the Endangered Species Act," *William and Mary Law Review* 38 (1997).
Saint-Paul, G. "The 'New Political Economy': Recent Books by Allen Drazen and by Torsten Persson and Guido Tabellini," *Journal of Economic Literature* 38 (2000): 915–25.
Samuelson, P. A. "The General Theory," 1st edn 1946, reprinted in vol. ii of *The Collected Scientific Papers of Paul A. Samuelson*, ed. Joseph F. Stiglitz (Cambridge MA: MIT Press, 1966).
Samuelson, P. A. *Foundations of Economic Analysis* (Cambridge MA: Harvard University Press, 1947).
Samuelson, P. A. *Economics: An Introductory Analysis* (New York: McGraw Hill, 1948).
Samuelson, P. A. "Evaluation of Real National Income," *Oxford Economic Papers 2* (1950): 1–29.

Samuelson, P. A. "The Pure Theory of Public Expenditures," *Review of Economics and Statistics* 36 (1954): 387–9.
Samuelson, P. A. *Economics: An Introductory Analysis* (New York: McGraw-Hill, 1958).
Samuelson, P. A. "Assumptions in Economic Theory," *American Economic Review, Papers and Proceedings* (1963).
Samuelson, P. A. "A Modern Theorist's Vindication of Adam Smith," *American Economic Review* 67 (1977): 42–9.
Samuelson, P. A. "The Canonical Classical Model of Political Economy", *Journal of Economic Literature* 16 (1978): 1415–34.
Samuelson, Paul A. and Solow, R. M. "Balanced Growth under Constant Returns to Scale," *Econometrica 21* (1953): 412–424, reprinted in *The Collected Scientific Papers of Paul Samuelson.*
Samuelson, P. A. and Solow, R. M. "Analytical Aspects of Anti-Inflation Policy," *American Economic Review 50* (1960): 177–194.
Sandoz, E. (ed.). *Political Sermons of the American Founding Era 1730–1805* (Indianapolis, 1991).
Sappington, D. F. M. "Incentives in Principal-Agent Relationships," *Journal of Economic Perspectives* 5 (1991): 45–66.
Sappington, D. F. M. and Stiglitz, J. F. "Privatization, Information and Incentives," *Journal of Policy Analysis and Management* 6 (1987): 567–82.
Sargent, T. J. "Rational Expectations, the Real Rate of Interest, and the Natural Rate of Unemployment," *Brookings Papers in Economic Activity* 2 (1973): 429–72.
Sargent, T. J. and Wallace, N. "Rational Expectations and the Theory of Economic Policy," *Journal of Monetary Economics* 2 (1976): 169–184.
Schmandt, Raymond H. "The Life and Work of Leo XIII," in Gargan, *Leo XIII and the Modern World* (New York: Sheed and Ward, 1961).
Schumpeter, J. A. *Capitalism, Socialism and Democracy*, 1st edn 1943, new edn (London: Unwin, 1987).
Schumpeter, J. A. *History of Economic Analysis* (London: Allen and Unwin, 1954).
Scitovsky, T. "A Note on Welfare Propositions in Economics," *Review of Economic Studies* 9 (1941): 77–88.
Sen A. K. "Terminal Capital and Optimal Savings," in *Socialism, Capitalism and Economic Growth: Essays Presented to Maurice Dobb*, ed. C. H. Feinstein (Cambridge: Cambridge University Press, 1967).
Senior, N. W. *An Outline of the Science of Political Economy* (London, 1836).
Senior, N. W. *Four Introductory Lectures on Political Economy, Delivered Before the University of Oxford* (London: Longmans, Brown, Green and Longmans, 1852).
Shelton, G. *Dean Tucker and Eighteenth Century Economic and Political Thought* (London: Macmillan, 1981).
Simon, H. A. "Theories of Decision Making in Economics and Behavioral Science," *American Economic Review* 49 (1959): 253–83.
Simon, H. A. "Rationality as Process and as Product of Thought," *American Economic Review* 68 (1978): 1–16.
Sisson, C. H. (ed.). *The English Sermon: an Anthology*, 3 vols, vol. ii 1650–1750 (Cheadle: 1976).
Skinner, Q. "Meaning and Understanding in the History of Ideas," *History and Theory* 8: (1969)3–53.

Skinner, Q. *The Foundations of Modern Political Thought*, 2 vols(Cambridge: Cambridge University Press, 1978).
Smith, A. *The Theory of Moral Sentiments*, 1st edn 1759, republished, ed E. G. West (Indianapolis: Liberty Classics, 1776).
Smith, A. *Lectures on Jurisprudence*, 1st edn 1762, republished, eds R. L. Meek, D. D. Raphael, P. G. Stein (Indianapolis: Liberty Classics, 1982).
Smith, A. *An Inquiry into the Nature and Causes of the Wealth of Nations*, 1st edn 1776, republished 1976, eds R. H. Campbell A. S. Skinner and W. B. Todd, 2 vols (Oxford: Clarendon Press, 1976).
Snooks, G. D. "Arbitrary Decree or Rational Calculation? The Contribution of the Domesday Book to Economic History and Economics," *Australian Economic History Review*, 30 (1990): 23–49.
Soderini, E. *The Pontificate of Leo XIII*, vol. i, transl. B. B. Carter (London: Burns, Oates and Washbourne, 1934)
Soderini, E. *Leo XIII Italy and France*, vol ii, transl. B. B. Carter (London: Burns, Oates and Washbourne, 1935)
Soloway, R. A. *Prelates and People: Ecclesiastical Social Thought in England, 1783–1852* (London: Routledge and Kegan Paul, 1969).
Sousberghe, L. de "'Propriété de droit naturel,' thèse néo-scolastique et tradition scolastique," *Nouvelle révue theologique* (1950): 580–607.
Spencer, H. *First Principles* (London: Williams & Norgate, 1862)
Sraffa, P. "The Laws of Return under Competitive Conditions," *Economic Journal* 35 (1926): 535–550.
Stamp, J. *The Christian Ethic as an Economic Factor* (London: Epworth, 1926).
Stamp, J. *Criticism and Other Addresses* (London: Benn 1931).
Stamp, J. *Motive and Method in a Christian Order* (London: Epworth 1936).
Stamp, J. *Christianity and Economics* (London. Macmillan, 1939)
Stangeland, C. E. *Pre-Malthusian Doctrines of Population: A Study in the History of Economic Theory* (New York: Columbia University Press, 1904).
Starkey, Thomas. *A Dialogue between Pole and Lupsett*, 1st edn 1538? Republished ed T. F. Mayer (London: Royal Historical Society 1989) Camden Miscellany 4th series, vol. 37.
Stephen, L. *History of English Thought in the Eighteenth Century*, 2nd edn, 2 vols (London: 1881).
Steuart, J. *An Inquiry into the Principles of Political Economy*, 2 vols, 1st ed 1767, ed. A. S. Skinner (Oliver and Boyd for the Scottish Economic Society, 1966).
Stewart, D. *Account of the Life and Writings of Adam Smith, LLD*, 1st edn 1793, republished in Adam Smith, *Essays on Philosophical Subjects*, eds W. P. D. Wightman, J. C. Bryce, and I. S. Ross (Oxford: Oxford University Press, 1980).
Stewart, D. *The Collected Works of Dugald Stewart, Esq., etc.*, 11 vols, ed. Hamilton William (Edinburgh: Constable, 1854–60).
Stigler, G. J. "The Ricardian Theory of Value and Distribution", *Journal of Political Economy* 60 (1952) 3:187–207.
Stigler, G. J. "The Economics of Information," *Journal of Political Economy* 69 (1961): 213–25.
Stigler, G. J. "The Theory of Economic Regulation," *Bell Journal of Economics and Management Science* 2 (1971): 3–31.
Stigler, G. J. *The Economist as Preacher* (Oxford: Blackwell, 1982).
Stiglitz, J. F. *Economics of the Public Sector* (New York: Norton, 1986).

Stiglitz, J. F. *Whither Socialism?* (Cambridge: MIT Press, 1994).
Swan, T. W. "Economic Control in a Dependent Economy," *Economic Record 36* (1960): 51–66.
Sweezy, P. "Economics and the Crisis of Capitalism," *Economic Forum* 3 (1935).
Swift, J. *Irish Tracts, 1720–1723 and Sermons*, vol ix of *The Prose Works of Jonathan Swift*, reprint of 1948 ed., eds Herbert Davis and Louis Landa (Oxford: Blackwell, 1963).
Sumner, J. B. *An Essay, Tending to Show that the Prophecies, Now Accomplishing, Are an Evidence of the Truth of the Christian Religion, Hulsean Prize Essay 1801* (Cambridge: Cambridge University Press, 1801).
[Sumner, J. B.] *Apostolical Preaching Considered, in an Examination of St. Paul's Epistles* (London: Hatchard 1815).
Sumner, J. B. *A Treatise on the Records of the Creation: with Particular Reference to Jewish History, and the Consistency of the Principle of Population with the Wisdom and Goodness of the Deity*, 2 vols, (London: Hatchard, 1816).
Sumner, J. B. *The Evidence of Christianity Derived from Its Nature and Reception* (London: Hatchard, 1824).
Sykes, N. W. *Church and State in England in the XVIIIth Century* (Cambridge: Cambridge University Press, 1934).
Sykes, N. W. *From Sheldon to Secker: Aspects of English Church History, 1660–1768* (Cambridge: Cambridge University Press 1959).
Tawney, R. H. *The Agrarian Problem of the Sixteenth Century*, 1st edn 1912, (New York: Harper Torchback, 1967).
Tawney, R. H. *Religion and the Rise of Capitalism*, 1st edn 1926 (London: Murray, 1936).
Tawney, R. H. *R. H. Tawney's Commonplace Book*, ed. J. M. Winter and D. M. Joslin. (New York: Cambridge University Press, 1972).
Taylor, F. M. "The Guidance of Production in the Socialist State," *American Economic Review* 19 (1929): 1–8.
Teichgraeber, R. F *Free Trade and Moral Philosophy: Rethinking the Sources of Adam Smith's Wealth of Nations* (Durham NC: Duke University Press, 1986).
Temple, W. *Christianity and Social Order.* 1st edn 1942, new edn, foreword by the Right Honourable Edward Heath, MBE, MP Introduction by the Reverend Canon Ronald H. Preston (London: SPCK, 1976).
Theocharis, R. D. *Early Developments in Mathematical Economics* (London: Macmillan, 1961).
Thünen, J. H. von *Der isolierte Staat in Beziehung auf Landwirtschafi und National* (Œkonomie (Hamburg: Perthes, 1826).
Tiemstra, J. *Reforming Economics: Calvinist Studies on Method and Institutions* (Lewiston: Mellen, 1990).
Tinbergen, J. "Quantitative Fragen der Konjunkturpolitik," *Weltwirtschaftliches Archiv* 42 (1935): 316–99.
Tillich, P. *A History of Christian Thought: From Its Judaic and Hellenistic Origins to Existentialism* (New York: Simon and Schuster, 1967).
Toland, J. *Christianity not Mysterious: or, A Treatise Shewing that There Is Nothing in the Gospel contrary to Reason nor Above It: and that no Christian Doctrine can be properly call'd A Mystery*, Ist edn 1696, facsimile reprint (New York: Garland, 1978).
Torrens, R. *An Essay on the External Corn Trade* (London, 1815).

Tracts for the Times No. 87, *On Reserve in Communicating Religious Knowledge*, part 4, (London: 1840).

Tuan, M-L. *Simon de Sismondi as an Economist* (New York: Columbia University Press, 1927).

Tuck, R. *Natural Rights Theories: Their Origin and Development* (Cambridge: Cambridge University Press, 1979).

Tucker, J. *A Brief Essay on the Advantages and Disadvantages which Respectively Attend France and Great Britain, with Regard to Trade. With Some Proposals for Removing the Principal Disadvantages of Great Britain, in a New Method*, facsimile reprint of 3rd edn 1749, vol. i in J. Tucker, *The Collected Works of Josiah Tucker*, 6 vols, with a new introduction by Jeffrey Stern (London: Routledge/Thoemmes, 1993).

Tucker, J. *The Elements of Commerce and Theory of Taxes*, privately printed by Tucker in 1755, first published 1931, reprinted from Schuler's 1931 edn, vol. iii in J. Tucker; *The Collected Works of Josiah Tucker*, 6 vols, with a new introduction by Jeffrey Stem (London: Routledge/Thoemmes, 1993).

Tucker, J. *Instructions for Travellers*, facsimile reprint of 2nd edn 1757, in J. Tucker, *The Collected Works of Josiah Tucker*, with a new introduction by Jeffery Stern (London: Routledge/Thoemmes, 1993).

Tullock, G. *The Politics of Bureaucracy* (Washington: Public Affairs Press, 1965).

Tully, J. *A Discourse on Property: John Locke and His Adversaries* (Cambridge, 1980).

Turgot, A. R. J. "Observations sur le mémoire de M. de Saint-Péravy en faveur de l'impôt indirect," 1st edn 1768, in A. R. J. Turgot, *Ecrits Economiques*, ed. B. Cazes (Paris: Calman-Lévy: 1970).

Ullman, W. *The Growth of Papal Government in the Middles Ages: A Study in the Ideological Relation of Clerical to Lay Power* (London: Methuen, 1970).

Vance, N. *The Sinews of the Spirit: the Ideal of Christian Manliness in Victorian Literature and Religious Thought* (New York: Cambridge University Press, 1985).

Varian, H. R. "Distributive Justice, Welfare Economics, and the Theory of Fairness," *Philosophy and Public Affairs (1974–75)*, 223–47. Reprinted in Hahn and Hollis, *Philosophy and Economic Theory* (Oxford: Oxford University Press, 1979).

Vaughn, K. "Economic Calculation under Socialism," *Economic Inquiry* 18 (1980): 535–554.

Vaughn, K. *Austrian Economics in America: The Migration of a Tradition* (Cambridge: Cambridge University Press, 1994).

Veblen, T. *The Engineers and the Price System*, 1st edn 1921, republished (New York: Kelley, 1965).

Vickers, D. *Man in the Maelstrom of Modern Thought: An Essay in Theological Perspectives* (Nutley, NJ: Presbyterian and Reformed Publishing Company, 1975).

Vickers, D. *Economics and Man: Prelude to a Christian Critique* (Nutley NJ: Craig, 1976).

Villeneuve-Bargemont, A. de. *Economic Politique Chrétienne ou Recherches sur la Nature et les Causes du Pauperisme, en France et sur les Moyens de le Soulager et de le Prévenir*, 3 vols (Paris: Poulin, 1834).

Villey, D. "The Market Economy and Roman Catholic Thought," *International Economic Papers 9* (1959): 93–124.

Viner, J. 1958. "*Adam Smith and Laissez* Faire," in *The Long View and the Short: Studies in Economic Theory and Policy* (Glencoe IL: Free Press, 1958).

Viner, J. " 'Possessive Individualism' as Original Sin," *Canadian Journal of Economics and Political Science* (1963): 548–59.

Viner, J. *The Role of Providence in the Social Order: An Essay in Intellectual History* (Philadelphia: American Philosophical Society, 1972).

Viner, J. *Religious Thought and Economic Society, 4 chs of an Unfinished Work*, eds Melitz and Winch (Durham, NC: Duke University Press, 1978).

Voltaire, F. M. A. "An Important Study by Lord Bolingbroke, or the Fall of Fanaticism", in *Voltaire on Religion: Selected Writings*, trans. and ed. Kenneth W. Applegate (New York: Ungar, 1975).

Voltaire, F. M. *A Philosophical Dictionary* (New York: Carlton House, 1940).

Wakefield, G. *Memoirs of the Life of G. W., B.A., Written by Himself* (London, 1792).

Wallace, R. *Ignorance and Superstition a Source of Violence and Cruelty, and in Particular, the Cause of the Present Rebellion. A Sermon Preached in the High Church of Edinburgh, Monday, January 6, 1745–6 Upon the Occasion of the Anniversary Meeting of the Society* in *Scotland for propagating Christian Knowledge* (Edinburgh: Fleming & Co. 1746) facsimile in Reprints of Economic Classics (New York: Augustus M. Kelley, 1969).

Wallace, R. *Various Prospects of Mankind, Nature and Providence* (London: A. Millar, 1761).

Walras, L. *Elements of Pure Economics*, 1st edn 1874, republished, trans. W. Jaffé (London: Allen & Unwin, 1954).

Walsh, V. C. *Scarcity and Evil* (Englewood Cliffs NJ: Prentice Hall, 1961).

Walsh, V. and Gram, H. *Classical and Neoclassical Theories of General Equilibrium: Historical Origins and Mathematical Structure* (New York: Oxford University Press, 1980).

Waterland, D. *The Works of the Rev. Daniel Waterland, D. D., formerly Master of Magdalene College, Cambridge, Canon of Windsor, and Archdeacon of Middlesex. To which is Prefixed a Review of the Author's Life and Writings by William van Mildert, D. D., Lord Bishop of Durham*, 2nd edn, 6 vols. (Oxford: Oxford University Press, 1843).

Waterman, A. M. C. "The Lord's Day in a Secular Society: a Historical Comment on the Canadian Lord's Day Act of 1906," *Canadian Journal of Theology* 11 (1965): 108–23.

Waterman, A. M. C. "Some Footnotes to the 'Swan Diagram:' or, How Dependent is a Dependent Economy?" *Economic Record* 42 (1966): 447–64.

Waterman, A. M. C. "Property Rights in John Locke and in Christian Social Teaching," *Review of Social Economy* 9 (1982): 97–115.

Waterman, A. M. C. "The Ideological Alliance of Political Economy and Christian Theology, 1798–1833," *Journal of Ecclesiastical History* 34 (1983): 231–44.

Waterman, A. M. C. "Malthus as a Theologian: the 'First Essay' and the Relation between Political Economy and Christian Theology," in *Malthus: Past and Present*, eds J. Dupâquier, A. Fauve-Chamoux and J. Grebenik (London: Academic Press, 1983).

Waterman, A. M. C. "On the Malthusian Theory of Long Swings", *Canadian Journal of Economics* 20 (1987): 257–70.

Waterman, A. M. C. "Economists on the Relation between Political Economy and Christian Theology: A Preliminary Survey," *International Journal of Social Economics* 14 (1987): 46–68.

Waterman, A. M. C. "Can 'Economic Policy' be 'Christian?,' " *Review of Social Economy*, 46 (1988): 203:11.

Waterman, A. M. C. "Hume, Malthus and the Stability of Equilibrium," *History of Political Economy*, 20 (1988): 85–94.

Waterman, A. M. C. "Denys Munby on Economics and Christianity," *Theology*, March/April (1990): 108–16.

Waterman, A. M. C. "A Cambridge 'Via Media' in Late Georgian Anglicanism", *The Journal of Ecclesiastical History* 42(1991a): 419–36.

Waterman, A. M. C. "The Intellectual Context of Rerum Novarum," *Review of Social Economy* 49 (1991b): 465–82.

Waterman, A. M. C. *Revolution, Economics and Religion: Christian Political Economy, 1798–1833* (Cambridge: Cambridge University Press, 1991c).

Waterman, A. M. C. "The 'Canonical Classical Model' in 1808 as viewed from 1825: Thomas Chalmers on the National Resources," *History of Political Economy* 23 (1991d): 221–4.

Waterman, A. M. C. "Analysis and Ideology in Malthus's Essay on Population," *Australian Economic Papers*, 31 (1992): 203–17.

Waterman, A. M. C. " 'The Grand Scheme of Subordination': The Intellectual Foundations of Tory Doctrine," *The Australian Journal of Politics and History* 40 (1994): 121–33.

Waterman, A. M. C. "Whately, Senior and the Methodology of Classical Economics," in HG. Brennan and A. M. C. Waterman (eds), *Economics and Religion: Are They Distinct*? (Boston: Kluwer, 1994).

Waterman, A. M. C. "Why William Paley was 'The First of the Cambridge Economists,' " *Cambridge Journal of Economics* 20 (1996): 3–86.

Waterman, A. M. C. "The Nexus between Theology and Political Doctrine in Church and Dissent," in Haakonssen, *Enlightenment and Religion: Rational Dissent in Eighteenth-Century Britain* (Cambridge: Cambridge University Press, 1996).

Waterman, A. M. C. "Recycling Old Ideas: Economics among the Humanities," *Research in the History of Economic Thought and Methodology* 15 (1997): 237–49.

Waterman, A. M. C. " 'Pure and Rational Religion:' A theological reading of Adam Smith's Wealth of Nations," unpublished paper presented to Southern Economic Association, Atlanta, Georgia, November 1997.

Waterman, A. M. C. "Reappraisal of 'Malthus the Economist,' 1933–97," *History of Political Economy* 30 (1998): 293–334.

Waterman, A. M. C. "Malthus, Mathematics and the Mythology of Coherence," *History of Political Economy* 30 (1998): 571–99.

Waterman, A. M. C. "Hollander on the 'Canonical Classical Growth Model:' A Comment," *Journal of the History of Economic Thought* 21 (1999): 311–13.

Waterman, A. M. C. "Enlightenment Religion: The Exclusion of Enthusiasm and Superstition," mimeograph, University of Manitoba Institute for the Humanities, October 1999.

Waterman, A. M. C. "Notes Towards an Un-Canonical, Pre-Classical Model of Political Œconomy", in *Reflections on the Classical Canon in Economics*, eds. F. L. Forget and S. Peart (London: Routledge, 2001).

Waterman, A. M. C. "The 'Sussex School' and the History of Economic Thought: British Intellectual History, 1750–1950," *European Journal of the History of Economic Thought* 9 (2002): 452–63.
Waterman, A. M. C. " 'New Political Economies' Now and Then: Economic Theory and the Mutation of Political Doctrine," *American Journal of Economics and Sociology* 61(2002): 13–51.
Waterman, A. M. C. "Economics as Theology: Adam Smith's Wealth of Nations," *Southern Economics Journal* 68 (2002): 907–21.
Waterman, A. M. C. "Mathematical Modeling as an Exegetical Tool: Rational Reconstruction," in *A Companion to the History of Economic Thought*, eds. W. J. Samuels, J. E. Biddle, J. B. Davis (Malden, MA: Blackwell, 2003).
Watts, Michael P. *The Dissenters* (Oxford: Clarendon, 1978).
Watson, R. *Anecdotes of the Life of Richard Watson, Bishop Llandaff*, 2 vols (London, 1817).
Watson, R. *An Apology for Christianity, in a Series of Letters Addressed to Edward Gibbon, Esquire* (Cambridge, 1776).
Watson, R. *A Collection of Theological Tracts*, 6 vols. (Cambridge: Cambridge University Press, 1785).
Werhahn, P. H. *The Entrepreneur: His Economic Function and Social Responsibility, Ordo Socialis*, no. 4 (Trier: Paulinus, 1990).
West, E. *Essay on the Application of Capital to Land; with Observation Shewing the Impolicy of Any Great Restriction on the Importation of Corn. . ., by a Fellow of University College, Oxford*. (London: Underwood, 1815).
Westminister Review
[Whately, Richard] *Historic Doubts Relative to Napoleon Buonaparte* (London: Parker, 1819).
Whately, E. J. (ed.). *Life and Correspondence of Richard Whately, D. D., Late Archbishop of Dublin*, 2 vols, (London: Longman Greens, 1866).
Whately, R. *Elements of Logic, Comprising the Substance of the Article in the Encyclopedia Metropolitana, with Additions, etc.* (London: Mawman 1826).
[Whately, R.] "Oxford Lectures on Political Economy," *Edinburgh Review*, September (1828) 170–84.
Whately, R. *Introductory Lectures in Political Economy*, 2nd edn, (London: Fellowes, 1832).
Whelan, Ruth. "The Wages of Sin is Orthodoxy: The Confessions of St Augustine in Bayle's Dictionnare," *Journal of the History of Philosophy 26* (1988): 195–206.
Whelan, Ruth. *The Anatomy of Superstition: A Study of the Historical Theory and Practice of Pierre Bayle. Studies on Voltaire and the Eighteenth Century* 259 (Oxford: The Voltaire Foundation, 1989).
Wicksell, K. *Lectures on Political Economy*, 2 vols, 1st edn 1901, trans. F. Classen, ed. L. Robbins (London: Routledge, 1935).
Wicksteed, P. H. *Essay on the Coordination of the Laws of Distribution* (London: Macmillan, 1894).
Wieser, F. von. *Natural Value*, 1st edn 1893, republished (New York: Kelley, 1971).
Wilberforce, W. *A Practical View of the Prevailing Religious System of Professed Christians, in the Higher and Middle Classes of This Country, Contrasted with Real Christianity* (London: Cadell, 1797).
Willey, B. *The Seventeenth Century Background: Studies of the Thought of the Age in Relation to Poetry and Religion* (London: Chatto & Windus, 1934).

Willey, B. *The Eighteenth-Century Background: Studies of the Idea of Nature in the Thought of the Period* (London: Chatto & Windus, 1940).
Willey, B. *Nineteenth Century Studies: Coleridge to Matthew Arnold* (New York: Columbia University Press, 1949).
Williams, N. P. *The Idea of the Fall and of Original Sin*, Bampton Lectures 1924 (London: Longman, 1927).
Williamson, O. F. *Markets and Hierarchies: Analysis and Antitrust Implications* (New York: Free Press. 1975).
Williamson, O. F. *The Economic Institutions of Capitalism: Firms, Markets, Relational Contracting* (New York: Free Press, 1985).
Williamson, O. F. *Economic Organization: Firms, Markets and Policy Control* (New York: New York University Press, 1986).
Winch, D. *Classical Political Economy and Colonies* (Cambridge MA: Harvard University Press, 1965).
Winch, D. *Adam Smith's Politics: An Essay in Historiographic Revision* (Cambridge: Cambridge University Press, 1978).
Winch, D. Collini, S. and Burrow, J. (eds) *That Noble Science of Politics: A Study in Nineteenth Century Intellectual History* (New York: Cambridge University Press, 1983).
Winch, D. *Malthus* (Oxford: Oxford University Press, 1987).
Winch, D. N. "Robert Malthus: Christian Moral Scientist, Arch-Demoraliser or Implicit Secular Utilitarian?," *Utilitas* 5 (1993): 239–53.
Winch, D. *Riches and Poverty. An Intellectual History of Political Economy in Britain, 1750–1834* (Cambridge: Cambridge University Press, 1996).
Winstanley, D. A. *Unreformed Cambridge: A Study of Certain Aspects of Its Development in the Eighteenth Century* (Cambridge: Cambridge University Press, 1935).
Wolfe, A. *One Nation After All!. What Middle-class Americans Really Think About* (New York: Viking Penguin, 1998).
Wolfe, A. *The Transformation of American Religion: How We Actually Live Our Faith* (New York: Free Press, 2003).
Wollstonecraft, M. *A Vindication of the Rights of Man, in a Letter to the Right Honourable Edmund Burke: Occasioned by His Reflections on the Revolution in France*, 2nd edn (London: Johnson, 1790).
Wong, Stanley "The 'F-Twist' and the Methodology of Paul Samuelson," *American Economic Review* (1973): 312–25.
Woodforde, James *The Diary of a Country Parson*, ed. John Beresford, 5 vols (Oxford: Oxford University Press, 1924–31).
Wordsworth, W. *Poetical Works*, ed. T. Hutchinson, new edn, revised E. de Selincourt (Oxford: Oxford University Press, 1950).
Worland, S. T. "Social Economy and the Theory of Justice: Two New Directions," *Review of Social Economy* (1977): 345–59.
Wrigley, E. A. and Schofield, R. S. *The Population History of England, 1541–1871 A Reconstruction* (London: Amold, 1982).
Young, A. A. "Increasing Returns and Economic Progress," *Economic Journal* 38 (1928): 527–42.
Young, B. W. *Religion and Enlightenment in Eighteenth-Century England: Theological Debate from Locke to Burke* (Oxford: Clarendon, 1998).

Index